THE INTERNAL SOLAR ANGULAR VELOCITY

ASTROPHYSICS AND SPACE SCIENCE LIBRARY

A SERIES OF BOOKS ON THE RECENT DEVELOPMENTS
OF SPACE SCIENCE AND OF GENERAL GEOPHYSICS AND ASTROPHYSICS
PUBLISHED IN CONNECTION WITH THE JOURNAL
SPACE SCIENCE REVIEWS

VOLUME 137

PROCEEDINGS

THE INTERNAL SOLAR ANGULAR VELOCITY

Theory, Observations and Relationship to Solar Magnetic Fields

PROCEEDINGS OF THE 8TH
NATIONAL SOLAR OBSERVATORY SUMMER SYMPOSIUM,
HELD IN SUNSPOT, MEXICO, AUGUST 11–14, 1986

Edited by

BERNARD R. DURNEY
National Optical Astronomy Observatories,
Tucson, Arizona, U.S.A.

and

SABATINO SOFIA
Yale University Observatory,
New Haven, Connecticut, U.S.A.

D. REIDEL PUBLISHING COMPANY

A MEMBER OF THE KLUWER ACADEMIC PUBLISHERS GROUP

DORDRECHT / BOSTON / LANCASTER / TOKYO

Library of Congress Cataloging in Publication Data

CP

National Solar Observatory (U.S.). Summer Symposium
(8th : 1986 : Sunspot, N.M.)
The internal solar angular velocity.

(Astrophysics and space science library; v. 137)
Includes index.
1. Sun–Rotation–Congresses. 2. Stars–Rotation–Congresses.
3. Sun–magnetic field–Congresses. I. Durney, Bernard. II. Sofia, Sabatino.
III. Title. IV. Title: Solar angular velocity. V. Series.
QB523.N38 1986 523.7′3 87–9703

ISBN-13: 978-94-010-8233-4 e-ISBN-13: 978-94-009-3903-5
DOI: 10.1007/978-94-009-3903-5

Published by D. Reidel Publishing Company,
P.O. Box 17, 3300 AA Dordrecht, Holland.

Sold and distributed in the U.S.A. and Canada
by Kluwer Academic Publishers,
101 Philip Drive, Assinippi Park, Norwell, MA 02061, U.S.A.

In all other countries, sold and distributed
by Kluwer Academic Publishers Group,
P.O. Box 322, 3300 AH Dordrecht, Holland.

Softcover reprint of the hardcover 1st edition 1987

TABLE OF CONTENTS

PART 3/THEORY: RADIATIVE ZONE

PART 4/THEORY: CONVECTIVE ZONE

Preface

It is clear that the discovery of solar eigenmodes and the resulting possibility of probing the solar interior is an event of primary importance for solar physics in general and for theories of the inner solar angular velocity in particular. While these theories are basic for the understanding of the solar spin down, differential rotation, dynamo and activity, they are however, extremely complex, and in all likelihood only limited further progress could have been achieved without the guidance of observations. Until recently and in spite of the scant observational basis the theoretical work has moved forward as the perusal of this book shows. There cannot be any doubt, however, that the present, rapidly expanding, worldwide observational program will lead ultimately to a vigorous theoretical development of the field.

It appeared to the organizers that a meeting centered on theories of the inner solar angular velocity, comprising presentations of the main research areas by the involved scientists, would significantly foster this development since it would help to clarify the basic ideas of the subject.

The meeting, held at the National Solar Observatory/Sacramento Peak, from August 11 to August 14, was the eighth in a series of summer symposia at Sacramento Peak.

The unqualified success of the meeting could not have been possible without the unlimited devotion of the staff at Sacramento Peak, Ray Smartt, Frank and Pat Hegwer, Ramona Elrod in particular. We are also grateful to the members of the organizing committee - Pierre Demarque, Tom Duvall, Bob Howard and Ray Smartt for their invaluable assistance and organization of logistical support. Dr. Robert F. Howard, National Solar Observatory director, kindly made partial funding and housing available at Sacramento Peak.

Finally, special thanks are owed to NASA. Without their financial support, as part of a grant awarded to Sacramento Peak, the meeting could not have taken place.

BERNARD DURNEY SABATINO SOFIA

LIST OF PARTICIPANTS

R. C. Altrock, AFGL/Sacramento Peak, Sunspot, New Mexico USA

I. Bernstein, Yale University, New Haven, Connecticut, USA

T. M. Brown, High Altitude Observatory, Boulder, Colorado, USA

A. Cannon, University of Sydney, New South Wales, Australia

F. Cattaneo, JILA/University of Colorado, Boulder, Colorado, USA

K. L. Chan, NASA/Goddard Space Flight Center, Greenbelt, Maryland, USA

S. M. Chitre, Tata Institute of Fundamental Research, Bombay, India

M. Clement, University of Toronto, Ontario, Canada

P. Demarque, Yale University Observatory, New Haven, Connecticut, USA

B. R. Durney, NOAO/National Solar Observatory, Tucson, Arizona, USA

T. L. Duvall, Jr., NASA/Goddard Space Flight Center, S. W. Solar Station, National Solar Observatory, Tucson, Arizona, USA

A. Endal, Applied Research Corporation, Landover, Maryland, USA

P. Fox, Yale University, New Haven, Connecticut, USA

G. A. Glatzmaier, Los Alamos National Laboratory, Los Alamos, New Mexico, USA

D. Gough, Institute for Astronomy, Cambridge, England

D. Guenther, Yale University Observatory, New Haven, Connecticut, USA

D. Haber, JILA/University of Colorado, Boulder, Colorado, USA

J. W. Harvey, NOAO/National Solar Observatory, Tucson, Arizona, USA

D. H. Hathaway, NASA/Marshall Space Flight Center, Huntsville, Alabama, USA

J. Herring, National Center for Atmospheric Research, Boulder, Colorado, USA

F. Hill, NOAO/National Solar Observatory, Tucson, Arizona, USA

R. F. Howard, NOAO/National Solar Observatory, Tucson, Arizona, USA

D. Hughes, JILA/University of Colorado, Boulder, Colorado, USA

H. P. Jones, NASA/Goddard Space Flight Center, S.W. Solar Station, National

Solar Observatory, Tucson, Arizona, USA

J. Kaufman, California Institute of Technology, Pasadena, California, USA

S. D. Kawaler, Yale University, New Haven, Connecticut, USA

S. L. Keil, AFGL/Sacramento Peak, Sunspot, New Mexico, USA

E. Knobloch, University of California at Berkeley, Berkeley, California, USA

J. R. Kuhn, Michigan State University, East Lansing, Michigan, USA

K. Libbrecht, Big Bear Solar Observatory, Cal Tech, Pasadena, California, USA

C. Marmolino, AFGL/Sacramento Peak, Sunspot, New Mexico, USA

B. Merryfield, JILA/University of Colorado, Boulder, Colorado, USA

D. F. Neidig, AFGL/Sacramento Peak, Sunspot, New Mexico, USA

L. J. November, NOAO/National Solar Observatory, Sacramento Peak Sunspot, New Mexico, USA

P. L. Palle, Instituto Astrofisico de las Islas Canarias, Tenerife, Spain

E. N. Parker, Enrico Fermi Institute, University of Chicago, Chicago, Illinois, USA

E. Pilger, Institute for Astronomy, University of Hawaii, Honolulu, Hawaii, USA

M. Pinonneaulp, Yale University, New Haven, Connecticut, USA

B. Popp, California Institute of Technology, Pasadena, California, USA

R. R. Radick, AFGL/Sacramento Peak, Sunspot, New Mexico, USA

E. J. Rhodes, Jr., University of Southern California, Los Angeles, California, USA

E. Ribes, Observatoire de Meudon, Meudon, France

I. W. Roxburgh, Queen Mary College, University of London, London, England

E. Schatzman, Observatoire de Nice, Nice, France

M. Schussler, Kiepenheuer Institut, Freiburg, West Germany

G. W. Simon, AFGL/Sacramento Peak, Sunspot, New Mexico, USA

A. Skumanich, High Altitude Observatory, Boulder, Colorado, USA

R. N. Smartt, NOAO/National Solar Observatory, Sacramento Peak, Sunspot, New Mexico, USA

S. Sofia, Yale University Observatory, New Haven, Connecticut, USA

E. A. Spiegel, Columbia University, New York, New York, USA

H. Spruit, Max Planck Institut, Garching, West Germany

S. R. Sreenivasan, University of Calgary, Calgary, Alberta, Canada

M. Stix, Kiepenheuer Institut, Freiburg, West Germany

J. Thomas, University of Rochester, Rochester, New York, USA

J. Toomre, JILA/University of Colorado, Boulder, Colorado, USA

I. Tuominen, University of Helsinki, Tahtitorninmaki, Helsinki, Finland

W. Unno, Kinki University, Higashi-Osaka, Japan

O. von der Luhe, NOAO/National Solar Observatory, Sacramento Peak, Sunspot, New Mexico, USA

H. Yoshimura, University of Tokyo, Department of Astronomy, Tokyo, Japan

J. P. Zahn, Observatoire du Pic du Midi, Bagneres-de-Bigorre, France

PROBLEMS OF THE SOLAR INTERIOR

IAN W. ROXBURGH
Queen Mary College
Mile End Road, London E1 4NS

ABSTRACT. The sun provides a testing ground for theories of stellar structure and evolution, the substantial increase in information afforded by observations on solar oscillations and the measurement of the solar neutrino flux suggests that we may not as yet have a full understanding of stellar interiors.

THE "STANDARD MODEL"

At the simplest level we observe the present luminosity $L_\odot$, radius $R_\odot$, mass $M_\odot$ and surface composition Z_i/X, but have no direct measure of helium abundance Y; we deduce the solar age $t_\odot$ from that of the solar system. If one asks if we can produce evolved models of the sun that fit these observations the answer is yes; such models have inhomogeneous evolution from an initially homogeneous model whose composition was (e.g. Bahcall et al 1982)

$$X = 0.72 \pm 0.02 \ , \quad Y = 0.26 \pm 0.02 \ , \quad Z = 0.018 \pm 0.004 \ .$$

Different authors find slightly different initial abundances – this in itself is a comment on the ignorance of the details of the physics of stellar interiors and perhaps on numerical accuracy. One should however emphasise that such "standard models" are adjusted to fit the observations – by adjusting the initial helium abundance and the "mixing length" models are obtained that have the Solar Luminosity and Solar Radius – 2 adjustable parameters to fit 2 observations!

Even these adjusted models sometimes smooth over what may be important phases of evolution at early times, the initial burning of ^{3}He and ^{12}C may produce a central convective core which may be sustained for some time by convective overshooting – g modes of oscillation may be excited by the ^{3}He and ^{12}C profiles which may also affect evolution. These early phases of nuclear burning may drive a convective core which may generate a central magnetic field even if an original field was dragged to the surface during an initial fully convective phase.

I am not here arguing that these effects took place – we don't know if they did – but I do wish to emphasise that even if they did we could still produce a model of the present sun that had the observed luminosity and radius by adjusting the initial helium abundance and mixing length.

B. R. Durney and S. Sofia (eds.), The Internal Solar Angular Velocity, 1–5.

SOLAR NEUTRINO PROBLEM

The observed flux of electron neutrinos by Davis and collaborators (Davis 1985) is about a factor of 3 above that predicted by standard models; this remains a problem suggesting that either the models are incorrect, the central temperature and central helium abundance being too high, or that our present understanding of neutrino physics is incorrect.

One way to lower the central temperature is to appeal to some mixing process in the solar interior; a fully homogeneous solar model predicts a neutrino flux below the observed value although it encounters difficulties in matching the solar oscillation data. Moreover it is not easy to mix a star – there has to be a supply of energy at a rate sufficient to overcome the stabilising effect of chemical evolution which produces helium rich material in the central regions. This may be estimated by noting that on a nuclear time scale $t_N \sim 10^{10}$ys the gravitational energy E_g about doubles, so to mix the sun there has to be a supply of energy at a rate E_g/t_N (Roxburgh 1984). If the mixing is on a shorter time scale then the rate of supply of energy is even greater. This places severe limits on the possible mechanisms for mixing, rotation and/or magnetic fields being too weak. Mixing driven by thermonuclear sources is energetically possible, since the rate of production of energy is some 10^3 times E_g/t_N. One possible candidate mechanism is the ^{3}He driven instability discovered by Gough and colleagues. (Dilke and Gough 1972, Christensen Dalsgaard et al 1974).

Another possibility is that our understanding of the physics of stellar interiors is badly in error, were our estimates of opacity or energy generation badly wrong in the central regions, or were some esoteric particles in part responsible for energy transport then models could be – and have been – produced that satisfy the observed solar neutrino constraint. (c.f. Faulkner et al 1986, Däppen et al 1986).

It would be of considerable help in analysing these possibilities if we had reliable estimates of the limiting g mode period and of p modes that penetrate the core.

THE ^{3}He INSTABILITY

I would like to emphasise the result obtained by Gough et al and confirmed by other workers, that inhomogeneous evolved models of the Sun become unstable to low order g modes at an age of about 2.10^8 ys. More recently Gavryusev and Gavryuseva (1987) have found higher order instabilities in models with mixed cores.

These instabilities have been mostly ignored by those calculating solar and stellar models, this does not seem a reasonable approach to the present author. It is the case that if it is assumed that the instability has no effect then solar evolutionary models return to stability after 3.10^9 ys due to the build up of ^{4}He in the central regions – but if the instability causes mixing the evolutionary sequence ignoring the instability is irrelevant! The growth rate of these instabilities are long $\sim 4.10^7$ years, but still large enough for large amplitude oscillations to have developed before chemical inhomogeneity suppresses the instability. Until the effect of these instabilities is understood I am reluctant to accept "standard models" as a correct description of the evolution of the sun – or of other solar type stars.

One major uncertainty is the effect of the solar convective zone on the instability, since g modes are evanescent in this zone. High order modes have low surface amplitudes and the effect of the convective overshoot region on the modes has not been properly analysed. Is it possible that such modes are currently excited in the sun but have negligible surface amplitude? If there were such finite amplitude oscillations in the interior – whether in the past or at the present time how would they affect the evolution of the sun. The original suggestion by Dilke and Gough was that the oscillations would trigger a finite amplitude instability leading to a mixing of the solar core, so producing a more homogeneous solar model. Even without such mixing large amplitude oscillations will affect the internal structure as they modify the energy production and energy transport. Since the relative temperature perturbation in a g–mode vanishes at the centre and reaches a maximum away from the centre this leads to enhanced energy production away from the centre, and therefore to a lower central temperature for a model of the present sun.

I have calculated such a model where the energy generation was modified by a factor $(1 + 6\ (\delta T/T)^2)$ and was able to reduce the neutrino flux to 3 SNU (Roxburgh 1985). This model is not correct but it demonstrates the possible importance of including the effects of the ^{3}He instability in solar evolutionary sequences.

Again it is data on oscillation modes that penetrate into the central core that will help resolve these problems.

CONVECTIVE OVERSHOOTING

It is reasonable to assume that convection driven in unstable regions penetrates into the surrounding stable layers and modifies their structure, the question is how important is this effect? For the sun it could be significant at the base of the outer convective zone, and thereby possibly influence the properties of normal modes of oscillation, and also in the early phase of solar evolution were there to have been a ^{3}He or ^{12}C driven convective core.

Attempts to model the effect of such overshooting suggests that there is a slightly sub adiabatic extension of the super adiabatic unstable region in which the convective overshooting is still effective in transporting energy, this may result in an inward transport of energy, the energy carried by radiation then exceeding the total flux. This almost adiabatic overshoot region then adjusts to the surrounding stable layer in a thin transition zone. The extent of the overshoot region cannot be reliably estimated – my simple model gives approximately half a pressure scale height but this is not reliable (c.f. Roxburgh 1985).

If such overshooting is included in a model of the sun it does not produce much change in the depth of the outer convective zone – from 0.27 to 0.3 $R_\odot$ in my models – this is because the model is normalised to give a solar radius by changing the adiabatic constant (or the mixing length) within the unstable region.

Very substantial overshooting will produce a sharp change in the gradient of sound speed in the solar interior – whether or not this can be determined from oscillation data has yet to be explored.

INTERNAL ROTATION

The solar oscillation data suggests an almost constant angular velocity in the solar interior — with perhaps a more rapidly spinning core and a less rapidly spinning shell between 0.2 and 0.3 $R_\odot$ (Duvall et al 1984). If this drop is real it poses a difficult problem — could it be that there has been a recent turnover of the core bringing low angular momentum material out from the centre and taking higher angular momentum material in to the centre (Gough, private communication)? Or is the region between 0.2 and 0.3 $R_\odot$ kept mixed by some instability and so maintaining differential rotation as is believed to be the case in the outer convective zone (Roxburgh 1986)? Or will the problem disappear with more detailed observations?

If there is even a small magnetic field in the solar interior then one should expect essentially uniform rotation, some differential rotation may be compatible with very weak fields but it is difficult to see how the interior could be differentially rotating if the internal field is 10^{-3} G or more. This need not be true if there were some mixed core or shell in the interior which produced an oscillating dynamo field and so separated the interior into different magnetic and rotation zones.

If there were no internal field there are a range of instabilities driven by differential rotation, but it is unlikely that these are effective in equalising angular velocity except just beneath the outer convective zone. Essentially this is because even a mild composition gradient requires a substantial energy supply to lift up heavy material and there is in general not enough energy in differential rotation to do this. This is also true if the ABCD instability of Spruit, Knoblock and Roxburgh (1983) which probably results in mild almost horizontal mixing.

However it is probably unreasonable to assume that the sun has no internal field; the decay time of a large scale field is of the same order as the solar life time so there should be a remnant of an initial field. If such an initial field were dragged to the surface by the outer convective zone there still remains the early evolutionary phase when the sun had a ^{3}He and ^{12}C driven convective core — if this produced a dynamo field the large scale components would have diffused out to thread the whole interior, and establishing constant rotation throughout the interior.

CONCLUSIONS

We do not as yet have a detailed understanding of the solar interior nor of its past evolution, but with increasingly accurate oscillation data and advances in inversion techniques we should have severe empirical restrictions to our possible solar models.

REFERENCES

Bahcall, J.N., Huebner, W.F., Lubow, S.H., Parker, P.D., and Ulrich, R.K., 1982, *Rev.Mod.Phys.* 54, 767.

Christensen-Dalsgaard, J., Dilke, F.W.W. and Gough, D.O., 1974, *Monthly Notices Roy.Astron.Soc.* 169, 429.

Davis, R., 1985 in M.L.Cherry, W.A.Fowler and K.Lande (eds.) *Solar Neutrinos and Neutrino Astronomy*, Am.Inst.Phys. p.1.

Däppen W., Gilliland, R.L. and Christensen-Dalsgaard J., 1986, *Nature* 321, 229.

Dilke, F.W.W. and Gough, D.O., 1972, *Nature*, 240, 262.

Duvall, T.L., Dziembowski, W.A., Goode, P.R., Gough, D.O., Harvey, J.W. and Leibacher, J.W., 1984, *Nature*, 310, 22.

Faulkner, J., Gough, D.O. and Vahia, M.N., 1986, *Nature*, 321, 226.

Gavryusev, V.G. and Gavryuseva, E.A., 1987 - to be published.

Roxburgh, I.W., 1984, in A.Meader and A.Renzini (eds.), *Observational Tests of the Stellar Evolution*, D.Reidel Publ.Co., Dordrecht, Holland, p.519.

Roxburgh I.W., 1985, *Solar Physics* 100, p.21.

Roxburgh I.W., 1986 in D.O.Gough (ed), *Seismology of the Sun and Distant Stars*, Reidel, P.249.

Spruit, H.C., Knobloch, E.W., and Roxburgh, I.W., 1983, *Nature*, 304, 320.

OBSERVATIONS OF SOLAR P-MODE ROTATIONAL SPLITTINGS

Timothy M. Brown and **Cherilynn A. Morrow**
High Altitude Observatory/National Center for Atmospheric Research*
P.O. Box 3000
Boulder, CO 80307
U.S.A.

ABSTRACT. In the last two years, several groups of workers have observed the splitting of solar p-mode frequencies that arises from the Sun's differential rotation. Early results suggested that throughout much of the convection zone, the differential rotation was much different than that seen at the surface. However, Duvall *et al.* (1986) found splittings that are consistent with the differential rotation being independent of depth, even within the radiative interior. We describe here the results of a new set of observations obtained with the Fourier Tachometer at NSO/Sunspot (Brown and Morrow 1986). The rotational splittings from these observations fall between those of Brown (1985) and Duvall *et al.* (1986). They appear consistent with a model in which the differential rotation resembles that seen at the surface throughout the convection zone, but becomes constant on spherical surfaces within the radiative interior.

1. INTRODUCTION

Solar p-mode oscillations have become a powerful tool for probing the Sun's interior. Besides carrying information about the solar structure (*e.g.* the sound speed as a function of depth, Christensen-Dalsgaard *et al.* 1985), the way in which oscillation frequencies depend on the direction of propagation of the corresponding wave can tell us something about flows in the solar interior. In particular, the oscillation frequency ν of a p-mode with particular values of the radial order n, angular degree l, and azimuthal order m depends on the Sun's internal rotation (Lynden-Bell and Ostriker 1967, Hansen, *et al.* 1977, Gough 1981, Cuypers 1980). In the simplest approximation (Brown 1985) this dependence has the form

$$\nu(n,l,m) - \nu(n,l,0) = -m \frac{\int_{-1}^{1} \overline{\Omega}(\theta)\, [P_l^m(\cos\theta)]^2 d\cos\theta}{\int_{-1}^{1} [P_l^m(\cos\theta)]^2 d\cos\theta} . \tag{1}$$

where θ is the colatitude and $\overline{\Omega}(\theta)$ is an appropriate average of the stellar rotation over the

*The National Center for Atmospheric Research is sponsored by the National Science Foundation

B. R. Durney and S. Sofia (eds.), The Internal Solar Angular Velocity, 7–17.

depth range sampled by the p-mode in question. Improved observational techniques have allowed us to exploit this dependence, making some features of the Sun's internal rotation structure accessible to observation.

To estimate the Sun's internal rotational structure, one must evidently measure the way in which oscillation frequencies vary with m at constant n and l. It is convenient to expand the observed frequency variation with m as a power series in m/l. Then Equation (1) implies that solid body rotation by itself causes ν to be a linear function of m. Indeed, Ω can contribute only odd powers to the series: latitudinal differential rotation adds cubic and higher odd powers. Even powers may arise from pole-equator variations in the solar structure (for details, see Brown 1985,1986 and Duvall *et al.* 1986). The coefficients of these higher powers thus provide a measure of $\Omega(\theta)$, averaged over some range of depths. The depth range over which Ω is averaged is determined primarily by the degree l of the modes in question. The l dependence of the coefficients therefore contains information about the depth variation of the rotational structure. Unfortunately, the depth average that determines the frequency variations is always dominated by contributions from the surface layers of the Sun, so that it can be very difficult to distinguish $\Omega(r,\theta)$ distributions that resemble one another near the surface, but differ at depth. This problem will be illustrated in Section 3.

It proves useful to expand the observed $\nu(m)$ values not in a simple power series, but rather in a series of Legendre polynomials (Duvall *et al.* 1986):

$$\nu(n, l, m) - \nu(n, l, 0) = l\sum_{i=0}^{N} a_i P_i(-m/l) \quad . \tag{2}$$

Note that Duvall *et al.* chose to define their Legendre polynomials in terms of m/L, where $L \equiv \sqrt{l(l+1)}$. We have used m/l instead, because this choice allows easier comparison with theoretical results in section 5. The choice of Legendre polynomials as basis functions minimizes dependence between coefficients in the expansion of $\nu(m/l)$. In terms of this expansion, the observed surface differential rotation (as measured by tracking surface magnetic features) is described by $a_1 = 442.8$ nHz, $a_3 = 21.7$ nHz, $a_5 = -2.5$ nHz (Snodgrass 1983).

Brown (1985, 1986) published the first observations of the m dependence of oscillation frequencies These observations spanned 5 days (giving frequency resolution of about 2 μHz), and covered $8 \leq l \leq 50$. After correction for an error in the assumed solar radius, the resulting a_1 values agreed fairly well with the surface equatorial value. a_3, however, was typically only about one-third of the surface value, throughout the range of l. Since modes with $l = 50$ are confined almost entirely to the convection zone, this result suggested that the latitudinal differential rotation was very different at the bottom of the convection zone than at the top. Brown (1985) found a_2 to be slightly positive (3 nHz, averaged over $10 \leq l \leq 50$), but ascribed no physical significance to this result. Libbrecht (1986) used Doppler observations taken with a birefringent filter to obtain similar results for modes with $5 \leq l \leq 20$. However, Duvall *et al.* (1986), using observations obtained at the South Pole, found a_1, a_3, and a_5 to agree with values computed from the surface rotation, for $20 \leq l \leq 98$. Moreover, they found $a_2 = 8$ nHz, significantly larger than Brown's or Libbrecht's values. These observations had frequency resolution of only about 5.5 μHz, but did not suffer from the diurnal gaps afflicting the other observations. One must now ask what is the origin of the discrepancy between the results by Duvall *et al.* and those of the other workers. Are some observations right and others wrong, or has the Sun changed in the interval separating the various observations? If the former, what are the correct splittings and what is the origin of the apparent differences? If the latter, what can possibly be happening inside the Sun? We believe that the best means of answering these questions is by examining more observational data, and by looking carefully at the analysis methods leading from the raw data to the final results. To this end, we will describe a new set of observations spanning 15 days (frequency resolution 0.8 μHz), obtained and analyzed by means similar to those used by Brown (1985, 1986). These observations are treated in more detail by Brown and Morrow (1986). In addition, we will discuss a source of systematic error in low-degree splitting measurements derived from cross-correlation analyses, and give linear splitting results from a much simpler (but less error-prone) analysis for degrees between 3 and 7.

2. OBSERVATIONS AND ANALYSIS

We used the Fourier Tachometer (Brown 1984, Evans 1980) at the Sacramento Peak Observatory Big Dome to obtain 2-dimensional, spatially resolved (100 x 100 pixels) Doppler images of the entire solar disk. We recorded images at a cadence of one per 60 s, for typically 7.6 hr per day on 13 of the 15 days between 28 Oct. 1984 and 11 Nov. 1984. The two days 5-6 Nov. were lost to bad weather. We observed in the 667.8 nm line of Fe I, attaining typical velocity noise of 3 m s^{-1} for single pixels in the center of the solar image in a 60 s integration.

The analysis procedure was very similar to that described by Brown (1985, 1986). Each velocity image was processed to remove instrumental and uninteresting physical velocity fields, and the residual velocity images were decomposed into spherical harmonics. We constructed time series of these spherical harmonic components, and produced power spectra from the time series by standard FFT techniques. We grouped the power spectra into sets with identical values of the nominal degree l_0, and arranged the spectra in each set into 2-dimensional $m - \nu$ diagrams. Finally, we extracted $\nu(m)$ relations from each $m - \nu$ diagram by a cross-correlation method similar to that used by Brown (1985, 1986) and Duvall *et al.* (1986).

There were a few significant differences between this analysis and that of Brown (1985, 1986). The most important of these were:
(1) We computed spherical harmonic coefficients for nominal degrees between 2 and 99. However, to economize on computing time, we computed values for only about half of the degrees between $l = 50$ and $l = 99$.
(2) We used an iterative technique to compute an average spectrum for each degree that was consistent with the frequency splittings derived from the cross-correlation analysis for that degree. In this respect, the analysis is similar to that of Duvall *et al.* (1986).
(3) We expanded the observed $\nu(m)$ relation in Legendre polynomials (*cf.* Eq. 2), rather than simple polynomials in m. In addition, we carried the expansion out to polynomials of 5th degree, instead of 3rd as in Brown (1985, 1986).

3. INTERMEDIATE DEGREE SPLITTING RESULTS

The Legendre polynomial coefficients a_i obtained from these fits are shown in Figs. 1 and 2, and as averages over small l ranges in Table 1. The dashed lines on the plots of odd-indexed coefficients indicate the expected values if the surface latitudinal differential rotation prevailed throughout the Sun. a_1 is in fair agreement with previous measurements derived from p-modes. Our values are smaller than those of Duvall *et al.* (1986) by 5.6 nHz, or about 1.3%, and smaller than those of Brown (1986) (corrected for a solar radius error) by 2.5 nHz. Our a_3 values, however, are intermediate between those of Brown 1985 and those of Duvall *et al.* 1986 and are probably incompatible with both.

The errors indicated in Figs. 1 and 2 are 1σ standard errors of the polynomial fits, derived from the scatter remaining about the 5th order fit. The regularity of the data suggests either that these errors are overestimated, particularly at large l, or that the errors are well correlated over a substantial range in l. In the Table, the quoted errors are 1σ errors derived from the scatter over the range of l values included in the averages, and are thus more representative of the apparent smoothness of the a_i.

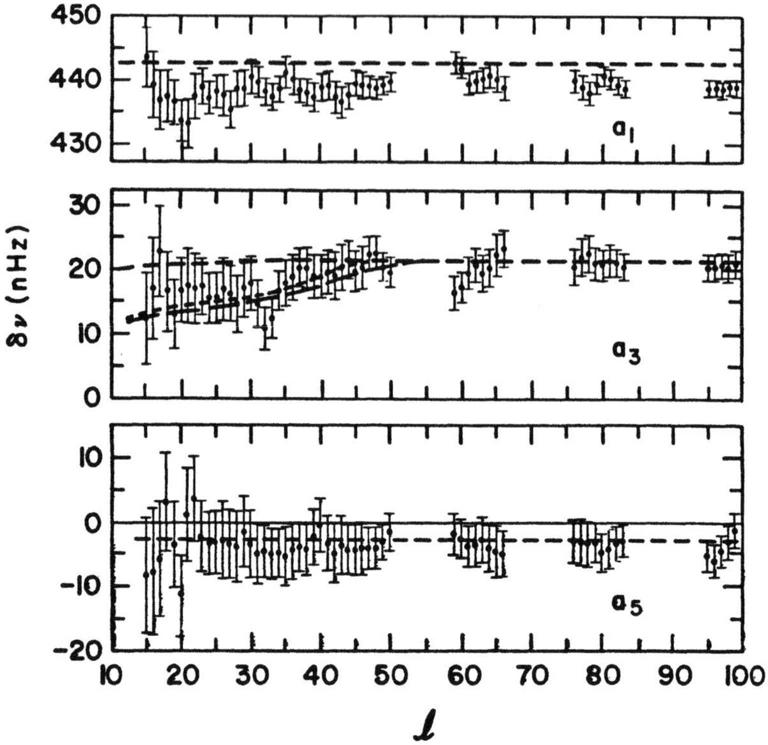

Figure 1. Frequency splitting coefficients a_i for $i = 1, 3, 5$, in sidereal nHz.

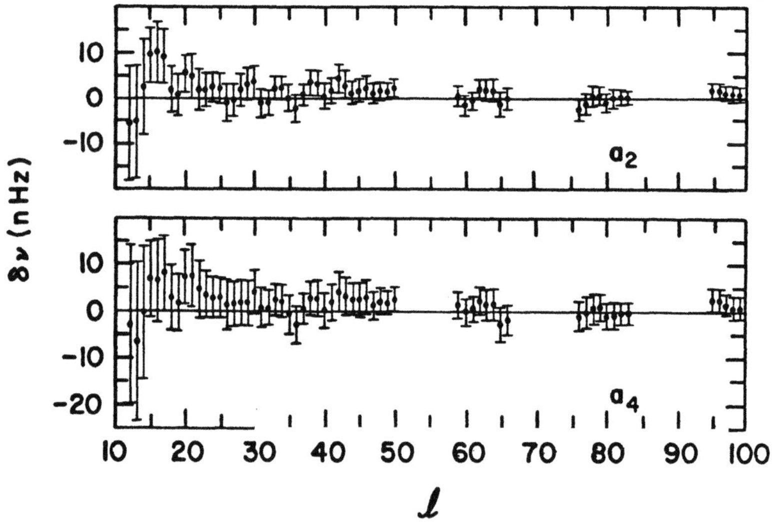

Figure 2. Same as Figure 1, but for the structure-sensitive coefficients a_2, a_4.

Besides various random noise sources, there may be significant systematic effects in these data. As mentioned above, earlier work with a 5-day run of similar observations gave values for a_3 that were significantly smaller than those in Fig. 1. An analysis of the first 5 days of the current data set yielded similarly small a_3 values, as did an erroneous analysis of the entire 15-day run, in which there was incorrect phasing between the three 5-day subsections. This suggests that low a_3 values may arise in some way from inadequate frequency resolution in the power spectra. (Why Duvall *et al.* obtain *larger* a_3 values with a 2-day observing run remains mysterious, however.) If poor frequency resolution is at fault, then one must ask whether the current resolution is in fact adequate, and whether the decrease in a_3 for degrees less than 40 may be simply a lesser form of the same problem. Further observations with better frequency resolution will probably be required to give a reliable answer to this question. Nevertheless, the good agreement with surface differential rotation results at large degree encourages us to believe that the trends illustrated in Fig. 1 are close to correct.

l range	a_1	a_2	a_3	a_4	a_5
15-18	439.3±3.0	7.8±3.9	17.2±4.3	6.1±2.2	-4.6±5.3
19-22	435.3±2.1	3.3±2.3	16.1±2.1	5.3±2.7	-2.5±6.6
23-26	438.1±0.7	1.4±1.6	16.4±0.9	2.6±0.9	-2.8±0.5
27-30	438.4±2.1	2.2±1.8	16.5±1.4	2.3±1.2	-2.9±1.0
31-34	438.6±1.0	0.7±1.9	13.7±2.6	1.5±0.9	-4.6±0.2
35-38	439.6±1.4	0.7±2.4	19.4±1.2	0.0±2.3	-4.3±0.6
39-42	438.4±1.0	2.6±1.7	19.2±0.3	2.3±1.6	-2.6±1.8
43-46	438.5±1.4	2.1±0.7	20.5±0.7	2.8±0.3	-3.9±0.4
47-50	439.4±0.5	1.7±0.5	21.3±1.5	1.9±0.5	-3.0±1.2
59-62	441.3±1.5	0.3±1.5	18.5±2.0	1.0±1.0	-2.7±0.8
63-66	440.4±0.8	0.7±1.5	21.4±1.8	-0.4±2.2	-3.8±1.0
76-79	439.5±0.8	-0.5±1.5	21.6±0.9	0.0±0.9	-2.9±0.2
80-83	440.2±0.9	0.1±0.5	21.0±0.4	-0.7±0.3	-3.5±0.8
95-99	439.5±0.2	1.7±0.5	20.6±0.2	1.8±0.9	-4.4±1.3

Table 1. Average a_i coefficients (nHz sidereal) with standard deviations. These standard deviations are computed from the scatter of the individual coefficients about the mean for each l range.

4. LOW DEGREE SPLITTING RESULTS

For degrees smaller than 15 or so, the rotational splittings given by the cross-correlation analysis become more and more unreliable. This is true not only in the sense that they suffer from large amounts of noise, but also in that they start to show the effects of systematic errors. These difficulties are unfortunate, since the low-degree modes provide the only currently available information on the solar rotation in the deep interior. Earlier work (Claverie *et al.* 1981, Duvall and Harvey 1984) has suggested that the Sun's equatorial rotation rate may rise significantly within about 0.2 $R_\odot$ of the center; one might hope that the current data set could throw further light on this important question. In this section, we will first discuss the source of one kind of systematic error in low-degree cross-correlation splitting measurements, and then describe the results of a different analysis for degrees between 3 and 7.

To understand why a cross-correlation analysis is likely to be in error for small degrees, recall that spectra obtained with a spatial filter tuned to degree l_0 will in fact respond to a range of degrees in the neighborhood of l_0. For multiplets with $l \neq l_0$, not only is the assumed degree incorrect, but so is the assumed range of azimuthal order m : for $l < l_0$ one produces more spectra than there are m values, while for $l > l_0$ there are more m values than spectra. In all cases, however, the temporal freqency range within any multiplet is preserved in the power spectra. The result of this mode confusion is illustrated schematically in Fig. 3. Evidently for multiplets with $l < l_0$, $d\nu/dm$ is underestimated because of the presence of spurious m values. For $l > l_0$, on the other hand, the existing modes are crammed into too few m values, leading to an overestimate of $d\nu/dm$. Since $d\nu/dm$ is just the linear component of the rotational frequency splitting, errors of this sort can have a direct effect on one's estimate of the solar rotation.

The magnitude of the error just described is typically about a factor of $1 \pm 1/l$, since one is usually sensitive primarily to modes with $l = l_0 \pm 2$, and the response weakens as the difference in l increases. Thus, for large degrees, one expects the errors to be intrinsically small, and moreover the effects from degrees greater and less than l_0 should nearly cancel. Notice, however, that the cross-correlation technique does something like a power-weighted mean of $d\nu/dm$, and that if all modes are equally excited, the total power in each multiplet would be proportional to $2l + 1$. Thus, for low degrees the multiplets with $l > l_0$ should dominate, leading to systematic overestimates of $d\nu/dm$. This tendency is particularly unpleasant in data (like the current set) that have had low-degree signals fitted out, decreasing the instrumental response to very small values for $l \leq 3$. In this case (as illustrated in Fig. 4), it seems likely that splittings measured with the cross-correlation method will be totally unreliable for degrees less than 5, and seriously affected for degrees less than 10.

To avoid low-degree problems of the sort just described, we have returned to the oldest and most basic means of measuring frequency splittings: locating power spectrum peaks by eye on a suitable graphics device (*e.g.* Duvall and Harvey 1984). We did this by showing $m-\nu$ diagrams for nominal degrees between 2 and 10 simultaneously as suitably normalized gray-scale plots on a Grinnel display. To get the best frequency resolution, and to minimize the dynamic range required in the display, we examined all of the spectra in four 250 μHz segments, covering the frequency range between 2.5 and 3.5 mHz. We then used the known frequencies for low-degree oscillations (Duvall and Harvey 1983) as a guide in identifying features in the $m-\nu$ diagrams. Displaying a range of nominal degrees facilitated the task of identifying multiplets, by allowing one to see which features were most likely to be associated with a particular value of l, and which ones arose as leakage from neighboring l values. Once the multiplets were identified, we estimated the total frequency splitting $\nu(l, m=l) - \nu(l, m=-l)$ by superposing a cursor on the high- and low-frequency ends of the multiplet and measuring the frequency difference.

This process is very direct, but since it makes no explicit use of all of the spectra for which $|m| < l$, it is comparatively imprecise. Moreover, the resulting splitting values contain no information about differential rotation: $[\nu(l,l) - \nu(l,-l)]/2l$ corresponds to the sum of the odd a_i in Eq. (2), and may be interpreted roughly as measuring the equatorial rotation rate. Inaccurate and incomplete as they are, these splittings are nonetheless interesting. The lowest degrees probe the rotation in the deepest parts of the Sun, allowing one to form at least some idea of the rotational profile in and near the energy-producing core. There have been several suggestions (*e.g.* Dicke, 1976) that the Sun's deep interior may rotate significantly faster than its surface. In addition, observations of $l = 1$ modes by Claverie *et al.* (1981) and of low-degree modes by Duvall and Harvey (1984) indicated that rotational frequency splittings may go through a minimum at about $l = 7$, and then increase to perhaps 600 nHz at $l = 1$. Any further information about this range of degrees is therefore of interest.

Table 2 shows the measured splittings (effectively the sum of all the odd a_i) for degrees between 3 and 7, in sidereal nHz, along with estimated errors and the number of multiplets contributing to each splitting measurement. Multiplet identifications proved impossible for degrees less than 3, and were difficult and somewhat tentative for $l = 3$. For degrees higher

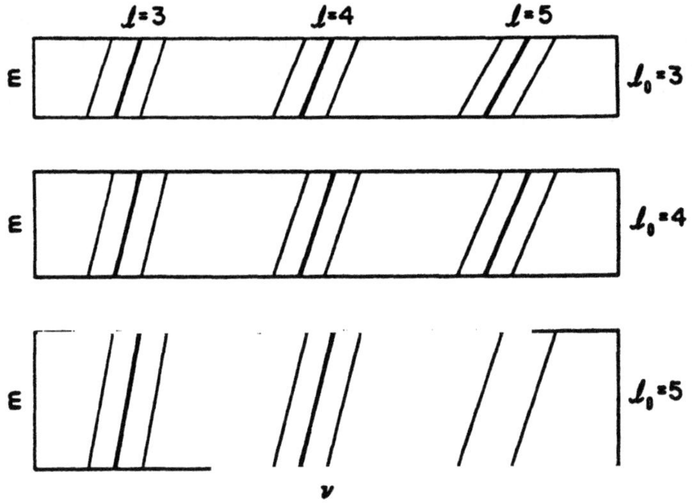

Figure 3. Schematic illustration of $m - \nu$ diagrams, showing effects of leakage from degrees near the nominal degree. Multiplet frequencies (heavy lines) are shown as a function of azimuthal order m, for a range of assumed degrees l and nominal degrees l_0. The light lines flanking each multiplet represent 1/day sidelobes, caused by diurnal gaps in the data.

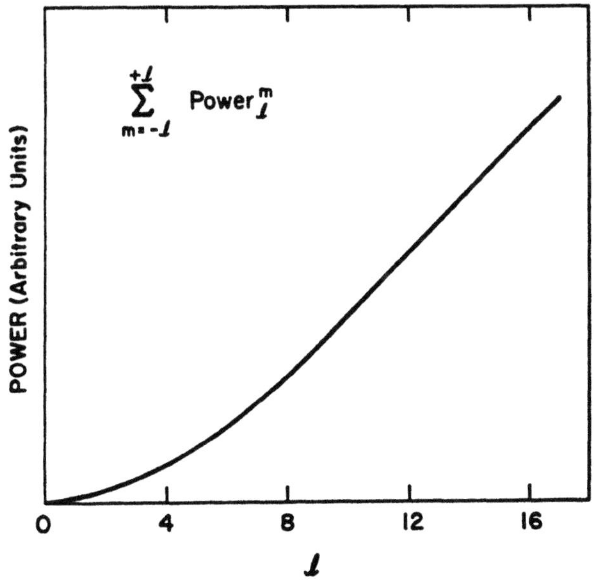

Figure 4. Expected total power per degree plotted against nominal degree, assuming that mode excitation is independent of both l and m, and assuming that the data processing filters out modes with degree less than about 2.

than 3, the identifications appear secure, and several multiplets with accompanying 1/day sidelobes were measured for each degree. All of these splittings are consistent with the surface equatorial rotation rate of 461 nHz (corresponding to magnetic features, Snodgrass 1983), except for $l = 7$, which is significantly low. However, they are also consistent with the idea that the splittings reach a minimum near $l = 7$, and then increase with decreasing degree. Given the large observational errors and the absence of splitting values for $l = 1$ and 2, one certainly cannot rule out the rapid central rotation suggested by earlier observations.

degree	splitting	standard error	number of multiplets
3	399	78	2
4	465	36	4
5	479	22	6
6	462	19	4
7	432	7	6

Table 2. Rotational splittings $\frac{[\nu(l,l)-\nu(l,-l)]}{2l}$ (nHz sidereal), with standard error and number of multiplets measured for each degree.

5. CONCLUSIONS

It will be necessary to perform a suitable inversion using the a_i coefficients in order to determine precisely what $\Omega(r,\theta)$ is most consistent with the a_i, and with what choices for $\Omega(r,\theta)$ they can be reconciled. However, one can obtain a very good preliminary idea by comparing the observed a_3 coefficients with those computed by Christensen-Dalsgaard (1986) assuming various models of the rotation profile. The dotted line on Fig. 2 shows the expected values of a_3 on the assumption that the rotation profile is given by

$$\Omega(r,\theta) = \begin{cases} \Omega_s(\theta) & 0.73R_\odot \le r \le R_\odot \\ \Omega_e & 0 \le r \le 0.73R_\odot \end{cases} \tag{3}$$

where Ω_s denotes the latitudinal variation of rotation observed at the solar surface, and Ω_e is the surface equatorial rotation rate. For this model, therefore, the Sun's radiative interior rotates as a solid body at the surface equatorial rate, and the convection zone has the surface latitudinal differential rotation, independent of depth.

The long-dashed line on Fig. 1 shows the expected a_3 values for a model in which

$$\Omega(r,\theta) = \begin{cases} \Omega_s(r\ \sin\theta) & 0.73R_\odot \le r \le R_\odot \\ \Omega_e & 0 \le r \le 0.73R_\odot \end{cases} \tag{4}$$

In this model, the radiative interior still rotates as a solid body, but within the convection zone Ω depends only on the perpendicular distance from the rotation axis (*i.e.*, rotation is constant on cylindrical surfaces). This model also matches the observed surface rotation at $r = R_\odot$.

It is evident that both of these models fit the observed a_3 values better than does a model in which $\Omega(\theta)$ is independent of radius (the upper dashed line on Fig. 1). Moreover, the a_3 data are clearly inadequate to distinguish between the two models with no latitudinal differential rotation below the convection zone. This inability to distinguish between very different models arises because the response of the a_i to Ω is weighted by the energy density in the oscillation modes, which is strongly peaked near the solar surface. Thus, models that are forced to agree in the neighborhood of the surface always give similar values for the a_i.

The a_5 values are noisier than those for a_3, but they are broadly in agreement with those expected from the surface rotation, at least for $l \geq 20$. This agreement is especially good for $l \geq 80$, *i.e.* for modes that are confined to approximately the upper third of the convection zone, and for which the data are particularly good. This suggests that at least in the upper part of the convection zone, the differential rotation matches that seen at the surface.

Based on these indications, we conclude that
(1) Latitudinal differential rotation is much smaller within most of the radiative interior than at the solar surface.
(2) The latitudinal differential rotation within the convection zone is qualitatively like that seen at the surface, but its detailed variation with depth remains unknown.

The observed a_1 appears to be nearly constant with l, while $a_1 + a_3 + a_5$ increases slightly with increasing l, for $l < 40$. These facts tend to confirm the conclusion of Duvall and Harvey (1984) and Duvall, *et al.* (1984) that the equatorial rotation rate is slightly smaller in the radiative interior than at the surface. They also suggest that the polar rotation rate at depth is larger than the surface value, keeping the latitudinal average nearly constant. One plausible model to explain these results is simply that the convection zone rotates differentially with latitude, and the radiative interior rotates as a solid body at a rate that is intermediate between the surface equatorial and polar rates. However, a more quantitative analysis will be required in order to decide whether the data are consistent with this model, or whether they demand a variation of Ω with radius below the convection zone. Another consideration is that Coriolis forces may modify the oscillation frequencies, opposing the frequency splitting due to advection. A simple estimate of such effects (Brown, *et al.* 1986) suggests that they could decrease a_1 by as much as 30 nHz, for degrees near 10. Theoretical work is needed to relate this process to the apparent dip in a_1 at low degrees.

Finally, we consider the implications of the even-indexed coefficients a_2, a_4. Nonzero values for these coefficients would result from pole-equator asymmetries in the solar structure. In the absence of depth variations, one would expect most such asymmetries to cause frequency shifts at $m = \pm l$ that are independent of l (in contrast to the rotationally-induced frequency shifts, which are proportional to l). Since the a_2 and a_4 coefficients in Fig. 2 are scaled to remove the rotational l-dependence, a true solar structural asymmetry should cause a_2 and a_4 to vary roughly as l^{-1}. Because this functional form reduces the importance of the noise at low degrees, we have extended the l coverage of Fig. 2 to show degrees as low as 11.

Duvall *et al.* (1986) found that, averaged over $20 \leq l \leq 98$, $a_2 = 8$ nHz, a value that they thought to be significantly different from zero. We too find a_2 and a_4 to be systematically positive, though by amounts substantially smaller than those reported by Duvall *et al.* (1986). Averaged over $11 \leq l \leq 99$, $a_2 = 1.5 \pm 0.4$ nHz, and $a_4 = 1.7 \pm 0.3$ nHz. There appears to be some tendency to increasing values at small l: least-squares fits to l^{-1} yield $a_2 = (58 \pm 27) l^{-1}$ and $a_4 = (64 \pm 36) l^{-1}$. It is important to note that both the averages and the l^{-1} fits are heavily influenced by the three points at $l = 15, 16, 17$; removing just these three points leads to substantially different results. For this reason, we believe that the observed a_2 and a_4 coefficients are best interpreted as upper limits, and probably reflect systematic errors in the measurement process.

There are several possible explanations for nonzero values for these coefficients, but for values as large as a few nHz, none of them are very satisfactory. Since modes with $l = 0$ travel over the solar poles, whereas those with $m = \pm l$ travel around the equator, all of these explanations require that the solar structure be different at low and high latitudes. Duvall *et al.* suggested that their observations might reflect some systematic difference in the character of convection near the solar equator, relative to the poles. A distortion of the solar figure could make a_2 nonzero, but for its sign to be positive, the Sun would have to be prolate. Finally, a properly oriented magnetic field could influence a_2, but to give a value as big as 8 nHz, large field strengths are required.

Based on our observations, one can place some limits on the possible magnitudes of these pole-equator inhomogeneities. The spatially-dependent convection hypothesis of Duvall *et al.* (1986) cannot be evaluated without a more specific model. The quadratic part of the relative frequency shift caused by a distortion of the solar shape is, within about an order of magnitude, equal to the relative difference between the polar and equatorial radii. The exact relation depends on how the geometric distortion affects the thermodynamic structure. For example, one can use relations from Gough and Taylor (1984) to estimate this relative frequency shift for a uniformly rotating Sun, with the result that for the interesting range of degrees, $\frac{\delta\nu}{\nu} \approx 0.30\frac{R_e - R_p}{R_e}$, where R_e and R_p are the equatorial and polar radii. If we take the quadratic frequency shift to be less than 55 nHz, so that $a_2 = 5$ nHz at $l = 11$, then the maximum allowed ellipticity of the Sun is about 6×10^{-5} (in the sense of a prolate Sun).

Last, we consider the allowable magnetic field magnitude implied by a_2. As in Brown (1985, 1986), we assume a model similar to that proposed by Dicke (1982), in which there is a large toroidal magnetic field confined within a radius R_c of the solar center. Then one may once again use the results of Gough and Taylor (1984) to estimate the quadratic frequency splitting. The resulting maximum allowed field is $8\text{–}20\times10^6$ G, depending on whether R_c is $0.7R_\odot$ or $0.3R_\odot$. These are not restrictive limits, but it may be possible to set lower bounds by considering other magnetic field topologies than the one chosen by Dicke, or magnetic fields that penetrate into the convection zone.

We thank Kim Streander, Lou Gilliam, and the staff of the Sacramento Peak Observatory Big Dome for their help in obtaining the data, and Peter Bandurian and Dan Haynes for assistance in reducing it.

REFERENCES

Brown, T.M. 1984, in *Proc. Snowmass Conf. on Solar Seismology from Space,* R. Ulrich (ed.), Jet Propulsion Laboratory, Pasadena, p. 157.

Brown, T.M. 1985, *Nature* **317,** 591.

Brown, T.M. 1986, in *Seismology of the Sun and the Distant Stars,* D.O. Gough (ed.), D. Reidel, Dordrecht, p. 199.

Brown, T.M., Mihalas, B.W. and Rhodes, E.J. Jr. 1986, in *Physics of the Sun, Vol I,* P.A. Sturrock (ed.), D. Reidel, Dordrecht, p. 177.

Brown, T.M. and Morrow, C.A. 1986, *Astrophys. J. Letters* (in press).

Claverie, A., Isaak, G.R., McLeod, C.P., van der Raay, H.B., and Roca Cortes, T. 1981, *Nature* **293,** 443.

Christensen-Dalsgaard, J. 1986, in *Cool Stars, Stellar Systems, and the Sun,* M. Zeilik and D.M. Gibson (eds.), Springer-Verlag, Heidelberg, p. 145.

Christensen-Dalsgaard, J., Duvall, T.L., Gough, D.O., Harvey, J.W. and Rhodes, E.J. 1985, *Nature* **315,** 378.

Cuypers, J. 1980, *A. Ap.* **89,** 207.

Dicke, R.H. 1976, *Solar Phys.* **47,** 475.

Dicke, R.H. 1982, *Solar Phys.* **78,** 475.

Duvall, T.L., Dziembowski, W., Goode, P.R., Gough, D.O., Harvey, J.W. and Leibacher, J.W. 1984, *Nature* **310,** 22.

Duvall, T.L. and Harvey, J.W. 1984, *Nature* **310,** 19.

Duvall, T.L., Harvey, J.W. and Pomerantz, M.A. 1986, *Nature* **321,** 500.
Evans, J.W. 1980, in *Solar Instrumentation: What's Next?*, R. Dunn (ed.), Sacramento Peak Observatory, Sunspot, p. 155.
Gough, D.O. 1981, *M.N.R.A.S.* **196,** 731.
Gough, D.O. and Taylor, P.P. 1984, in *Oscillations as a Probe of the Sun's Interior,* L. Belvedere and L. Paterno, (eds.), Memoire della Societa Astronomica Italiana **55,** 69.
Hansen, C.J., Cox, J.P. and Van Horn, H.M. 1977, *Ap. J.* **217,** 151.
Libbrecht, K.G. 1986, *Nature* **319,** 753.
Lynden-Bell, D. and Ostriker, J.P. 1967, *M.N.R.A.S.* **136,** 293.
Snodgrass, H.B. 1983, *Ap. J.* **270,** 288.
Snodgrass, H.B. 1985, *Solar Phys.* **94,** 13.

LATITUDE AND DEPTH VARIATION OF SOLAR ROTATION

T. L. Duvall, Jr.*
Laboratory for Astronomy and Solar Physics
NASA Goddard Space Flight Center
Greenbelt, MD 20771 U.S.A.

J. W. Harvey
National Solar Observatory
National Optical Astronomy Observatories†
P.O. Box 26732
Tucson, AZ 85726 U.S.A.

M. A. Pomerantz
Bartol Research Foundation
University of Delaware
Newark, DE 19711 U.S.A.

ABSTRACT. Spectra of solar intensity oscillations in the degree range ℓ = 20-98 were obtained with a 92% duty cycle over a 50 hour period from the geographic South Pole. The spectra have been analyzed for frequency shifts caused by solar internal rotation as functions of latitude and depth. The results have been published elsewhere and some of the intermediate steps in producing the results are illustrated here.

1. INTRODUCTION

The scientific results that we presented to this meeting are the same as published by Duvall, Harvey and Pomerantz (1986a). In summary, spectra of solar intensity oscillations were obtained from observations made at the geographic South Pole in December 1981. The observations covered 50 hours with 92% continuity. For a given degree, ℓ, between 20 and 98, spectral frequency shifts of p modes as a function of m were determined by a cross-correlation technique. These shifts were then expressed as coefficients of terms in an orthogonal basis series of Legendre polynomials. The coefficients were compared with those derived from observations of surface rotation with the conclusion that latitudinal differential

* Visiting Astronomer, National Solar Observatory.
† Operated by the Association of Universities for Research in Astronomy, Inc., under contract with the National Science Foundation.

B. R. Durney and S. Sofia (eds.), The Internal Solar Angular Velocity, 19–22.

rotation appears to persist at least to the bottom of the convection zone in a form similar to that of the surface. In this extended abstract we illustrate some of the key intermediate steps in producing the results.

2. AN OBSERVATIONAL EXAMPLE

The observations were full-disk images obtained with a 0.6 nm filter centered on the CaII K line and a solid-state detector array having a resolution of ~10 arc s. In the first step of the reduction, raw images were corrected for pixel-to-pixel variations of bias and gain. A few dead pixels were replaced by interpolated values. Figure 1 is one image from near the beginning of the analyzed observing sequence. Each image, such as this one, was the sum of nine 10-s exposures and has a noise level at a typical pixel of about 1 part in 10000.

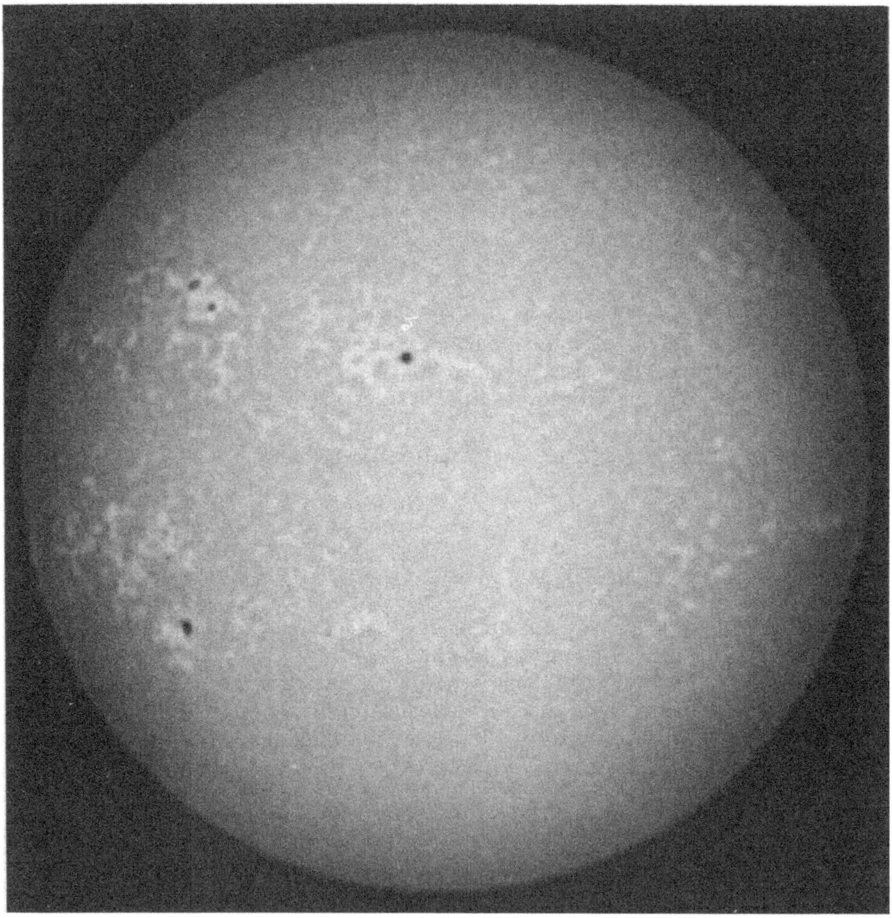

Figure 1. Ca II K-line image obtained on December 24, 1981. This is a 90-s sum of images and has been corrected for instrumental artifacts. North is at the top.

In the next step of the reduction, a low-degree fitting routine removed limb darkening, instrumental variations, and (unfortunately) low-spatial-frequency oscillations from the images. The images were interpolated onto a longitude - sine latitude grid and spherical harmonic coefficients were estimated in the manner pioneered by Brown (1985). Temporal power spectra were estimated in the normal way.

3. AZIMUTHALLY-AVERAGED POWER SPECTRUM

For each spherical harmonic, ℓ, the spectra for the different azimuthal orders, m, were combined with an assumed value of solar rotation removed. This procedure substantially reduced the noise level and revealed for the first time the discrete nature of solar p-mode oscillations in an intermediate degree, ℓ-ν spectrum. The clean appearance of this spectrum (figure 2) is due to the absence of spectral sidelobes that results from the absence of nighttime gaps in the data and to the high signal-to-noise ratio resulting from the azimuthal averaging. The columns of this picture were used in the cross-correlation procedure that produced the rotational splitting results of Duvall, Harvey and Pomerantz (1986a). The same azimuthally-averaged spectrum was also used by Duvall, Harvey and Pomerantz (1986b) to determine accurate mode frequencies.

Figure 2. Azimuthally-averaged ℓ-ν spectrum of p-mode oscillations. The abscissa is ℓ from 6 to 250 (left to right) and the ordinate is frequency from 2238 to 4679 μHz (bottom to top). The frequency resolution of 5 μHz is too coarse to resolve modes with ℓ>100.

4. AZIMUTHAL ORDER - FREQUENCY SPECTRA

Rotation is revealed by frequency shifts of p-modes as a function of m. These frequency shifts are visible in ν-m spectra of the sort shown in figure 3 for a selection of ℓ values. Values of the frequency shifts for a given value of ℓ were measured by cross correlating the rows of the ν-m spectra with the azimuthally-averaged spectrum. The results are published in Duvall, Harvey and Pomerantz (1986a).

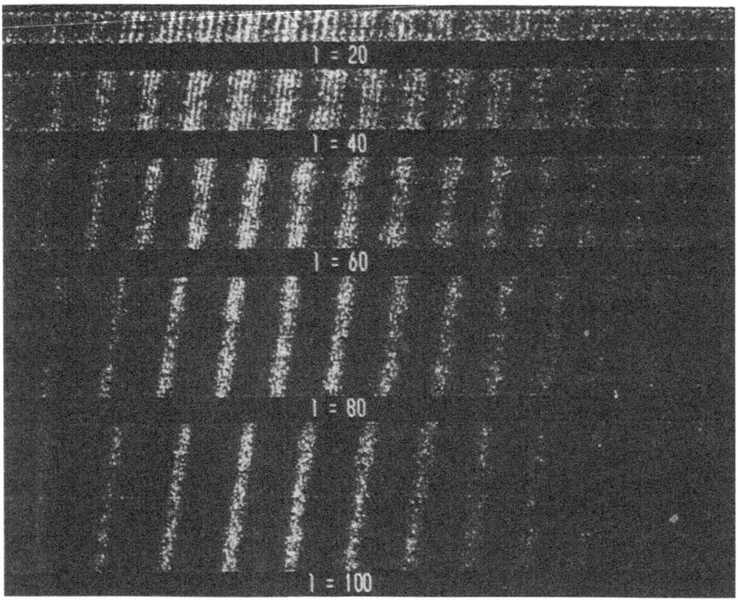

Figure 3. A selection of ν-m spectra for ℓ values of 20, 40, 60, 80, and 100. The ordinate is m and the abscissa is ν from 2273 to 5045 μHz.

5. ACKNOWLEDGEMENT

This work was supported in part by NSF grant DPP81-19627.

REFERENCES

Brown, T. M.: 1985, *Nature* **317**, 591-594.

Duvall, T. L. Jr., Harvey, J. W. and Pomerantz, M. A.: 1986a, *Nature* **321**, 500-501.

Duvall, T. L. Jr., Harvey, J. W. and Pomerantz, M. A.: 1986b, in *IAU Symposium 123, Advances in Helio- and Asteroseismology,* ed. J. Christensen-Dalsgaard and S. Frandsen, (Dordrecht, Reidel), in press.

OBSERVATIONS OF SURFACE VELOCITY FIELDS

Robert F. Howard
National Solar Observatory
National Optical Astronomy Observatories*
Post Office Box 26732
Tucson, Arizona 85726-6732 USA

ABSTRACT - This brief review discusses several recent papers on the motions of individual sunspots and sunspot groups. Measurements of spot areas and positions from Mount Wilson white-light, full-disk plates were carried out for an interval covering most of this century. Rotation and meridional motions were derived from these measures. In addition motions of individual spots within sunspot groups were analyzed. A number of results have emerged which may shed some light on the problem of the linkage of sunspot magnetic flux tubes to subsurface layers and to each other.

The Mount Wilson full-disk, white-light solar plate collection has been measured and these measurements have been analyzed (Howard, Gilman and Gilman 1984) for information about areas and daily surface motions (in longitude and latitude) of individual sunspot umbrae. The interval covered in the original analysis was 1921 through 1982. Later work extended that interval to 1917 through 1984 and included more than 100,000 spot determinations of solar rotation from more than 20,000 plates. The observations continue, at least through the present time, so there is hope of extending the analysis into the future. These measurements, which required about two years of effort, constitute a valuable set of data for the study of solar surface motions. Spot area measures were made at the same time, along with precise diameter determinations, but these results have not been analyzed in detail yet.

The purpose of this program has been to study the daily positions of individual sunspots in order to examine in detail the properties of longitude (rotation) and latitude (meridional) motions of these spots. In the past, nearly all sunspot rotation or meridional flow studies have dealt with sunspot groups instead of individual sunspots. While the group information is certainly important, another important possible tracer in the search for clues about surface dynamics of the sun is the individual sunspots. In principal they may give different results than one obtains from the groups because they may be linked in a subtly different way to the surface gas. They have the advantage that there are more of them than there are spot groups, but, of course, they are harder to measure because they sometimes have irregular shapes and, in addition, their short lifetimes do not lead to ease in making daily position measurements for rotation.

*Operated by the Association of Universities for Research in Astronomy, Inc. under contract with the National Science Foudation.

B. R. Durney and S. Sofia (eds.), The Internal Solar Angular Velocity, 23–26.

There are plans to extend and complement the data set by adding measurements of at least one other long, high quality solar plate collection. This will enable us to verify some results that are close to the noise limit, such as the variation of the rotation rate of individual sunspots with phase in the activity cycle. Also the use of data originating in a different geographical location will enable us to fill in many of the gaps in the existing data due to cloudy weather at Mount Wilson and also to add a large number of small sunspots whose lifetimes are less than one day. Altogether the data set that will result from this combined effort should be significantly more complete and longer than any such set now available from any one site.

From the data analyzed so far a number of results have been derived. Differential rotation, of course, is clearly seen in the results. The overall group results are quite similar to those of Newton and Nunn (1951). Individual spots rotate faster than groups by about 1%. The largest spots rotate slower than small spots by about 2%. This appears to be a real effect, although some authors attribute this to a real difference (decrease) in rotation rate with spot age (Balthasar, Schüssler, and Wöhl 1982). The discrepancy in the spot and group rates results from the way in which the group positions are determined. We use the area-weighted position for each day to determine the daily rotation motion. Thus the larger, slower moving spots dominate in these position determinations, and the groups tend to have a rotation rate slower than the average of the individual sunspots.

A correlation of individual sunspot rotation rate with the solar cycle phase was noted in the results after correcting for the effects of variations due to latitude drift of the spot zone during each activity cycle (Gilman and Howard 1984a). On the average the rotation rate of sunspots increases by about 1% near sunspot minimum. A smaller increase near the maximum phase is also observed. A roughly similar pattern is seen for the Mount Wlson Doppler rotation results in the interval for which they are available for comparison (1967 - 1982). This agreement raises questions regarding the relationship between the subsurface rotation of the plasma and the fields. For example, do the changes in the rotation rate of the plasma and the fields take place at the same depth, or are the plasma changes more shallow, and does a shallow plasma effect then tend to sweep the fields along with it? Overall, because compensating changes at high latitude are not observed in the Doppler data, these variations must represent a transfer of angular momentum between the surface and deeper layers.

We examined separately the rotation rates of leader and follower sunspots -- defined as spots west or east respectively of the area-weighted, east-west centroid of the group. Leader spots on average rotate faster than follower spots by about 0.1% (Gilman and Howard 1985). This difference appears to be independent of phase in the activity cycle. The variations in rotation rate from group to group are greater or comparable to this figure, so this implies a physical connection between the leading and following spots that is greater than that between neighboring groups. These results came as no surprise to anyone. The separation of the leading and following parts of sunspot groups during the evolution of the group is a well-known phenomenon (e. g. Brunner 1913) and cannot be separated phenomenologically from pure rotation.

The nature of the connection of the sunspot magnetic flux tubes to the subsurface layers is also addressed by the results of a study of the rotation rates of individual spots within a group (Gilman and Howard 1986) using the same data set. In this investigation we looked at the differential rotation of spots, and found that within a group the spots obey a rotation relationship with latitude that closely approximates that of the groups themselves. This suggests that if the spot flux tubes are connected to a common flux tube bundle below the surface, then their surface rotation properties are strongly influenced by that of the surrounding plasma. An additional result from this study was that the primary relative motions of leader and follower spots in a group can be characterized as a simple expansion of the group along its major axis. We detected little systematic rotation of the pattern about group center. Although on rare occasions rotation of groups about a vertical axis at a substantial rate is observed for some time period, this is not an effect that shows any systematic properties. On average spot groups do not rotate about their own axes.

The daily meridional motions of individual sunspots were also studied (Howard and Gilman 1986). This is a much more difficult observational problem than the rotation determinations because the amplitude of these motions is around three orders of magnitude smaller than that of the rotation rate, so that small errors of measurement have much more influence on the results. In general the individual measurement accuracy in this study is lower than that in most previous works, but the number of spots measured is significantly greater. The meridional motions averaged over both the north and south hemispheres shows a smooth progression, with poleward motions at the highest latitudes and equatorward motions at lower latitudes. This is in general agreement with several previous results (Richardson and Schwarzschild 1953; Tuominen 1976). Viewed separately in each hemisphere, however, there is a mid-latitude northward flow in each hemisphere with an amplitude of a few hundredths of a degree per day. Leader and follower sunspots seen separately show substantially the same behavior, which lends some credence to the result, although a northward flow in both hemispheres is a rather strange phenomenon, which does not link with any other solar behavior that we are aware of. It is difficult to imagine any sort of instrumental effect that might cause such a result. In general such effects, or perspective effects such as the tilt of the solar equatorial plane to the ecliptic, should cause false patterns that are symmetric about the equator in long-term averages, not antisymmetric. Thus the source of the apparent mid-latitude northward flow remains unknown.

Since we have both longitudinal and latitudinal motion data, we are able to examine the correlation between these two quantities (Gilman and Howard 1984b). For spot groups our covariances between these quantities are similar to Ward's (1965). But for individual spots the covariance is reduced by a factor of about 3 from the Ward value. The correlation for spots is still in the same direction and still significant, however. We conclude that the correlation between longitudinal motions of individual sunspots is real and implies that there is angular momentum transport toward the equator as inferred by Ward. The unpublished discussion at the time of the Ward paper between Ward and Leighton concerning the role that the known expansion of spot groups along an inclined axis plays in the correlation result is now settled in part in favor of Leighton. This is a measurable effect. But the residual seen in individual spots is still a real effect. The amplitude we have found implies that it would take 6 months to a year to build up the observed equatorial acceleration; hence, whatever is braking differential rotation can be no more vigorous than that.

This study of sunspot motions has proved to be a very valuable project. Further work with an enhanced data set should provide still more information about solar motions and perhaps more clues about the large-scale dynamics of the solar interior.

REFERENCES

Balthasar, H., Schüssler, M., and Wöhl, H., 1982, *Solar Phys.*, **76**, 21.

Brunner, W., 1913, *Publ. Zurich Obs.*, **5**, 1.

Gilman, P. A., and Howard, R., 1984a, *Astrophys. J.*, **283**, 385.

Gilman, P. A., and Howard, R., 1984b, *Solar Phys.*, **93**, 171.

Gilman, P. A., and Howard, R., 1985, *Astrophys. J.*, **295**, 233.

Gilman, P. A., and Howard, R., 1986, *Astrophys. J.*, **303**, 480.

Howard, R., and Gilman, P. A., 1986, *Astrophys. J.*, **307**, 389.

Howard, R., Gilman, P. A., and Gilman, P. I., 1984, *Astrophys. J.*, **283**, 373.

Newton, H. W., and Nunn, M. L., 1951, *Mon. Not. Roy. Astron. Soc.*, **111**, 413.

Richardson, R. S., and Schwarzschild, M., 1953, *Problemi della Fisica Solare*, (Rome: Accademia Nazionale dei Lincei), p. 228.

Tuominen, J., 1976, *Solar Phys.*, **47**, 541.

Ward, F., 1965, *Astrophys. J.*, **141**, 534.

SPACELAB EXPERIMENTS ON CONVECTION IN A ROTATING SPHERICAL SHELL WITH RADIAL GRAVITY

Juri Toomre and **John E. Hart**
Department of Astrophysical, Planetary and Atmospheric Sciences,
and Joint Institute for Laboratory Astrophysics,
University of Colorado, Boulder, Colorado 80309-0440, USA

Gary A. Glatzmaier
Earth and Space Sciences, ESS5, Los Alamos National Laboratory,
Los Alamos, New Mexico 87545, USA

ABSTRACT — Experiments on thermal convection in a rotating, differentially-heated hemispherical shell of fluid with a radial gravity field were carried out in the microgravity environment of Spacelab 3 which was flown on the space shuttle Challenger in May 1985. Such global-scale convection interacting with rotation has bearing on flows within the Sun and the giant planets, all of which rotate and possess deep shells of convection. The radial force of gravity is modelled by imposing a strong electric field across the shell, with dielectric polarization forces producing radial accelerations proportional to temperature. The thermally-driven circulations result from imposed radial and latitudinal temperature gradients on the bounding surfaces. When the rotation is rapid and its influence strong, the experiments with spherically-symmetric heating yield north-south oriented columnar convection ("banana cells") in equatorial and sub-equatorial regions. As the radial heating is increased, these roll-like cells interact with mid-latitude waves, ultimately being destroyed by more isotropic convection that moves down from the pole. When a significant equator-to-pole temperature difference is imposed, "spiral waves" develop on top of a strong meridional circulation. Intricate, nonaxisymmetric, convective patterns that propagate in longitude and evolve in time are observed, including some "triangular waves" that couple mid-latitude and equatorial disturbances. Schlieren visualizations of these laboratory flows are compared briefly to three-dimensional nonlinear simulations that can be conducted at the more modest heating rates.

1. INTRODUCTION

Global-scale flows within the convection zones of rotating stars like the Sun are substantially controlled by a combination of buoyancy and Coriolis forces. These forces must also have a significant role in shaping the circulations observed in the atmospheres, and probably extending deep into the gaseous interiors, of the giant planets Jupiter, Saturn and Uranus. Although the energy sources for driving convection within these planets differs from that of main-sequence stars, a property common to all these objects is that they are spherical and they rotate. As a consequence, the horizontal component of the Coriolis forces varies with latitude. This has long been thought to be a key ingredient in determining the differential rotation of the

B. R. Durney and S. Sofia (eds.), The Internal Solar Angular Velocity, 27–44.

convective envelope of the Sun, the structure of its large-scale convection cells and of possible magnetic dynamo action, the formation of jet streams and mid-latitude waves on Earth, and the strong zonal flows and cloud band orientations on the giant planets.

The dynamics of all these thermally-driven flows must respond both to the sphericity and to the latitudinal variation of Coriolis force. Yet such effects have been difficult to model properly in the terrestrial laboratory. In planets and stars the rotation ($\boldsymbol{\Omega}$) and gravitational acceleration ($\boldsymbol{g}$) vectors range from parallel at the pole to perpendicular at the equator, whereas in the laboratory these vectors must be parallel to avoid oscillatory gravity in the rotating reference frame. Previous laboratory experiments on thermally-driven global circulations have been confined either to studies of convection near the poles where $\boldsymbol{\Omega}$ and $\boldsymbol{g}$ are nearly parallel and the geometry is effectively flat (Rossby 1969), or to studies of motions close to the equator where $\boldsymbol{\Omega}$ and $\boldsymbol{g}$ are nearly perpendicular, with the effects of gravity being simulated by centrifugal buoyancy forces associated with very rapid rotation (Busse & Carrigan 1976, Carrigan & Busse 1983). The latter experiments have permitted study of the properties of convection in equatorial and mid-latitude regions of very rapidly rotating objects. In particular, such experiments confirmed that the preferred orientation of convection is along the axis of rotation in the form of "banana cells", as had been predicted earlier by theory (Durney 1970; Busse 1970, 1973; see also Gilman 1986). Yet the centrifugal convection experiment cannot consistently address interactions of polar and equatorial convection modes, nor be used to study the evolution of the global-scale convective structures that occur at low rotation rates and which span large extents in latitude, since all these processes depend crucially on the varying orientation of $\boldsymbol{g}$ and $\boldsymbol{\Omega}$ as one moves from equator to pole.

Our laboratory experiments with electroconvection permit us to consistently simulate the full latitudinal dependence of Coriolis forces on spherical surfaces. We have studied convection within a rotating spherical shell of fluid, using electrohydrodynamic polarization forces to generate the required *radial* buoyancy forces. Within a dielectric fluid with an imposed electric field and thermal gradients, there will be temperature-dependent polarization forces present in addition to normal buoyancy. However, the electrostatic "radial gravity" resulting from the polarization forces is relatively weak; the properties of available dielectric fluids limit the polarization forces to be typically one-tenth of the value of terrestrial gravity, g. Therefore experiments that are not to be contaminated by effects of the unidirectional g need to be carried out in a microgravity laboratory. We conducted such electroconvection experiments using our instrument, the "Geophysical Fluid Flow Cell" (or GFFC), during the flight of the Spacelab 3 microgravity laboratory onboard the space shuttle Challenger in May 1985.

Our scientific goal with the GFFC experiment, in conjuction with numerical simulations, has been to address some of the more basic questions about convection in simple spherical cavities. The objective is to study global-scale convection interacting with rotation, for differential heating distributions varying with both radius and latitude. Namely, how does convection redistribute angular momentum within shells of fluid which are unstably stratified? Does the resulting differential rotation have varying forms depending upon the magnitude of imposed latitudinal gradients in temperature? What are the cell shapes of large-scale convection as the rotational constraint and the differential heating are varied to span both high and low rotation rates, while considering both spherically-symmetric and latitudinally-varying differential heating? Such basic questions are central to understanding motions within both stars like the Sun and the giant planets.

2. THE EXPERIMENT

The motion of an incompressible fluid is governed by the vorticity equation, obtained by taking the curl of the momentum equations. Within a dielectric liquid subject to imposed electric field gradients, the non-conservative part of the polarization forces contribute to vorticity generation, much as do the normal buoyancy forces. As shown in Hart (1976), this generation is proportional to

$$\text{curl (body force)} = \alpha \nabla T \times \nabla \Phi + \frac{\varepsilon\gamma}{2} \nabla T \times \nabla \boldsymbol{E}^2, \tag{1}$$

with α the thermal expansivity, T the temperature, Φ the geopotential, $\boldsymbol{E}$ the electric field, ε the average dielectric constant of the liquid, and γ the temperature coefficient of the dielectric constant. In usual laboratory experiments $\nabla\Phi$, or g, is unidirectional, but here the $\nabla \boldsymbol{E}^2$ term can be used to impose the desired radial geometry of the gravity field. Indeed, a simulation of thermal buoyancy in a radial geopotential can be achieved when an electric field $\boldsymbol{E} \simeq r^{-2}\hat{\mathbf{r}}$ is generated in the liquid by applying a uniform electric potential across two concentric spherical surfaces. The resulting electrostatic "gravity" , which is proportional to $\nabla \boldsymbol{E}^2$, falls off like r^{-5} instead of r^{-2} as one might prefer for geophysical modelling. This issue has been addressed by Hart & Toomre (1977) by considering the linear convective instability of spherical shells with varying gravity distributions. For the aspect ratio of our experiment, the linear eigenfunctions are almost identical, suggesting that the qualitative structure of the observed convective motions achieved with the electrostatic gravity would carry over to an inverse square law gravity.

The GFFC experiment has been described briefly in Hart *et al.* (1986) and in more detail in Hart, Toomre & Glatzmaier (1986). As sketched in Fig. 1, the experiment is based on a differentially-heated hemispherical shell of dielectric fluid. The fluid, a silicone oil of low viscosity, is contained between two hemispherical shells that form part of a spherical capacitor. The inner bounding shell at radius $R_i = 2.402$ cm is made of polished nickel, and the outer at radius $R_o = 3.300$ cm is a transparent sapphire dome about 1 cm thick. The electrostatic radial gravity is achieved by imposing an alternating electrical potential V_o across the fluid shell. The field reverses in a period short compared to the charge relaxation time for the fluid, and this ensures an absence of electric currents. The cell assembly, including the high-voltage supply (to produce V_o as high as 10 kV rms) and thermal control circuitry, is mounted on a turntable that rotates at constant but adjustable rates Ω. At its most rapid rotation, the turntable makes one revolution in 2 seconds; faster rates are not chosen in order to ensure that centrifugal forces are negligible compared to buoyancy forces associated with the electrostatic gravity. The thermal boundary conditions are axisymmetric, but allow for variations of both inner (T_i) and outer (T_o) surface temperature with latitude θ. Although uniform differential

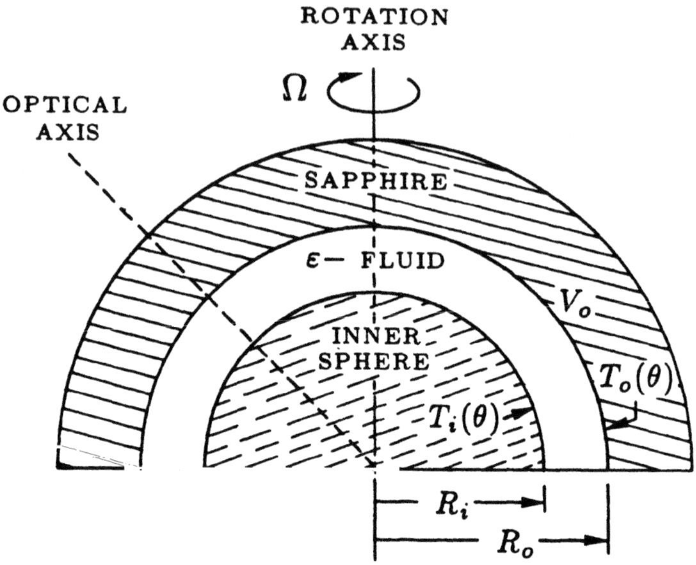

Figure 1. Cross-section of the experimental cell used to study global-scale convection under the effects of rotation Ω in a hemispherical shell of fluid in the presence of an electrostatic radial gravity field. That gravity is achieved by imposing an alternating electrical potential V_o across the fluid shell. The dielectric silicone working fluid is bounded on the outside by a transparent sapphire dome to permit optical probing of the flows and thermal structures that result from differential heating.

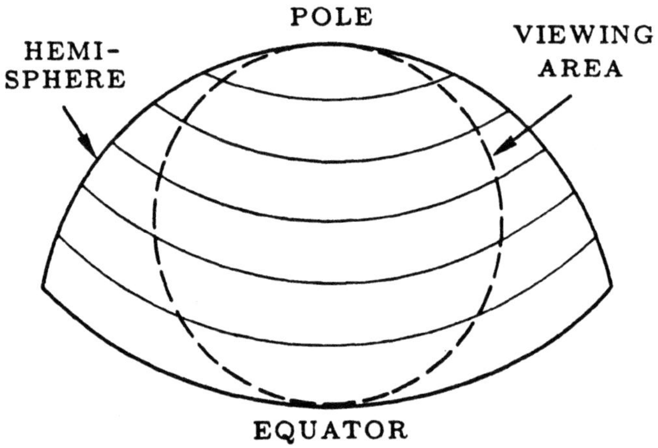

Figure 2. Sketch of the Schlieren viewing area with respect to the hemispherical shell. All the photographs to follow show a region extending from the pole (top) to the equator (bottom), and cover about a quadrant in longitude; the basic rotation is from right (west) to left (east). Lines of constant latitude are indicated in 15° increments; to help define the scale of the viewing area, that at 30° encompasses about 110° in longitude.

heating, with $T_i > T_o$ and independent of θ, is the simplest and most relevant case for modelling objects with spherically symmetric forcing, several experiments were run with a hotter pole. These zonally-symmetric boundary temperatures are maintained by computer-controlled heaters and coolers.

Since the hemispherical shell of fluid is bounded on the outside by a transparent sapphire dome, the convective flow patterns are readily visualized by back-focus Schlieren and by shadowgraph photography. A non-rotating optical train focuses light from a grating onto the inner sphere. Thermally induced refractive index variations in the fluid cause individual rays to bend. When the incident rays are compared with the reflected rays, various measures of the radially-averaged temperature field in the fluid are obtained. Because of space and weight limitations the optical system does not image the whole hemisphere. As indicated in Fig. 2, the data consists of photographic images recorded on 16mm color film which depict a circular view extending from equator to pole and covering about a quadrant in longitude; the control computers were programmed to provide various sequences of Schlieren and shadowgraph images that sampled different longitudes. The experiments yielded over 50,000 images during the 110 hours of operation of the GFFC on Spacelab 3.

There are several non-dimensional parameters that serve to define the convective flow regimes. The Prandtl number $P_r =$ (viscosity ν)/(thermal diffusivity κ) $= 8.4$ is fixed, as is the aspect ratio $R_i/R_o = 0.73$. One of the variable parameters is the Taylor number

$$T_a = \frac{4\Omega^2(R_o - R_i)^4}{\nu^2}, \tag{2}$$

which provides a measure of the rotation effects. For the most rapid rotation of the turntable (period of 2 s), $T_a = 6.3 \times 10^5$. Another variable is the Rayleigh number, which indicates the degree of thermal forcing with respect to dissipation, and is defined as

$$R_a = g_o \frac{\gamma \Delta T_r (R_o - R_i)^3}{\nu\kappa}, \tag{3}$$

where g_o is the equivalent electrostatic gravity at $r = R_o$ and $\Delta T_r \equiv T_i - T_o$ is the radial temperature difference across the fluid at the equator. Only convectively unstable cases with $T_i > T_o$ everywhere are considered here. The GFFC experiment permitted the study of a wide range of slowly and rapidly rotating convection, with Taylor numbers from 0 to 6×10^5 and Rayleigh numbers from zero to over 10^6. The latter value is well over one hundred times critical at Taylor numbers less than about 10^6.

We have investigated several cases with uniform radial differential heating, as well as others with combined radial and latitudinal differential heating. The latter permit us to assess some of the effects of latitudinal gradients on radially unstable stratifications. Such non-uniform heating is likely to be of relevance to rapidly rotating stars, giving rise to meridional circulations whose non-axisymmetric instabilities are not well understood. The various configurations can be distinguished by the dimensionless heating parameter

$$H = \frac{\Delta T_\theta}{\Delta T_r}, \tag{4}$$

which compares the imposed pole-to-equator temperature difference at the boundaries to the radial temperature difference. In our experiments H is either zero (spherically symmetric heating) or positive (with relatively warmer inner and outer poles). This latter preference was required by the design of the optics; a polar cooler would otherwise obscure the visualizations. Further, we thought that such hot inner-pole objects might be of interest in the study of Jovian dynamics, as suggested by the observed heat balance for Jupiter (Ingersol 1976). Indeed, theoretical studies have shown that the spatial patterns or planforms selected by global convective modes can be strongly influenced by latitudinal gradients in the applied heating (Hathaway, Toomre & Gilman 1980), and this may have some relevance to the structure of the cloudbands on Jupiter.

3. SUMMARY OF RESULTS

We have sampled a wide range of R_a, T_a and H with the GFFC experiment, attempting to get a sense of the planforms selected by the convection, and their apparent stability, as the boundary conditions are changed to drive progressively more nonlinear flows under various rotational constraints. In wandering through such multi-dimensional parameter space, we selected regions that appeared to be distinctive and promising based on the linear instability analyses and nonlinear simulations that we had carried out prior to programming the experiment scenarios for the flight. Our surveys were intended to dwell only briefly on the mildly nonlinear flows achieved near the onset of various classes of convection, and to then proceed systematically to the strongly nonlinear flow regimes for which our nonintrusive optical probing had been optimized. Some of the surveys were repeated, others reversed in the order in which the parameters were incremented, to get some feeling for the robustness of the resulting convective patterns and heat transports.

Typical scenarios occupied about 3 hours of experiment time, during which specific radial and latitudinal temperature gradients were maintained on the boundaries, while gradually incrementing the rotation rate Ω (and thus T_a) and the voltage V_o (and thus the radial gravity g_o, and so R_a), sampling of order 20 to 30 values during a given run. This approach generally yielded flows that had equilibrated before being sampled in detail by bursts of Schlieren images or shadowgraphs, for the adjustments to changes in Ω and V_o were much more rapid than the more gradual responses to increments in ΔT_r or ΔT_θ. Although the sampling in parameter space is quite sparse, we are pleased that most scenarios yielded flows of considerable diversity in which the rich changes in pattern selection could be readily traced. There also arose a number of dynamical features which to us were fully a surprise. We shall next display a brief sampling of the types of flows that were realized, and comment on what we have generally inferred from these. More detailed spectral analysis and interpretations of some of the convective flows, along with results obtained from numerical simulations, are presented in Hart, Glatzmaier & Toomre (1986).

3.1 Soccer balls and finer tessellations: slow rotation, $H \simeq 0$.

Figure 3 presents a sequence of Schlieren images obtained when imposing only a uniform radial temperature gradient ($H \simeq 0$) across the fluid shell. Here the

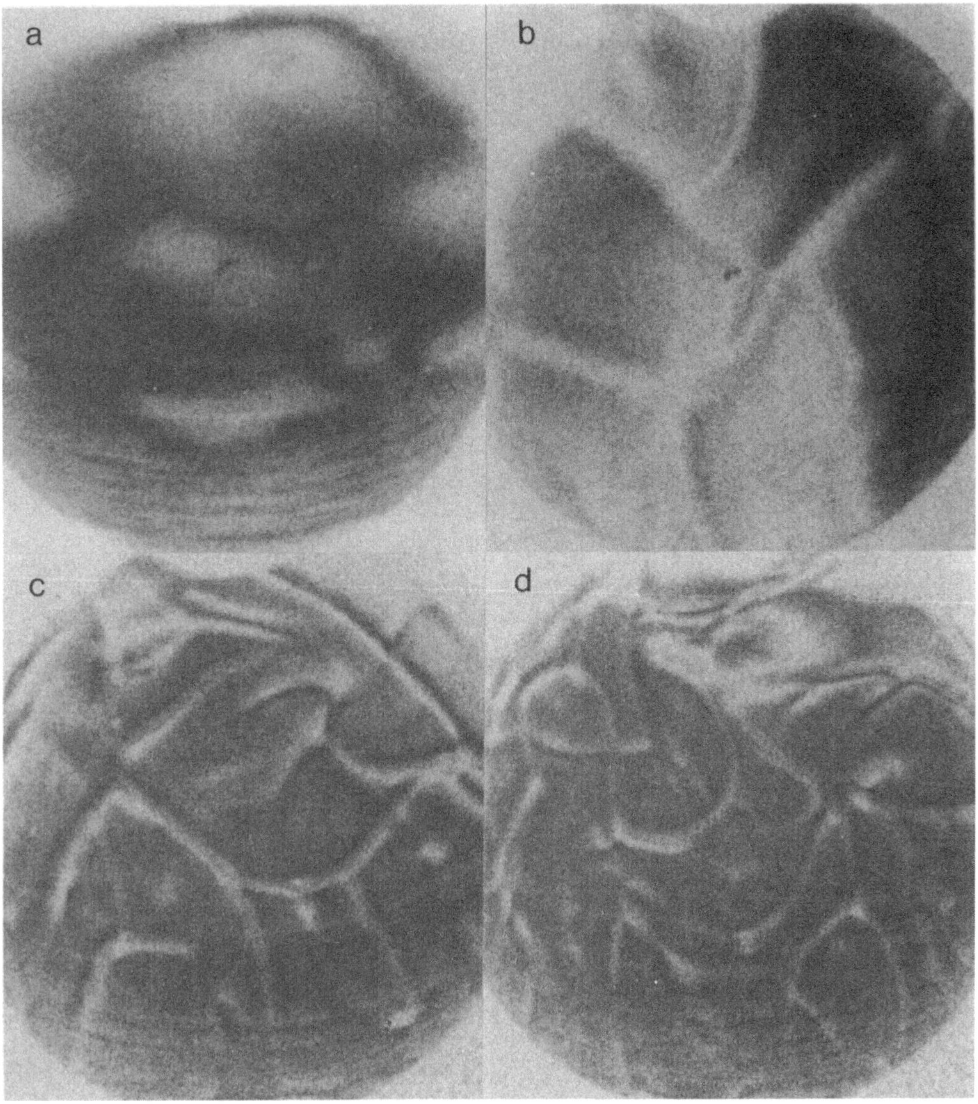

Figure 3. Planforms for convection as observed with Schlieren photography in experiments with a uniform radial temperature difference (about 10°C) imposed between the inner and outer spheres ($H \simeq 0$) and a weak rotational influence (48 s rotation period). The convection has a polygonal pattern with no preferred orientation, appearing at onset rather like a soccer ball. At higher R_a (achieved by increasing the voltage) the scale of the convection cells shrinks, their boundaries become sharper and finer tessellations appear, and the flows become vigorously time dependent. Here $T_a = 1.1 \times 10^3$ and $H \simeq 0$. (a) $R_a = 1.8 \times 10^3$, (b) $R_a = 9.4 \times 10^3$, (c) $R_a = 1.2 \times 10^5$, (d) $R_a = 2.5 \times 10^5$.

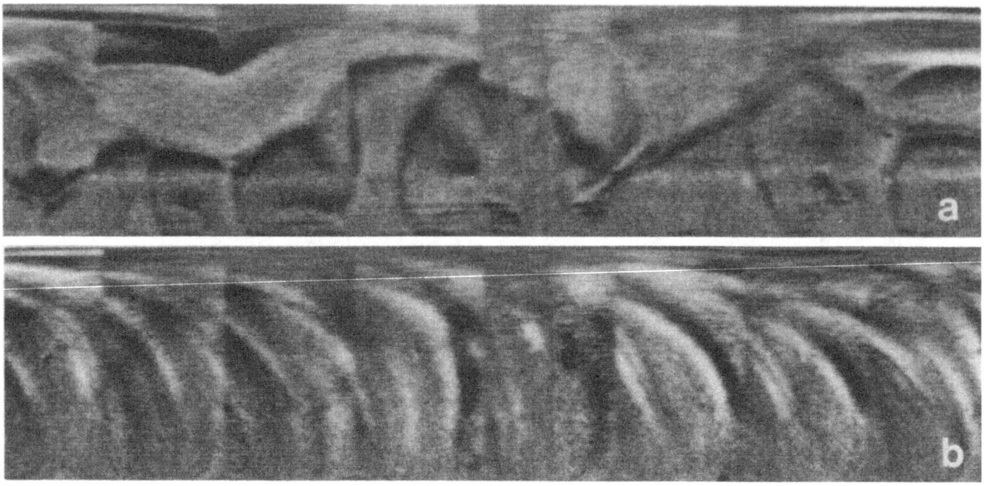

Figure 4. Unfolded Mercator projections of the Schlieren data combined from many frames to illustrate the planforms of convection over the entire hemisphere. The polar regions are at the top, the equatorial regions at the bottom, and a full 360° in longitude is spanned right (west) to left (east). The polygonal patterns in (a) are as in Fig. 3b, the spiral waves in (b) as in Fig. 8a. The perceptible breaks between image segments in such a projected montage arise because the Schlieren responds to gradients that are cartesian in the image plane, and these do not simply coincide with lines of latitude and longitude.

rotational influence is weak, with the slow rotation (48 s period) yielding a modest $T_a = 1.1 \times 10^3$; there is no preferred direction imposed on the convection cells. The very soft pattern of weak thermal fluctuations close to the onset of convection (Fig. 3a) is replaced by more distinctive cells (Fig. 3b) as R_a is increased. The slowly evolving tiling pattern consists of narrow updrafts of hot fluid that form sheets at the cell boundaries; broad regions of cool, sinking flow make up the cell interiors. Such polygonal patterns, possessing no preferred orientation for the cell boundaries, are reminiscent of "soccer balls", noting that the boundaries intersect in triplets. These polygonal cells are similar to those obtained by Busse (1975) from theoretical analysis of planform selection in nonrotating spherical shells. A global view of our soccer ball pattern is shown in Fig. 4a, which presents a Mercator projection of Schlieren data for the entire hemisphere by forming an unfolded composite of images like those in Fig. 3b.

We are presently uncertain what produces the pronounced asymmetry of narrow upflows and broad downflows. Such symmetry breaking may be a consequence of the electrostatic gravity being strongest near the inner boundary. Yet some of the nonlinear simulations of convection that we have carried out with both r^{-2} and r^{-5} gravity show similar asymmetry in the radial motions. This either suggests that the symmetry breaking is readily achieved with many functional forms, or possibly that it is intrinsic to nonlinear interactions within spherical shells. Further theoretical work is needed to resolve such basic questions about planform selection for these relatively simple nonlinear flows.

The convection becomes more complicated with increasing R_a. Figures 3c,d show that the horizontal scale of the cells has decreased sharply and that thermal gradients at the cell boundaries are more pronounced; the patterns now possess many finer tessellations. By examining sequences of such images in time, we find that the patterns change rapidly; what are displayed here are individual snapshots of quite turbulent flows for these higher R_a. The motions appear to be horizontally isotropic, with no evident pattern propagation with longitude.

3.2 Banana cells and high-latitude modes: rapid rotation, $H \simeq 0$.

Turning to more rapid rotation while still retaining only a radial temperature gradient ($H \simeq 0$), we observe that for moderate to high T_a the convection appears first as columnar cells aligned with the rotation axis, and that these cells propagate slowly in longitude. Increasing R_a leads to the variety of patterns shown in Figs. 5 and 6. These modes are roughly in accord with what is expected from the Taylor-Proudman constraint if the thermal forcing is not too large: the motion should be invariant in the direction of $\boldsymbol{\Omega}$. As seen in Fig. 5a, which is a Schlieren picture sensitive to nearly east-west gradients, the cells extend from the equator to somewhat north of the critical latitude of 43°. (At that latitude a cylinder tangent to the inner sphere and parallel to $\boldsymbol{\Omega}$ intersects the outer sphere.) These low-latitude disturbances are called "banana cells" because the columns have to bend to conform to the spherical boundaries. Such banana-cell patterns had been deduced from a variety of theoretical calculations (Durney 1970; Busse 1970, 1973; Gilman 1975; Glatzmaier & Gilman 1981) and observed in multi-mode numerical simulations (Gilman 1978; Glatzmaier 1983, 1985).

Extended sequences of shadowgraph images for the parameter values of Fig. 5a reveal that the banana cells propagate slowly in a prograde direction (at only about 0.5° per basic rotation). This sense of propagation for columnar disturbances at low latitudes is as has been argued by Busse & Cuong (1977) from considering how fluid columns, or vortex lines, parallel to $\boldsymbol{\Omega}$ should respond (while conserving angular momentum) when displaced radially and stretched or contracted in length by the geometry of the shell. They similarly deduced that disturbances poleward of the critical latitude should propagate instead in a retrograde sense. Indeed, as the Rayleigh number is increased, such a switch of the sense of propagation with latitude is apparent from movie sequences of our shadowgraphs. As is suggested by Figs. 5b,c,d, with increasing R_a there appear prominent mid-latitude and polar disturbances, suggesting that new classes of modes at higher latitudes have become unstable as well as the low-latitude columnar cells. These features at higher latitudes possess a weakly retrograde motion, while the banana cells in equatorial regions retain their prograde motion. The interaction between these different modes probably accounts for the wavering and vacillations now apparent in the banana cells, and progressive erosion of the northward tips of those columnar cells as R_a increases. The upwelling and downwelling flows within the banana cells appear to be increasingly concentrated into radial jets. The shadowgraph sequences also show distinct eddy-like disturbances from the polar regions propagating rather

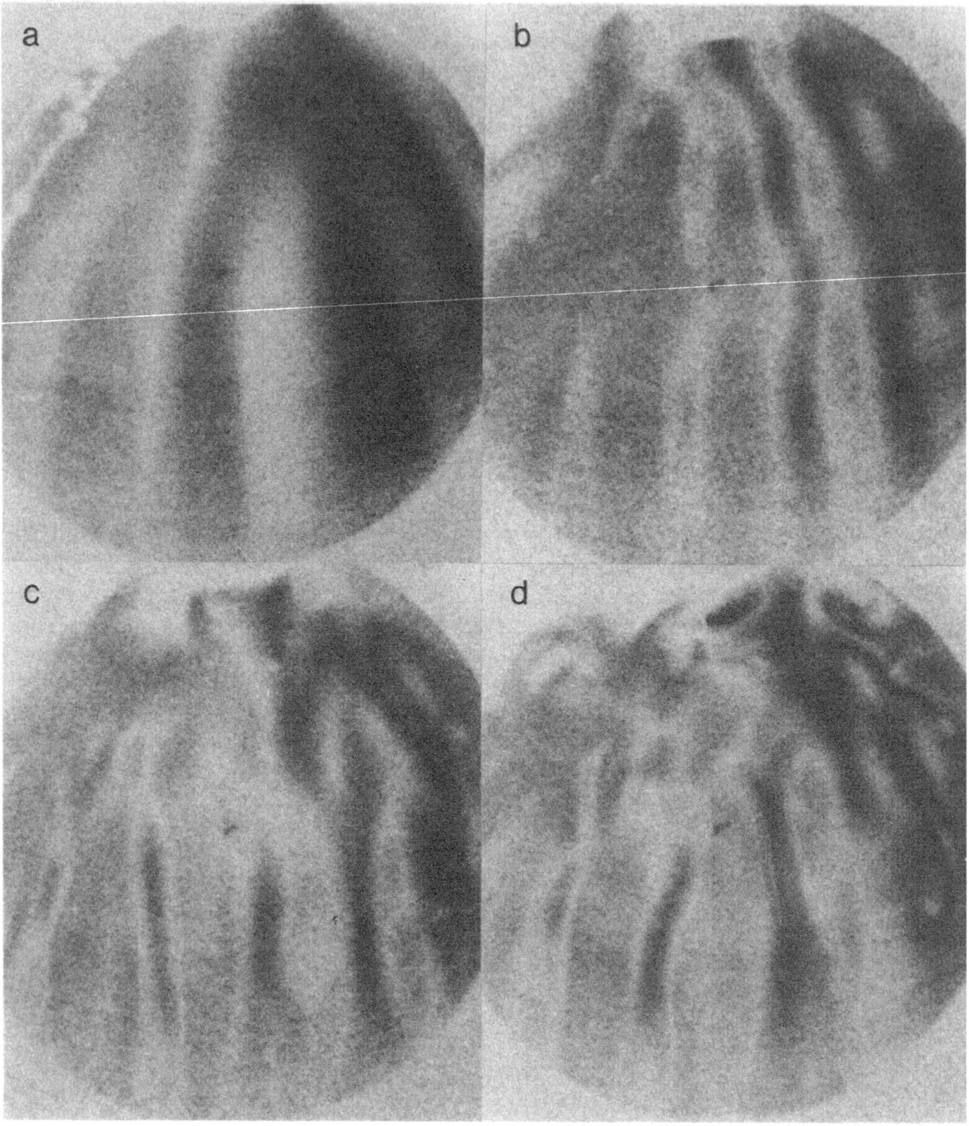

Figure 5. As in Fig. 3 with $H \simeq 0$, but rotational effects are emphasized, leading to motion in the form of slowly propagating columnar convection, or banana cells, aligned with the rotation axis. At first the cells extend from the equator poleward to somewhat north of the critical latitude of 43°; increasing R_a and T_a leads to a wavering of the banana cells, with erosion of their northward tips as polar modes become more prominent. (a) $R_a = 4.7 \times 10^3$, $T_a = 3.9 \times 10^4$ (8 s rotation period), (b) $R_a = 5.2 \times 10^4$, $T_a = 6.3 \times 10^5$ (2 s rotation period), (c) $R_a = 7.6 \times 10^4$, $T_a = 6.3 \times 10^5$, (d) $R_a = 8.6 \times 10^4$, $T_a = 6.3 \times 10^5$.

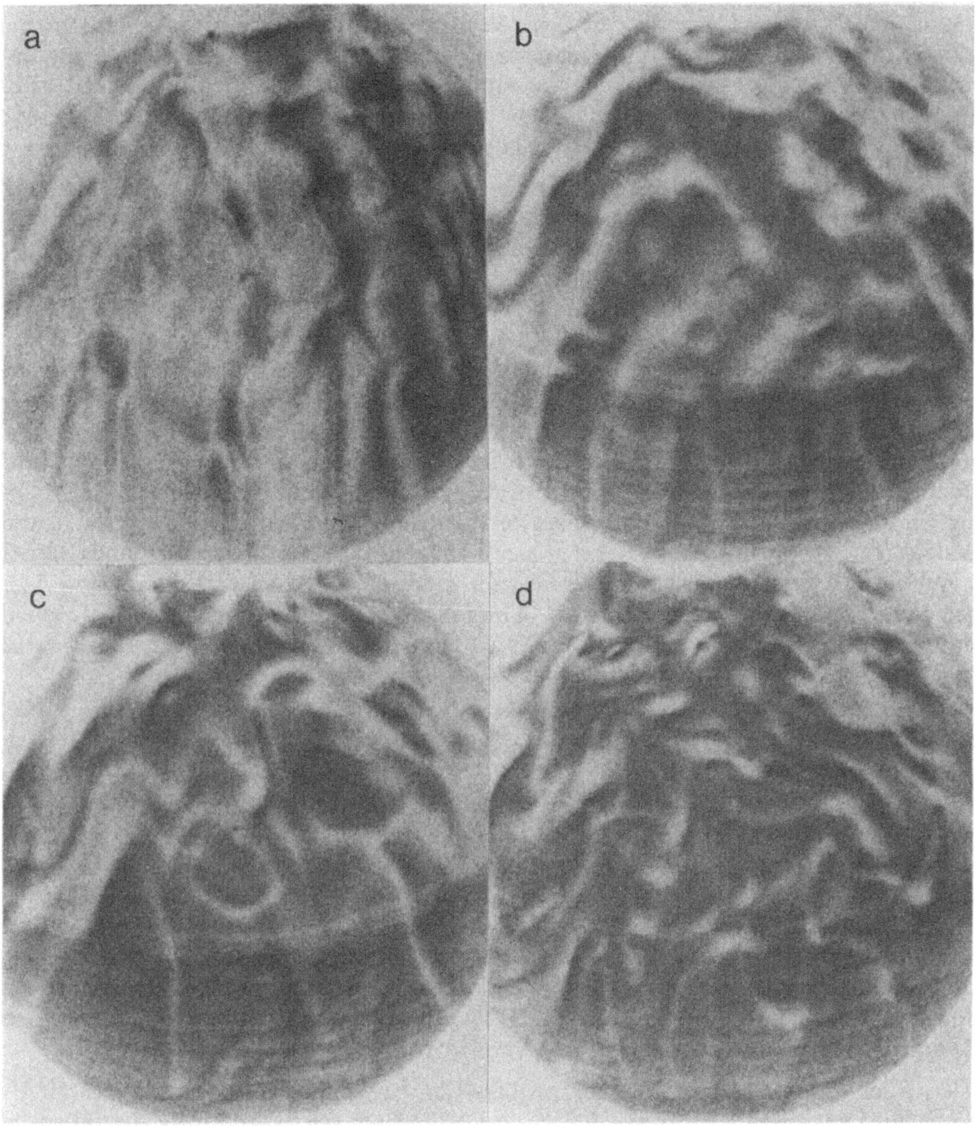

Figure 6. Continuation of Fig. 5 to higher R_a, showing that the banana cells are now limited to a narrow interval near the equator by the mid-latitude and polar convection modes. At the highest R_a the flow is extremely turbulent and the convective elements appear to be horizontally isotropic. Views (a) and (b) are nearly coincident images of the same site, separated in time by only one rotation period (2 s), with the Schlieren grating in (a) responding to nearly east-west gradients in the thermal field, in (b) to nearly north-south gradients. View (c) uses the latter grating, but is many rotations later, revealing that the convection is of the same general character, but no individual features are recognizable. Here $T_a = 6.3 \times 10^5$, and (a)–(c) have $R_a = 1.8 \times 10^5$, (d) $R_a = 3.8 \times 10^5$.

erratically, often equatorward. As Figs. 6a,b,c reveal, the competition between the various disturbances leads to the equatorial columns being limited to a narrow interval of about 20° in latitude near the equator, and this is further diminished with the higher R_a in Fig. 6d. The convection occupying the majority of the hemisphere is by then highly turbulent and nearly horizontally isotropic.

Figure 7 shows a snapshot from one of our numerical three-dimensional simulations carried out for the more modest values of R_a which are accessible with the resolution available on present supercomputers. Figure 7 presents the mid-depth radial velocity pattern, which is nearly the same as the radially-averaged thermal pattern, from the simulation that coincides with parameter values for Fig. 5a. Banana cells are the dominant feature in such simulations, and are joined by polar disturbances as R_a is increased. The convective rolls in the simulations are robust structures at lower latitudes, with the boundaries of given cells changing only gradually as they propagate in longitude. That phase speed is basically different at high and low latitudes, and this leads to the rolls shearing and reconnecting at mid latitudes. Guided by such behavior from the detailed velocity and thermal fields in the simulations, we could then recognize this also in the experimental data: there is slow retrograde drift above the critical latitude (though the disturbance signal in our images is weak until the mid-latitude modes strengthen with increasing R_a), and prograde drift at lower latitudes. Chance alignments can occasionally produce features, as in Fig. 5a,b and so too in Fig. 7, that appear to extend from the equator to nearly the pole, but they do not survive the shearing.

We should emphasize that the speeds of phase propagation are only a very small fraction of the peak speeds attained by the convection. Thus the slow drifts are almost an incidental feature, though they do influence our impressions of how persistent are the global-scale patterns. We found from the simulations first carried out with purely radial temperature gradients ($H = 0$) that the sense of phase propagation at low latitudes was retrograde, whereas the experiments indicated a propagation that was prograde. We later deduced that one of the equatorial coolers did not function on the GFFC during the flight, and thus there was a small latitudinal gradient ($H \lesssim 0.1$) present in all our experiments. In repeating the simulations with a weak latitudinal temperature gradient added, the convective flows were much like those in Fig. 7, but the phase propagation at low latitudes was now prograde, in agreement with the experiments.

Such subtleties about phase propagation bring into question the seemingly persuasive arguments by Busse & Cuong (1977) that are based on local analysis of vortex dynamics. However, as Fig. 6 should remind us, the convection at the higher R_a is very complicated, and whereas simple concepts of entities like banana cells or vortex columns do provide a framework for discussing the experimental results, they do not address the intricate nonlinear interactions that probably abound in such flows. We have not yet attempted numerical simulations at the high R_a in Fig. 6, for the computational resources needed to resolve the spatial scales adequately are still prohibitive. However, such modelling efforts may be required if we are to sort out many of the highly nonlinear processes that contribute to what is seen in Fig. 6, and it is not at all evident whether prograde phase propagation in the equatorial zone is, at large R_a, a detail that remains controlled by weak latitudinal gradients.

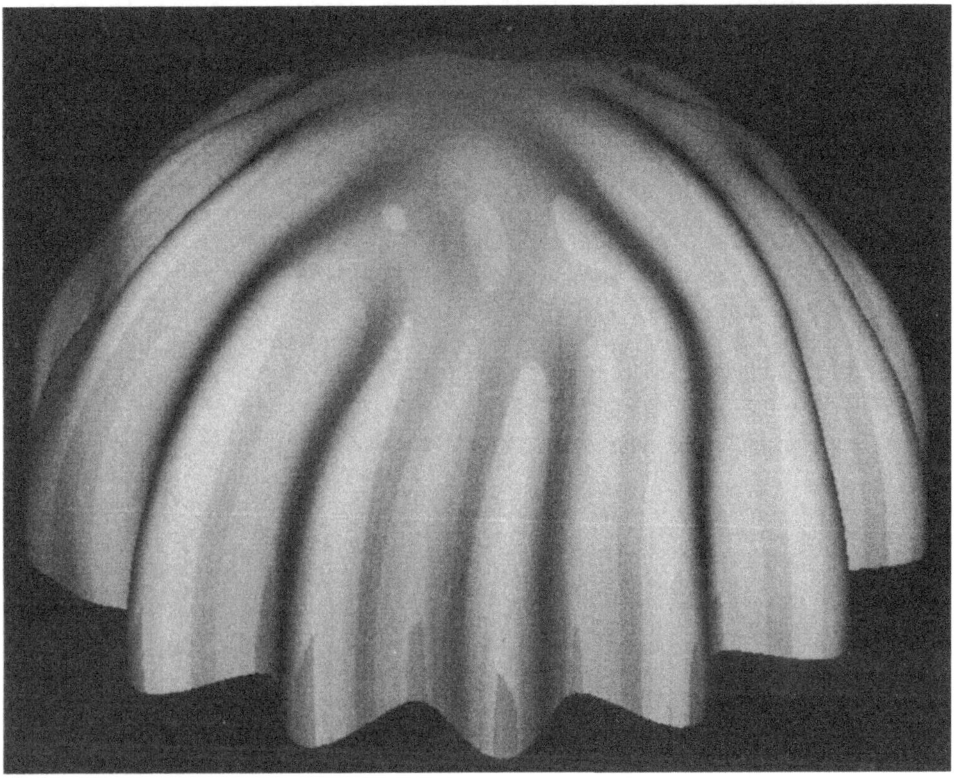

Figure 7. Perspective view from a three-dimensional numerical simulation of the radial velocity at mid-depth within the upper hemisphere at one instant in time. Elevation of surface represents upflow, depression downflow. This simulation studies the formation of banana cells, for conditions corresponding to those in Fig. 5a with rapid rotation and purely radial temperature gradient ($R_a = 4.7 \times 10^3$, $T_a = 3.9 \times 10^4$, $H = 0$).

Based on the results of these GFFC experiments, it is somewhat intimidating to guess how global-scale convection within the Sun might actually behave. Our work is in part motivated by the desire to provide experimental and theoretical underpinnings for predictions about convection deep within the Sun. Yet the substantial changes in the flows realized in Figs. 5 and 6 as the Rayleigh number is increased (by only a factor of 100 or so) suggests that we have hardly begun to understand nonlinear rotating convection for even the simplest type of thermal forcing within spherical shells. Solar conditions may well be far removed in parameter space, and thus one should be cautious in making inferences from our experimental data for the Sun, tempting as it may be.

3.3 Spiral waves and cross-branches: slow rotation, $H \neq 0$.

The richness of responses to changes in the thermal forcing can also be seen when a substantial latitudinal gradient (with $H \simeq 1$) is imposed in addition to the unstable radial gradient. Because the pole is hotter than the equator on each boundary, a meridional circulation must always exist, for there is no static state of pure conduction, and now even a slow basic rotation will produce substantial zonal flows. As Fig. 8a shows, the convective instability interacting with such axisymmetric motion leads to a spiral pattern of convection at low T_a. These convective rolls extend from the pole to mid-latitudes, possessing about a 45° tilt toward the west (right) as one traces an updraft towards the equator; the updrafts here too are narrow and the downdrafts broad. As R_a is increased at this low T_a, the rolls extend to lower latitudes (Fig. 8b) and form cross-branches as they make time-dependent interconnections. These secondary instabilities yield more complicated kinks in the irregular patterns at higher R_a (Figs. 8c,d), and the motion evolves chaotically. Yet the preferred average tilt of the cell boundaries is still quite evident. These "spiral wave" patterns of convection all propagate slowly in the prograde direction.

A full hemispheric view of the spiral waves in Fig. 8a is presented in the unfolded Mercator projection shown in Fig. 4b. Although this composite view indicates a characteristic tilted pattern nearly everywhere, we should comment that occasionally the spiral pattern is broken at certain longitudes by a convective roll extending from equator to pole without any appreciable tilt. Such remarkable defects in the pattern are fairly persistent, surviving for many rotations; we were first alerted to look for such pattern defects after noticing them in our numerical simulations for the parameter values of Fig. 8a. These experiments yielding spiral waves serve to emphasize that convective pattern selection is strongly influenced by meridional circulations and associated zonal flows, and that secondary instabilities once again add considerable subtlety to the overall process.

3.4 Triangular waves and mode interference: rapid rotation, $H \neq 0$.

We finally turn to convection under combined radial and latitudinal temperature gradients when the rotation is rapid, and the results are quite complicated. Figure 9a shows that with an 8 s rotation period and low R_a, quasi-axisymmetric bands occur near the pole, and these possess waves that propagate in a retrograde sense; weak banana cells can just be discerned in the equatorial regions. With increasing R_a and a strong north-south temperature gradient ($H \simeq 1.6$), more rapid rotation (2 s period) yields the vigorous high-latitude wavy spirals, and a relative absence of equatorial bananas, evident in Fig. 9b. Decreasing H to unity, while retaining nearly the same R_a and T_a, leads to reassertion of low-latitude banana cells, and they coexist with strong polar disturbances. Figures 9c,d show snapshots of the resulting turbulent flow, imbedded in which are persistent "triangular waves" at mid latitudes. Based on results from our numerical simulations, we believe that these striking triangular wave shapes are an interference pattern between the prograde-propagating equatorial columns and the retrograde-propagating polar disturbances. The triangular patterns also propagate slowly with a retrograde sense.

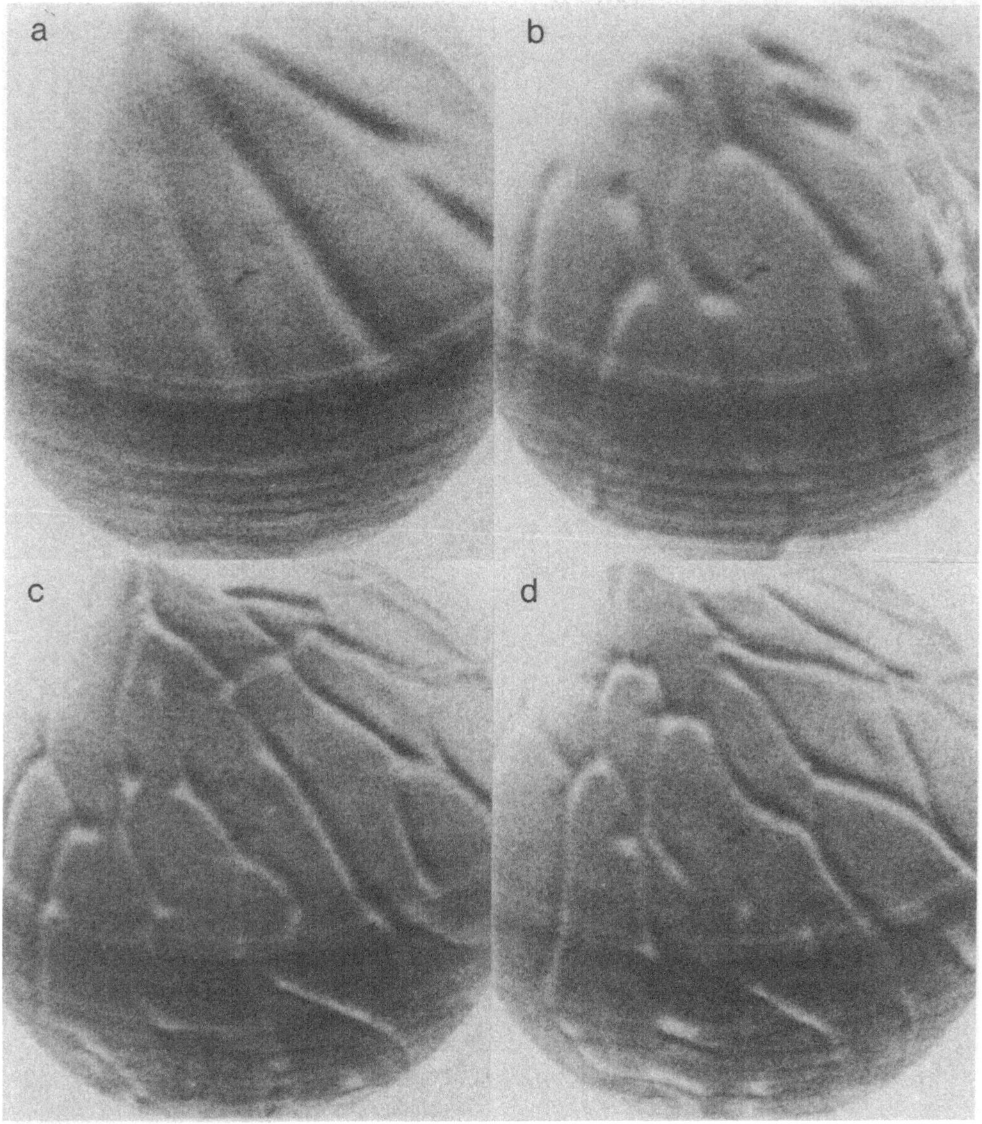

Figure 8. Planforms for convection under slow rotation when the temperature increases poleward on each spherical boundary in addition to a uniform radial temperature difference between the spheres ($H \simeq 1$), with $T_a = 1.1 \times 10^3$ (48 s rotation period). The latitudinal temperature gradient and the resulting meridional circulation lead to convection cells with a distinctive east-west tilt. These spiral waves experience secondary instabilities that produce cross-branches. (a) $R_a = 1.6 \times 10^4$, (b) $R_a = 5.0 \times 10^4$, (c) $R_a = 1.9 \times 10^5$, (d) $R_a = 2.1 \times 10^5$.

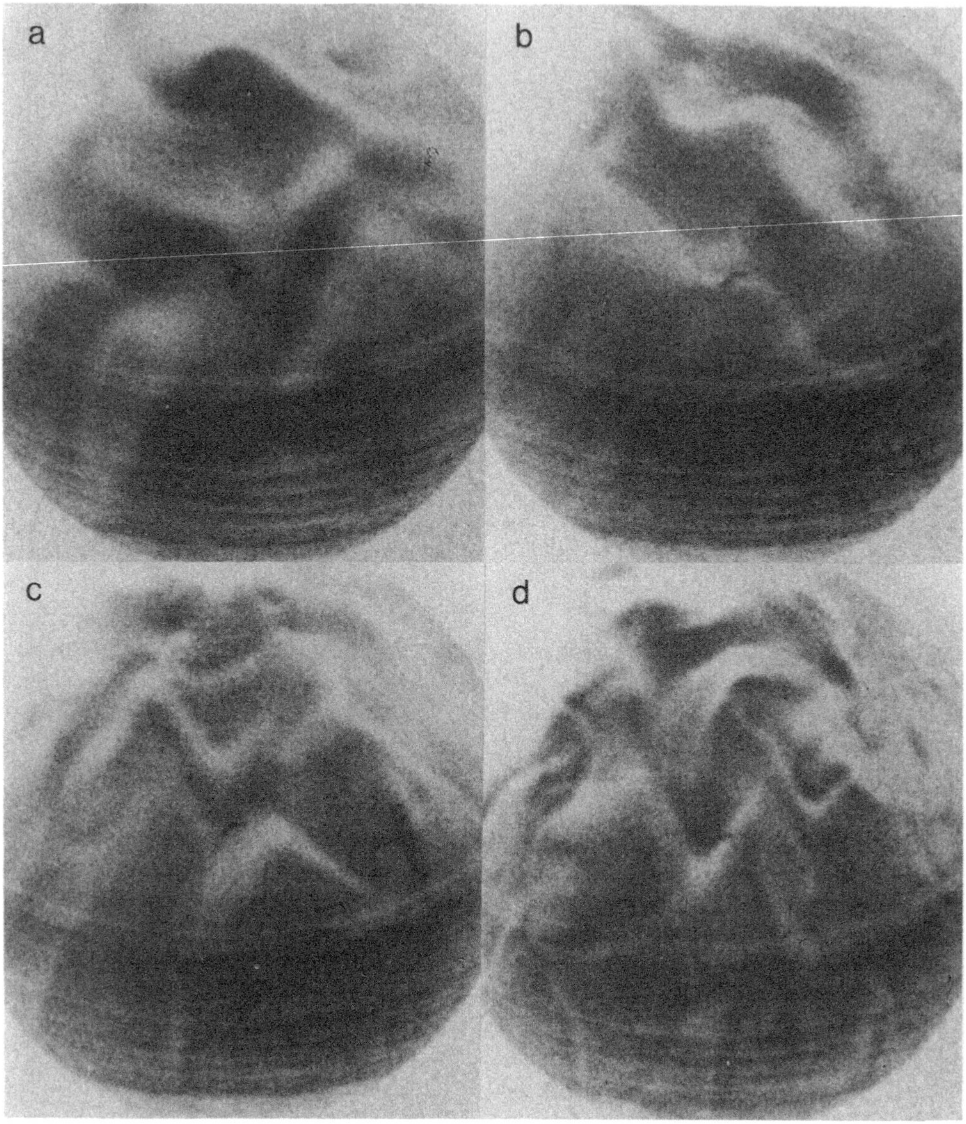

Figure 9. As in Fig. 6, but rotation is more rapid. Wavy disturbances at high latitudes interact with banana cells at low latitudes, yielding unusual triangularly-shaped waves at mid-latitudes. (a) $R_a = 1.7 \times 10^4$, $T_a = 3.9 \times 10^4$ (8 s rotation period), $H \simeq 1$, (b) $R_a = 1.4 \times 10^5$, $T_a = 6.3 \times 10^5$ (2 s rotation period), $H \simeq 1.6$, (c) $R_a = 1.6 \times 10^5$, $T_a = 6.3 \times 10^5$, $H \simeq 1$, (d) $R_a = 1.9 \times 10^5$, $T_a = 6.3 \times 10^5$, $H \simeq 1$.

4. CONCLUSIONS

Our data from the GFFC experiment on Spacelab 3 have illustrated the diverse forms of fluid motion that can occur in geometrically simple models of global convection with rotation and radial gravity. Several new types of convection were observed: mid-latitude waves interacting with the low-latitude columnar convection rolls or "banana cells", "spiral waves" near the poles when latitudinal heating gradients are present on the bounding surfaces, and "triangular waves" coupling mid-latitude and equatorial disturbances under similar differential heating. The richness of the patterns revealed from our brief surveys here in parameter space suggests that considerable care is required when trying to predict or interpret how convection should behave in real stars like the Sun or in the giant planets like Jupiter. Certainly concepts based on the Taylor-Proudman theorem have a central role in many of the flows realized in our rapidly rotating experiments when $H \simeq 0$, yet the resulting banana cells are subject to secondary instabilities that produce intricate time dependence and eventually turbulent flows in which only hints of the simpler patterns are recognizable. With $H \neq 0$ the convection is initiated as spiral waves at higher latitudes, and these tilted cells with multiple branchings move to lower latitudes with increasing R_a. Under more rapid rotation the polar disturbances interact with equatorial columns to form triangular waves at mid-latitudes, though here too the flows are turbulent. These results from the GFFC experiment emphasize once again that convection in rotating spherical shells has many intriguing features which present major challenges to theory.

Acknowledgments — This research is supported by the National Aeronautics and Space Administration through contract NAS-8-31958 to the University of Colorado. The efforts of our co-investigators have been most essential in making the GFFC into a working experiment: George Fichtl, Bill Fowlis and Fred Leslie have collaborated with us from NASA Marshall Space Flight Center, and Peter Gilman from the National Center for Atmospheric Research. Anil Deane and Neal Hurlburt from the University of Colorado participated in mission operations, and they with Bill Merryfield have assisted with the digital processing of the photographic data. The engineering and fabrication of the instrument was guided by Ray Kroger of Aerojet Electrosystems Company.

5. REFERENCES

Busse, F.H. 1970 Thermal instabilities in rotating systems. *J. Fluid Mech.* **44**, 441–460.

Busse, F.H. 1973 Differential rotation in stellar convection zones II. *Astron. & Astrophys.* **28**, 27–37.

Busse, F.H. 1983 A model of mean zonal flows in the major planets. *Geophys. & Astrophys. Fluid Dyn.* **23**, 153–174.

Busse, F.H., and Carrigan, C.R. 1976 Laboratory simulation of thermal convection in rotating planets and stars. *Science* **191**, 81–83.

Busse, F.H., and Cuong, P.G. 1977 Convection in rapidly rotating spherical fluid shells. *Geophys. Astrophys. Fluid Dyn.* **8**, 17–41.

Carrigan, C.R., and Busse, F.H. 1983 An experimental and theoretical investigation of the onset of convection in rotating spherical shells. *J. Fluid Mech.* **126**, 287–305.

Durney, B.R. 1970 Nonaxisymmetric convection in a rotating spherical shell. *Astrophys. J.* **161**, 1115–1127.

Gilman, P.A. 1975 Linear simulations of Boussinesq convection in a deep rotating spherical shell. *J. Atmos. Sci.* **32**, 1331–1352.

Gilman, P.A. 1978 Nonlinear dynamics of Boussinesq convection in a deep rotating spherical shell. III. Effects of velocity boundary conditions. *Geophys. Astrophys. Fluid Dyn.* **11**, 181–203.

Gilman, P.A. 1986 The solar dynamo: observations and theories of solar convection, global circulations and magnetic fields. Chapter 5 in *Physics of the Sun* (ed. P.A. Sturrock), Vol. 1, Reidel, pp. 95–160.

Glatzmaier, G.A. 1983 Numerical simulations of stellar convective dynamos. I. The model and method. *J. Comp. Phys.* **55**, 461–484.

Glatzmaier, G.A. 1985 Numerical simulations of stellar convective dynamos. II. Field propagation in the convection zone. *Astrophys. J.* **291**, 300–307.

Glatzmaier G.A., and Gilman, P.A. 1981 Compressible convection in a rotating spherical shell. II. A linear anelastic model. *Astrophys. J. Suppl.*, **45**, 351–380.

Hart, J.E. 1976 Studies of Earth simulation experiments. NASA Report CR-2753, 66 pp.

Hart, J.E., and Toomre, J. 1977 Geophysical fluid dynamics with dielectric polarization forces. Report 2, NASA Contract NAS8-31958, 65 pp. (Available from JEH or JT).

Hart, J.E., Toomre, J., Deane, A.E., Hurlburt, N.E., Glatzmaier, G.A., Fichtl, G.H., Leslie, F., Fowlis, W.W., and Gilman, P.A. 1986 Laboratory experiments on planetary and stellar convection performed on Spacelab 3. *Science* **234**, 61–64.

Hart, J.E., Glatzmaier, G.A., and Toomre, J. 1986 Space-laboratory and numerical simulations of thermal convection in a rotating hemispherical shell with radial gravity. *J. Fluid Mech.* **173**, 519–544.

Hathaway, D.H., Toomre, J., and Gilman P.A. 1980 Convective instability when the temperature gradient and rotation vector are oblique to gravity. II. Real fluids with effects of diffusion. *Geophys. Astrophys. Fluid Dyn.* **15**, 7–37.

Ingersol, A.P. 1976 Pioneer 10 and 11 observations and the dynamics of Jupiter's atmosphere. *Icarus* **29**, 245–251.

Rossby, H.T. 1969 A study of Benard convection with and without rotation. *J. Fluid Mech.* **36**, 309–335.

THE EQUATORIAL ROTATION RATE IN THE SOLAR CONVECTION ZONE

Frank Hill
National Solar Observatory
National Optical Astronomy Observatories*
Tucson, Arizona 85726 USA

ABSTRACT. The state of knowledge of the rotation rate in the solar convection zone as inferred from helioseismology is reviewed, and the implications for dynamo and convection theories are discussed.

1. INTRODUCTION

While we have much information on how the Sun behaves, we have as yet limited knowledge of why it acts the way it does. For example, we have known for 100 years that sunspots appear with an 11 year cycle, but a satisfactory theory of the mechanism that causes this is lacking. Since the spots are magnetic in nature, and because the decay time for magnetic fields is much shorter than the age of the Sun, it is generally felt that there must be some dynamo mechanism that regenerates the field. Apparently this mechanism must possess at least a 22 year polarity cycle. Attempts to numerically model this process have been numerous, and have met with varying degrees of success.

There are two basic models of the solar dynamo - kinematic and dynamical. Kinematic theories solve the magnetic induction equation using assumed velocity fields, while dynamical theories attempt to model both the magnetic and velocity field by solving approximations of the full set of MHD equations including the equations of motion. Kinematic models have been successful in explaining the cycle period and the butterfly diagram (e.g. Yoshimura 1978), but to achieve the correct direction of toroidal field migration, they require that the angular solar velocity, Ω, increase with depth, that is, $\partial\Omega/\partial r<0$, where r is the solar radius measured outward from the center. Dynamical theories (e.g. Glatzmaier and Gilman 1981), on the other hand, predict that the angular velocity should be constant upon cylinders aligned with the rotation axis, and since the surface differential rotation shows that $\Omega_{pole}<\Omega_{equator}$, Ω should decrease with depth ($\partial\Omega/\partial r>0$). While the dynamical theories have not yet included all of the essential physics, a reliable determination of $\partial\Omega/\partial r$ by helioseismology could help choose a valid model. However, Weiss (1986) points out that even if $\partial\Omega/\partial r$ turns out to be positive, kinematic theories can still be viable if the dynamo action occurs in a shell just below the base of the convection zone.

*Operated by the Association of Universities for Research in Astronomy, Inc. under contract with the National Science Foundation.

B. R. Durney and S. Sofia (eds.), The Internal Solar Angular Velocity, 45–50.

2. THE ROTATION RATE IN THE OUTER CONVECTION ZONE

Figure 1 summarizes our current knowledge of the sidereal equatorial rotation rate in the outer 30% of the convection zone. The published results from five helioseismological studies have been plotted, along with errors either published or inferred from the scatter of the published data.

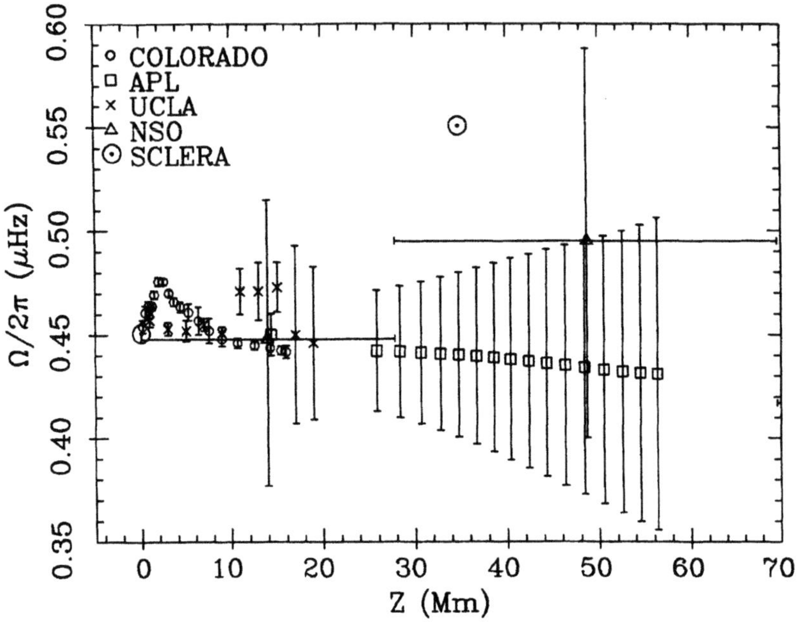

Figure 1: The equatorial sidereal rotation rate Ω as a function of depth Z in the outer solar convection zone, as inferred from five different studies using helioseismology. References and details are provided in the text.

The data of the Colorado group (F. Hill et al. 1986) were obtained using spectrographic measurements of the Doppler shift of the Fe I 5434 and Mg I 5173 lines over a long narrow observing region. The frequency splittings of 709 sectoral modes with spherical harmonic degree $80 \leq \ell \leq 995$ and order $0 \leq n \leq 8$ were combined using the Backus-Gilbert optimal averaging inversion technique. The splittings were measured from independent time strings spanning either 3 or 5 days. The plotted points are the average of the Colorado data sets, with the error bars representing the standard deviation of the average. A scale error apparently exists in the Colorado data, as the rotation rate at the shallowest point is substantially different from the latest published Mt. Wilson surface rate of 0.454 μHz (Howard 1984). This inconsistency may arise either from a misidentification of the spatial scale of the images since the apparent solar diameter was not measured in the course of the observations, or from a slow drift of the telescope. This error has been corrected in Figure 1 by offsetting the data points so that the rate at the shallowest depth is equal to the Mt. Wilson rate. When this correction has been made, the Colorado data becomes nearly continuous with the APL data at a depth of 16 Mm.

The APL data (F. Hill, Rust and Appourchaux 1986) were obtained from Fabry-Perot observations of the entire solar disk in the Ca I 6103 line. Again, Doppler images were obtained with the data set spanning 3 days. The frequency splittings of 79 sectoral modes with degree $80 \leq \ell \leq 150$ and order $3 \leq n \leq 7$ were combined using the Backus-Gilbert technique. Since the solar diameter was measured during the course of these observations, no scale correction has been applied to the points in Figure 1. The error bars represent the Backus-Gilbert estimate and are a weighted average of the errors present in the frequency splittings.

The UCLA data (Deubner, Ulrich and Rhodes 1979) were obtained in a manner similar to the Colorado observations, but using the Fe I 5576 line. Here, however, data from 3 individual days were analyzed, rather than a time string covering a number of consecutive days. Consequently, daily variations were seen that were most likely due to mode beating rather than a solar cause. Similar large daily variations were seen in an earlier Colorado study (F. Hill, Gough and Toomre 1984) which considered data obtained on individual days. The UCLA data was inverted using a linear approximation to the depth dependence of the rotation rate. Effective depths were assigned by first computing the theoretical frequency splittings from a solar model that included the hypothetical linear rate dependence, and then assuming that these theoretical splittings arose only from the rotation at a single depth. This procedure has been criticized by Gough (1978) on the grounds that this is only valid if the rotation rate is actually linear with depth, and if the modes are sensitive to conditions at just a single depth or over a narrow range of depths. Neither of these assumptions are valid in the Sun. Sectoral modes with degree $125 \leq \ell \leq 900$ and order $0 \leq n \leq 8$ were used in the analysis. In Figure 1, the published UCLA data for the 3 individual days have been averaged and converted from a velocity in ms^{-1} to the corresponding difference between the surface rate and the inferred rotation rate at the effective depth. The errors bars represent my calculation of the standard deviation of the average. Since guiding problems prevented the determination of the surface rate, the Mt. Wilson value has been added to the data.

The NSO data (Duvall et al. 1984, as displayed in Durney 1985), were obtained from spectrographic observations in the Ni I 6315 line. The time string spanned 17 days, and 180 sectoral modes with degree $1 \leq \ell \leq 100$ were used in the inversion. However, the frequency splittings at each ℓ were averaged over all available orders, n, resulting in a total of 37 splittings that already represented an average over extended depth ranges. Several types of inversion techniques were used; the procedure deemed most reliable, and the one used for the data shown in Figure 1, was a Monte Carlo piecewise constant approximation that minimized the scatter between the observations and theoretical splittings. The horizontal error bars show the width of the depth steps, and the vertical error bars are the published estimates.

The SCLERA data (H. Hill, Rabaey and Rosenwald 1986) were obtained from diameter measurements. A total of 30 modes was apparently used, with degree $1 \leq \ell \leq 2$ and order $5 \leq n \leq 22$. Additional constraints were provided by long period modes identified as g - modes. The inversion method was a modified Twomey procedure. No error bars were available for the inferred rotation rate.

The data in Figure 1 represent a wide variety of observational and techniques and inversion procedures. It is thus gratifying that the results shown are mostly consistent to within the inferred errors. The only region of discrepancy lies in the depth range of 1 to 5 Mm, where the Colorado and the UCLA results differ. However, the drawbacks of the UCLA inversion procedure, coupled with the lack of reliable spatial scale determinations in both data sets, may explain this discrepancy.

While this agreement provides confidence in the reality of the results, the data are really not yet good enough to provide us with the information that is needed to calibrate dynamo and convection theories. For instance, given the inferred errors, it is currently impossible to determine even the sign of $\partial\Omega/\partial r$, let alone its magnitude at depths greater than 16 Mm. However, at the very outer edge of the Sun the Colorado data do appear to be good enough to allow us to infer at least the change in the rotation rate. Evidently, there is a rise in Ω in the very outer edge of the convection zone above the H^- ionization zone. Below 2 Mm, $\partial\Omega/\partial r$ appears to be positive, at least down to a depth of 16 Mm.

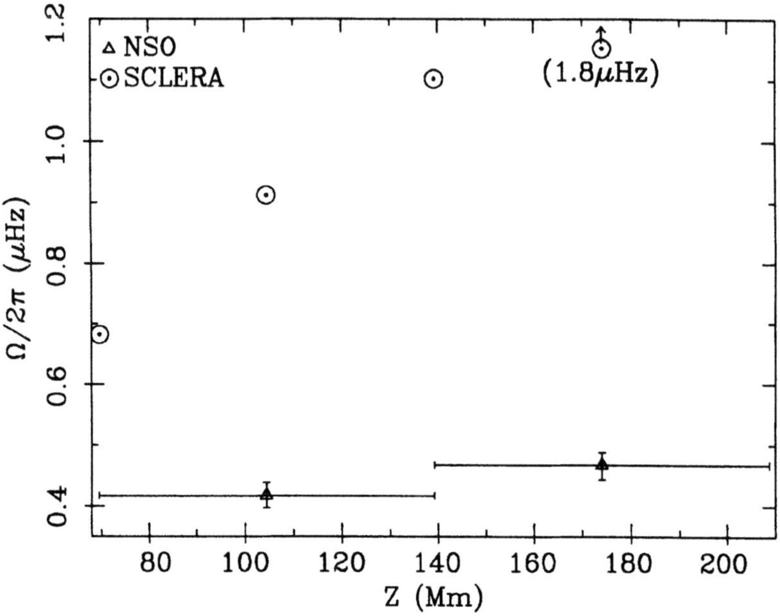

Figure 2: The equatorial sidereal rotation rate Ω as a function of depth Z in the inner solar convection zone.

3. THE ROTATION RATE IN THE INNER CONVECTION ZONE

Figure 2 shows the sidereal equatorial rotation rate in the lower 70% of the convection zone. To date, only two published studies have provided a measurement of Ω in this region. These studies are the NSO study (Duvall et al. 1984) and the SCLERA diameter measurements (H. Hill, Rabaey and Rosenwald 1986).

As can be seen in Figure 2, these measurements are clearly in disagreement. While the interpretation of the diameter measurements has been controversial, the discrepancy may indicate that the height dependence of the oscillation eigenfunctions in the solar atmosphere plays a more important role in the measurement of the frequency splittings than is currently believed. The results of both studies suggest that $\partial\Omega/\partial r$ appears to be negative in this region of the convection zone, but the measured values of $\partial\Omega/\partial r$ differ widely.

4. DISCUSSION AND CONCLUSIONS

The equatorial rotation rate in the solar convection zone has now been measured by several different helioseismological methods. The results are in qualitative agreement in the outer 30% of the convection zone, where it appears that $\partial\Omega/\partial r<0$ in the outer 2 Mm of the zone, and $\partial\Omega/\partial r>0$ over depths of 2 to 16 Mm. However, at depths of 16 to 70 Mm, it is currently not possible to determine the sign of $\partial\Omega/\partial r$. In the lower 70% of the convection zone, there is considerable disagreement among the few available results, although it appears that $\partial\Omega/\partial r<0$ between depths of 70 and 210 Mm.

The rise of Ω in the very outer edge of the convection zone should probably not be taken as evidence that the kinematic dynamo operates just below the photosphere. It is more likely that the rise represents a thin boundary layer within which conservation of angular momentum is the dominant process. This region also contains the H^- ionization zone which is believed to be the region of driving for the granulation. The rise may thus be a consequence of the turbulence generated in this region. Below 2 Mm, $\partial\Omega/\partial r>0$, suggesting that the dynamical models of the dynamo are valid. These results are consistent with the model of Gilman and Foukal (1979). However, since it is currently not possible to determine the sign of $\partial\Omega/\partial r$ below depths of 16 Mm, and since it is most likely that the bulk of the convection zone plays a role in the dynamo, the choice of a dynamo model must remain unresolved at this time. This is especially true if the dynamo action occurs in a narrow region just below the convection zone.

It is possible to quantitatively measure $\partial\Omega/\partial r$ over the outer 16 Mm of the convection zone using the Colorado data. Figure 1 shows three regions where $\partial\Omega/\partial r$ is roughly constant. In the first region covering depths of 0 to 2 Mm, $\partial\Omega/\partial r=(-8.2\pm1.9)\times10^{-11}$ rad s^{-1} km^{-1}. In the second region covering depths of 2 to 9 Mm, $\partial\Omega/\partial r=(+2.7\pm0.4)\times10^{-11}$ rad s^{-1} km^{-1}. Finally, in the third region covering depths of 9 to 16 Mm, $\partial\Omega/\partial r=(+0.55\pm0.55)\times10^{-11}$ rad s^{-1} km^{-1}. Note that below depths of 9 Mm, the data are consistent with $\partial\Omega/\partial r=0$. These values can be compared with the value of Foukal (1972), who obtained a value of $\partial\Omega/\partial r \approx -3\times10^{-11}$ rad s^{-1} km^{-1} by assuming that the observed higher rotation rate of sunspots reflected conditions at a depth of 5 Mm. The gradient in the outer 2 Mm inferred from the Colorado data is nearly three times this value, reflecting the shallower depth at which the rotation rate peaks. The magnitude of the rise is consistent with the observed higher sunspot rotation rate.

The value of $\partial\Omega/\partial r$ in the inner convection zone is also of interest as the seat of the dynamo as well as the driving of giant cell convection may be here. Turbulent models of the convection zone suggest that the lower regions should be rotating faster than the upper part (Durney 1987, these proceedings), thus these models predict that $\partial\Omega/\partial r<0$ in this region. Indeed, the data in Figure 2 are consistent with this prediction, with $\partial\Omega/\partial r=(-0.45\pm0.40)\times10^{-11}$ rad s^{-1} km^{-1} as determined from the NSO data, and $\partial\Omega/\partial r=-7.5\times10^{-11}$ rad s^{-1} km^{-1} as inferred from the SCLERA data. It would be desirable to obtain additional information about the rotation rate at these depths, since the sparse data currently available differ by a factor of 16 in the measured value of $\partial\Omega/\partial r$.

While helioseismology has provided some information on the equatorial rotation rate in the convection zone, the data are not yet to the point where dynamo and convection theorists can use the results as input to their models. This is because both the depth resolution and noise levels needed to provide reliable and useful results require the precise measurement of frequency splittings of modes with high degree. So far, most studies have concentrated on modes with moderate and low degrees using full disk measurements. While these measurements provide extremely valuable information on the rotation rate as a function of both depth and latitude (eg. Brown 1985, Libbrecht 1986, Duvall, Harvey and Pomerantz 1986, see also H. Hill et al. 1986), they are unable to finely resolve the details of the flows in the convection zone. At this point, due to limited time strings, poor spatial filter response, and unstable instrumentation, high degree measurements have not been sufficiently accurate to provide the desired quality of results. The GONG project will not normally provide the requisite data, since it will usually obtain information only on intermediate and low degree modes ($\ell \leq 300$). This data will allow the resolution of flows in the lower convection zone, but cannot tell us about the upper convection zone. The proposed high-resolution SOHO experiment would be able to probe the upper convection zone. SOHO data also holds the potential of mapping horizontal flows, such as might be due to giant convective cells, through the use of three-dimensional Fourier transforms and tomographic techniques.

ACKNOWLEDGEMENTS. This paper has profited from collaborations and scientific discussions with Juri Toomre, Douglas Gough, Dave Rust, Deborah Haber, Larry November, Thierry Appourchaux, Jack Harvey, Tom Duvall, Bernard Durney, John Leibacher, Tim Brown, Bill Jeffrey, Peter Gilman and Ed Rhodes.

REFERENCES

Brown, T.M. 1985, *Nature* **317**, 591.
Deubner, F.-L., Ulrich, R.K., and Rhodes, E.J. Jr. 1979, *Astron. Astrophys.* **72**, 177.
Durney, B.R. 1985, *Astrophys. J.* **297**, 787.
Duvall, T.L. Jr., Dziembowski, W., Goode, P.R., Gough, D.O., Harvey, J.W., and Leibacher, J.W. 1984, *Nature* **310**, 22.
Duvall, T.L. Jr., Harvey, J.W., and Pomerantz, M.A. 1986, *Nature* **321**, 500.
Foukal, P. 1972, *Astrophys. J.* **173**, 439.
Gilman, P.A., and Foukal, P. 1979, *Astrophys. J.* **229**, 1179.
Glatzmaier, G.A., and Gilman, P.A. 1981, *Astrophys. J. Supp.* **45**, 351.
Gough, D.O. 1978, in *Proc. Workshop on Solar Rotation*, ed. G. Belvedere and L. Paterno, Catania: Oss. Astrofis. Catania, 255.
Hill, F., Gough, D.O., and Toomre, J. 1984, *Mem. Soc. Astron. Ital.* **55**, 153.
Hill, F., Gough, D.O., Toomre, J., and Haber, D.A. 1986, in *Advances in Helio- and Asteroseismology, IAU Symp. 123*, ed. J. Christensen-Dalsgaard, Dordrecht: Reidel, in press.
Hill, F., Rust, D.M., and Appourchaux, T. 1986, in *Advances in Helio- and Asteroseismology, IAU Symp. 123*, ed. J. Christensen-Dalsgaard, Dordrecht: Reidel, in press.
Hill, H.A., Rabaey, G.F., and Rosenwald, R.D. 1986, in *Relativity in Celestial Mechanics and Astrometry, IAU Symp. 114*, ed. J. Kovalevsky and V.A. Brumberg, Dordrecht: Reidel, 345.
Hill, H.A., Rabaey, G.F., Yakowitz, D.S., and Rosenwald, R.D. 1986, *Astrophys. J.* **310**, 444.
Howard, R. 1984, *Ann. Rev. Astron. Astrophys.* **22**, 131.
Libbrecht, K.G. 1986, *Nature* **319**, 753.
Weiss, N.O. 1986, in *Seismology of the Sun and the Distant Stars*, ed. D.O. Gough, Dordrecht: Reidel, 253.
Yoshimura, H. 1978, *Astrophys. J.* **220**, 692.

MORE EVIDENCE FOR A SOLAR LATITUDE DEPENDENT LIMB TEMPERATURE VARIATION

J. R. Kuhn
Department of Physics and Astronomy
Michigan State University
E. Lansing, MI 48824-1116

ABSTRACT. A preliminary analysis of solar limb data obtained during 1983, 1984 and 1985 supports the observation that the sun has a 0.6°K effective limb temperature variation with a temperature minimum near 50° North and South latitude and no detectible solar cycle time scale variability.

I. INTRODUCTION

Von Zeipel's theorem assures us that a non-rotating, spherically symmetric star with no surface magnetic field or Reynolds stresses has a uniform surface brightness (or temperature). Of course these conditions are only approximately true for the sun. A smooth latitudinal temperature variation should be associated with the solar differential rotation. Qualitative arguments suggest a limb temperature modulation, ΔT, at the level of

$$\Delta T \sim (v^2/\phi)T \sim 0.1°K \tag{1}$$

where T is 5700°K and v^2/ϕ is the ratio of the rotational kinetic and potential energy density at the photosphere.

Since 1983 the Princeton Solar Distortion Telescope has operated during the summer observing season from Mt. Wilson collecting solar limb brightness data (cf. Dicke et al. 1986). From 1983 to 1985 about 20,000 5 min observations of the limb brightness in two broad-band colors with about 2° angular resolution (as measured from the center of the solar disk) have been obtained. The telescope was designed to accurately obtain the brightness of between (approximately) 5" and 25" of the solar limb. Although the data were primarily intended to provide an accurate measure of the shape of the sun they are nearly ideal for measuring limb brightness variations also (cf. Kuhn et al. 1985 denoted KLD). Here I will briefly review the evidence for a 0.6°K limb temperature variation observed in 1983 and present a simplified analysis of data from 1984 and 1985 that supports the earlier findings

B. R. Durney and S. Sofia (eds.), The Internal Solar Angular Velocity, 51–57.

and suggests that the temperature variation does not change with the solar cycle.

II. THE DATA

As described in KLD, limb intensity observations $I_{ij}(\theta)$ are obtained as:

$$I_{ij}(\theta) = \int_{r_j}^{\infty} F_i(\theta)dr \tag{2}$$

where j = 1-3 is the disk identification, r_j describes the amount of exposed limb, and i = 1,2 identifies the red and green color band data. The integral is over a 2° wide aperature in the solar image plane at position angle θ (measured clockwise from the northernmost point on the image), and extending from r_j to about 20 arcseconds beyond the visible limb of the sun. The intensity data are conveniently expressed in terms of the displacement of the limb (in arcseconds) that would produce an equivalent intensity measure by:

$$D_{ijk} = \{I_{ij}(\theta_k) - \bar{I}_{ij}\}/\bar{F}_j \ . \tag{3}$$

Here θ_k is obtained by signal averaging the data into 128 distinct angular bins (k = 1,128), $\bar{I}_{ij}$ is the rotationally averaged intensity signal from (1) and F_j is the mean brightness of the limb at the edge of the occulting disk (r_j). Thus D_{ijk} can be expressed in units of arcseconds (the limb displacement required to produce the observed intensity signal). To measure an effective limb temperature or color variation with latitude we consider the quantities

$$C_{jk} = D_{1jk} - D_{2jk} \tag{4}$$

formed by subtracting the red and green normalized data. Note that any limb shape signal will not appear in the "color" measure C_{jk}.

If a limb effective temperature variation exists in the data for limb exposure j as $\Delta T_j(\theta)$ then the red (i = 1) and green (i = 2) data must also satisfy

$$\frac{I_{1j}(\theta) - \bar{I}_{1j}}{\bar{I}_{1j}} = 3.1\ \frac{\Delta T_j(\theta)}{T} \text{ and } \frac{I_{2j}(\theta) - \bar{I}_{2j}}{I_{2j}} = 4.8\ \frac{\Delta T_j(\theta)}{T} \ .$$

Both black-body temperature variations and facular brightness contributions show this color dependence. Contamination of the limb brightness data by faculae make temperature variations of the background quiet photosphere difficult to distinguish, so that measurements of mean limb temperature variations over a time period of several days necessarily mix any photospheric effects with occasional resolved or unresolved faculae and spots. I've found a useful way of using data over a long period of time (which is necessary to achieve a temperature sensitivity better than 0.1°K) which allows a separation of the facular and photospheric brightness signals. This approach depends on knowledge of the empirical color distribution (as opposed to the first moment or mean) at each latitude bin j of the C_{jk} data in (4). Since any given day's measurement of C_{jk} (at fixed j and k) always samples the photosphere and only occasionally includes a contribution from faculae a simple and effective statistical model can be constructed (see KLD) that results in a measurement of the probability of a facular contribution (a distribution function that depends on solar latitude), the background color distribution, and the facular color distribution.

Figure 1 shows the observed distributions, C_{jk} (disk 1 color data), for observations taken during 1983, 1984 and 1985. The color distribution of data obtained during each season at different position angles along the solar limb are displaced vertically in each plot. Data obtained near the sun's equator are the central traces in these figures.

Color-dependent atmospheric and mirror limb distortions have been considered as possible systematic noise sources (see KLD). Note that atmospheric contributions will not be spatially synchronous with the solar rotation axis. The data has also been divided into halves and analyzed separately. The general features described below appear in both datasets, so that it is unlikely that the measurements are greatly affected by the slow seasonal geometric variation of the mean mirror orientations. The telescope mirror orientations were rotated on a daily basis. An analysis of the data obtained from a single mirror orientation shows no systematic differences from the combined data shown here. Further, we have no reason to expect any color dependent mirror optics related signal to be symmetric about the solar equator.

III. ANALYSIS

Several general conclusions are obvious from the 1983-85 datasets of Fig. 1. The traces near the equator, but displaced north and south of it, show a distribution with a tail in the negative (green) color direction. This is clearly the result of faculae in active latitude bands. Note that from 1983 to 1985 the facular contribution at these latitude bands declines with the solar cycle as it should.

It's also notable that the color distributions far from the equator have approximately the same form but show a systematic shift. The amount of the color displacement is symmetric about the equator. In addition there is no obvious variation in this pattern between 1983 and 1985.

In KLD I described a statistical model based on 4 unknown discrete functions (172 unknowns) that are obtained from 2048 measured values per limb exposure. Results of this analysis gave the probability of a facular contribution in a given angular bin (L_j), the color distribution of the background photosphere (B_i), the convolution of the

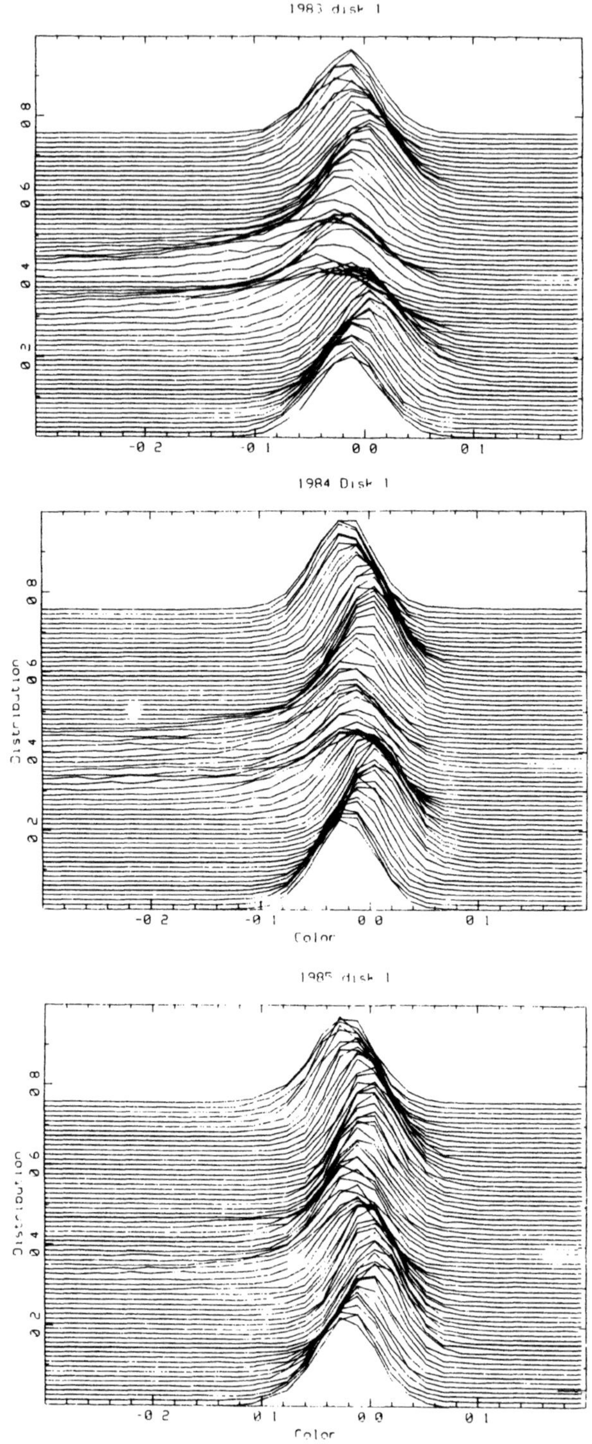

Figure 1: Disk 2 color distributions from 1983-1985.

facular and background color distribution W_i, and the color shift in the background distribution in each angular bin (S_j). Figure 2 displays the results of this analysis performed on disk 1 data taken in 1983. Figure 3 shows the corresponding temperature variation that would produce the color distribution data from all three limb exposures.

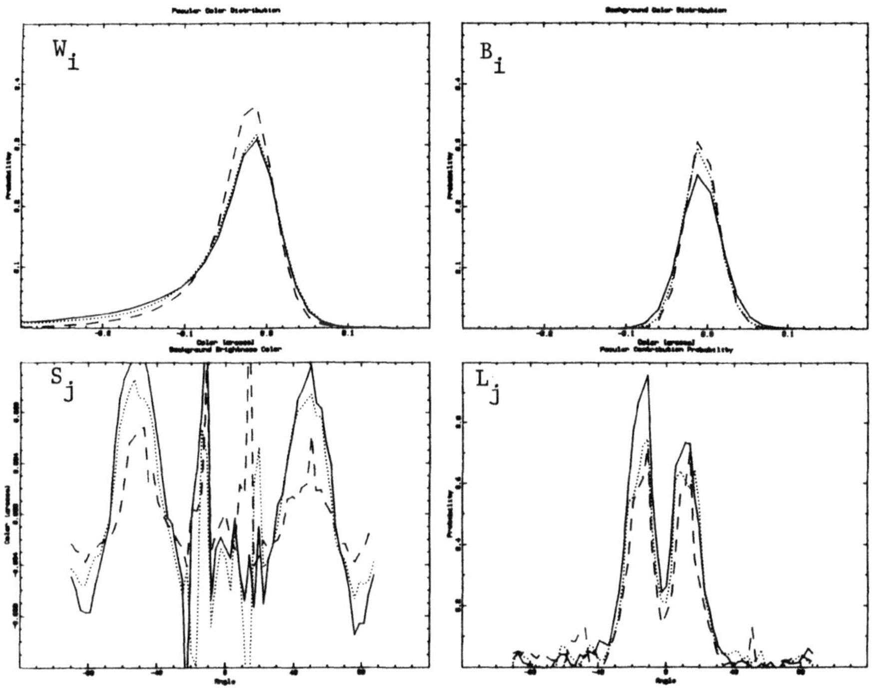

Figure 2: 1983 facular model description. Solid, dotted, and dashed curves are from disk 1, 2, and 3 data.

The latitude distributions displayed in Fig. 2 show that faculae are confined to the active latitude bands and have little effect on the color distribution outside of this region. In 1984 and 1985 there were even fewer faculae, consequently a simpler procedure follows if we assume $L_j = 0$ (no faculae). In the active latitude regions we expect to get unrealistically cooler temperature but the purpose of this analysis is to simply demonstrate that the magnitude and latitude distribution poleward of the active regions has not changed significantly during 1983-1985.

The data from the largest limb exposure (disk 1) has the largest temperature sensitivity and the shift function S_j is plotted for three years in Figure 4. The plus, cross and triangle symbols show data from 1983, 1984, and 1985. Where the facular signal is important data have not been plotted. From this preliminary analysis is it clear that neither the latitude of the temperature maximum nor its magnitude has declined dramatically during 1983-85.

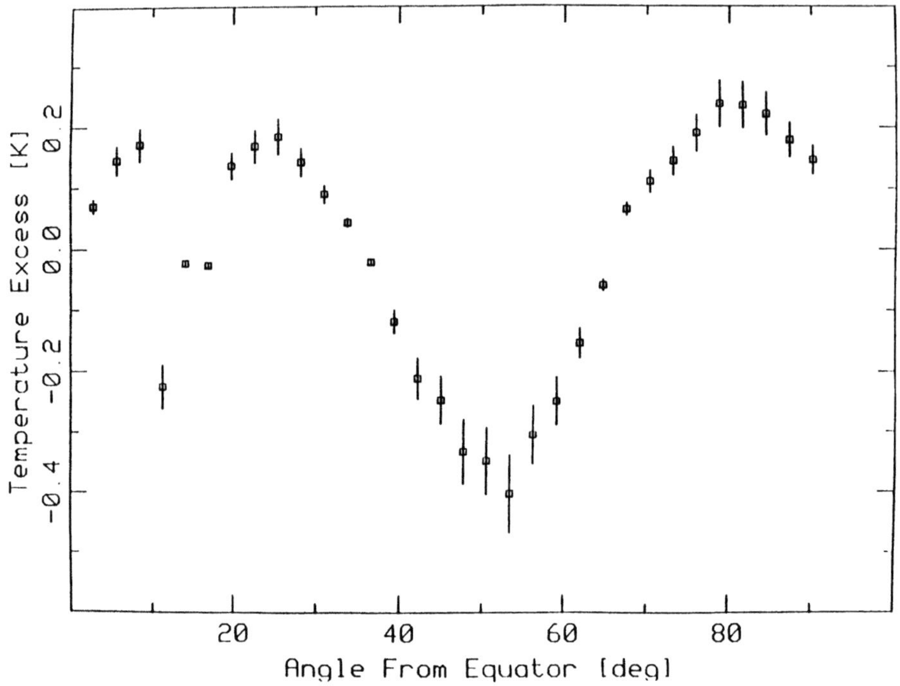

Figure 3: Limb temperature variation determined from 1983 observations.

CONCLUSIONS

The temperature perturbation is a puzzle. It shows no sign of having the time dependence of the solar cycle. This might be expected if it were directly related to the solar dynamo. Furthermore we see no evidence of a P_2 temperature variation. From the data in Fig. 3 we let $\Delta T(\theta) = \Delta T_2 P_2(\cos\theta) + \Delta T_4 P_4(\cos\theta)$ then $|\Delta T_2| < 0.16°K$ (2σ limit) and $\Delta T_4 = 0.21°K$ with significant higher order terms.

This research was supported in part by the NSF under grant no. ATM 8604400.

REFERENCES

Dicke, R.H., Kuhn, J.R. and Libbrecht, K.G. Ap. J. (1986) in the press.

Kuhn, J.R., Libbrecht, K.G. and Dicke, R.H. Ap. J. 290, 758 (1985) denoted KLD.

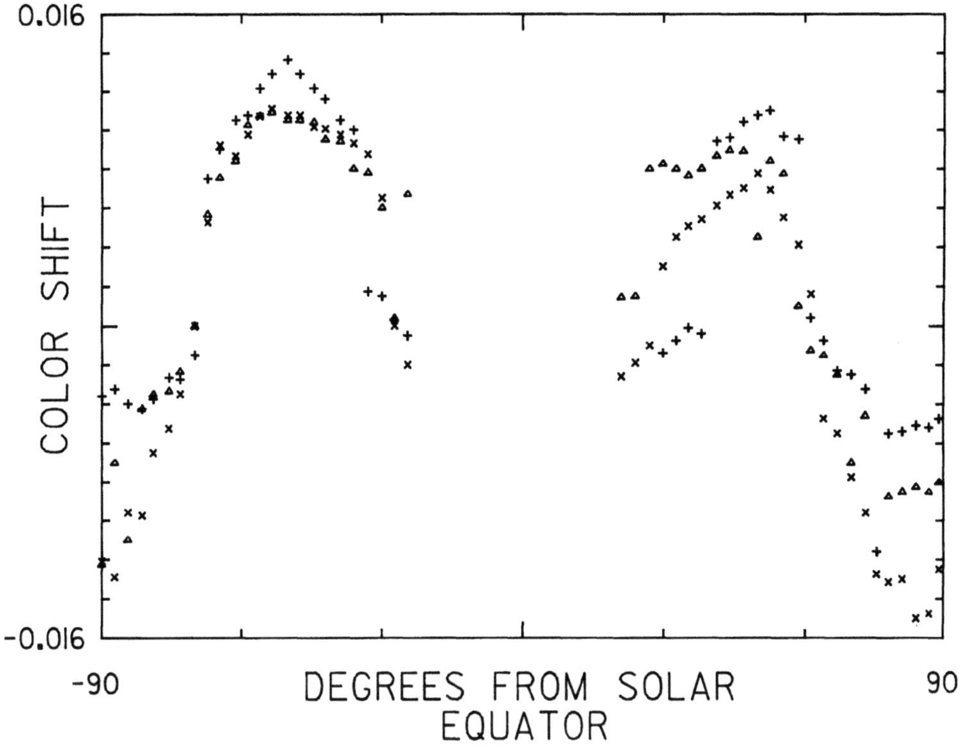

Figure 4: Disk 1 limb color variation during 1983-1985.

Fine Structure in Solar Oscillation Spectra

Ken G. Libbrecht
Solar Astronomy 264-33
California Institute of Technology
Pasadena, CA 91125

ABSTRACT. I discuss here two aspects of helioseismology which rely on precise measurements of the relative frequencies of solar oscillation modes: rotational and magnetic mode splittings, and solar cycle changes in the mode frequencies. Recent data are presented, along with speculation into the future of the measurements.

1. Rotational Splittings

The sun's differential rotation as a function of latitude is well known and can be accurately measured by both Doppler techniques and by following the rotation of sunspots. This differential rotation and the differential rotation as a function of radius are important inputs into any theory of the solar dynamo. Unfortunately the radial dependence of the sun's rotation has not been so easy to measure, nor is it possible to determine it theoretically.

Therefore a great deal of attention has been given to the prospect of measuring the sun's internal rotation using the frequency splitting of solar oscillation modes. Two questions are in the forefront: what is the rotation as a function of radius and latitude in the convection zone, where it drives the dynamo, and does the sun possess a rapidly rotating core.

The first measurements of the rotational splitting of the solar oscillation frequencies by Duvall and Harvey[1] showed that the sun's interior is not rotating at a vastly different rate than it's surface, but since the measurements included only sectoral modes nothing could be said about the sun's interior rotation as a function of latitude. Later measurements by Brown[2] and Libbrecht[3] improved the situation, and the latest measurements by Brown[4] and Duvall, Harvey, and Pomerantz[5] are good enough to begin distinguishing between different models for the solar rotation.

I have made some improvement to my earlier measurements at low l by combining data from 1985 and 1986. The two data sets were 12 and 14 days long, respectively, and both were obtained at Big Bear Solar Observatory using the dedicated helioseismology telescope there. The frequencies of individual modes were determined from the data using essentially the same techniques as I described earlier[3], since this provides some improvement over a simple cross-correlation of power spectra. The mode frequencies were fit to Legendre polynomials as described in Ref. 5, and the resulting fit coefficients are shown in Figure 1.

These results are consistent with previously published results, and the error bars are unfortunately not much smaller. Both the A_1 and A_3 terms are smaller

B. R. Durney and S. Sofia (eds.), The Internal Solar Angular Velocity, 59–62.

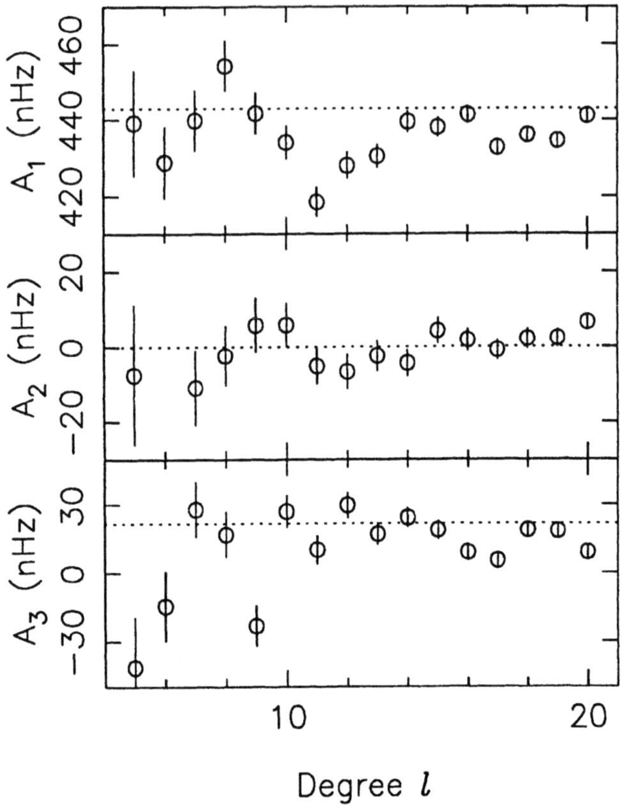

Figure 1. Rotational splitting of solar oscillation modes, in the form of Legendre polynomal fit coefficients[5] as a function of spherical harmonic degree l. The dotted lines show the coefficients for rotation at the surface rate.

than their surface values, and there does not appear to be significant evidence for a non-zero A_2 term at these low l-values.

Although these measurements and those cited above are headed in the right direction, higher accuracy must be achieved before a really good determination of the sun's interior rotation rate can be realized. Errors in the Legendre fit coefficients are currently at about ± 5 μHz at intermediate l-values, and are much worse for $l < 5$. Fortunately at intermediate l-values, which are the most important for measuring the rotation of the convection zone, there is still room for another quick order-of-magnitude gain in splitting sensitivity.

Figure 2 shows snips of power spectra from 1985 data for modes with a single l and n, but different order m. The rotational splitting is of course very clear in these data, but note also the mixing between the different m-states. Since we can only see a single hemisphere of the sun, the $Y_l^m(\theta,\phi)$ are not orthogonal, and so several different modes are included when a single Y_l^m is fit to a Doppler image. Since the different modes are excited to different amplitudes, it is difficult to accurately determine the frequency of a single oscillation mode.

With a timeseries greater than one month long it should be possible to clearly resolve the different m-states, and therefore determine the mode frequencies to a

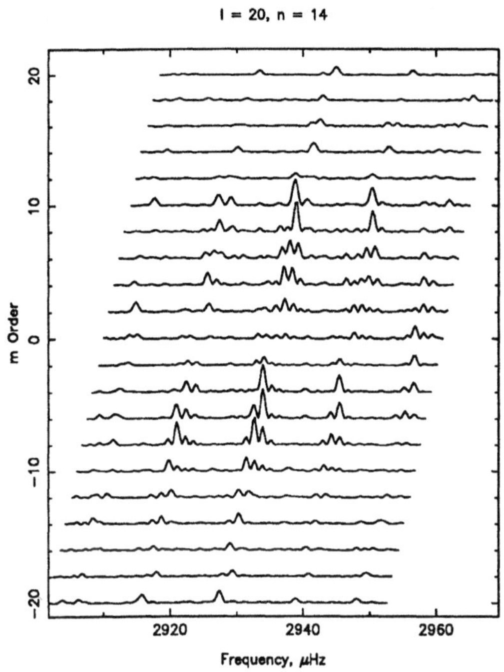

Figure 2. Power spectra snips from a 12-day timeseries, for modes with $l = 20$, $n = 14$, and even m. Multiple peaks in some of the spectra show overlap between the different m-states, which is a result of viewing only one hemisphere of the sun. The different m are separated by 0.8 μHz in frequency, so are just barely resolved in this figure.

much greater accuracy. Of course this will only work if the mode linewidths are less than the 0.8 μHz spacing between the modes. Although this is not the case near the power maximum at $\nu = 3$ mHz, it appears to be true for the lower frequency modes. Modeling of the data could be used if the linewidths are too large to be cleanly resolved. At Caltech we are currently analyzing a 4-month data set consisting of over 100 days of Doppler data, which should be more than long enough to do the job.

Measuring the splitting of the very low l modes, in particular $l = 1$, is important for determing the rotation rate of the solar core, but working at these low l continues to be a problem. It is worth noting that no helioseismology group has been able to produce usable $l = 1$ power spectra when imaging the entire solar disk. Tests were made at Big Bear using a simple DC light bulb shining on the camera in place of the sun, without any narrow bandpass filter. The low spatial frequency noise turned out to be the same as with real solar data, indicating that the noise originates either in the CCD camera or the electronics which digitizes the camera signal. I suspect the same is true for other groups. This will be a difficult (and expensive) problem to fix, so at Caltech we are trying to measure the low-l splitting using a different approach. A current project involves imaging the sun on a four-quadrant photocell, which we hope will reduce the noise enough to allow us to make a better measurement of the $l = 1$ splitting.

2. Frequency Shifts

Recently Woodard and Noyes[6] have presented data indicating that the solar oscillation mode frequencies change with the solar cycle. Since this is a very interesting result, I have also used the 1985 and 1986 data sets to look for shifts in the mode frequencies.

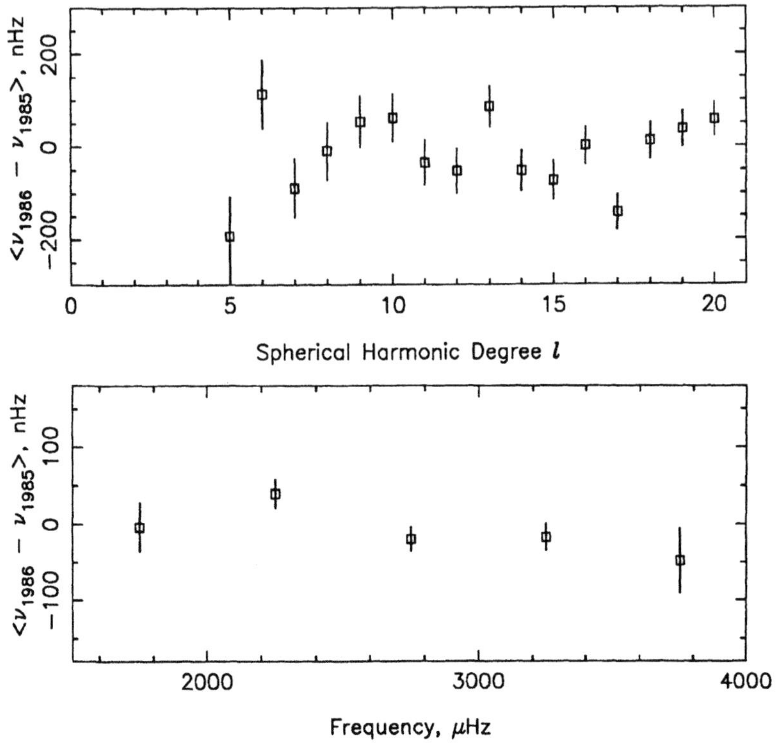

Figure 3. Shifts in solar oscillation mode frequencies between 1985 and 1986 as a function of ν and l. Both plots are consistent with zero frequency shift.

The results shown in Figure 3 indicate that the frequencies were constant between the summers of 1985 and 1986 at about the 0.02 μHz level. The errors in the present frequency comparison are much smaller than that presented by Woodard and Noyes, but the two results need not be inconsistent. The present comparison spans only one year, and that year is very close to solar minimum, so it is to be expected that the frequency shifts would be much less. Nevertheless one should note that we now have the machinery to measure frequency shifts with very high accuracy during the upcoming solar cycle.

References.

1. Duvall,T.L.Jr & Harvey,J.W. *Nature* **310,** 19 (1984).
2. Brown,T.M. *Nature* **317,** 591 (1985).
3. Libbrecht,K.G. *Nature* **319,** 753 (1986).
4. Brown,T.M. *These Poceedings.*
5. Duvall,T.L.Jr, Harvey,J.W. & Pomerantz,M.A. *Nature* **321,** 500 (1986).
6. Woodard,M.F. & Noyes,R.W. *Nature* **318,** 449 (1985).

CHROMOSPHERIC ACTIVITY IN OPEN CLUSTERS

E. J. Pilger
University of Hawaii
Institute for Astronomy
2680 Woodlawn Drive
Honolulu, HI 96822
USA

ABSTRACT. Spectra of Ca II H and K are being taken for stars of similar mass in the Hyades, the Pleiades, and NGC 752. These spectra will be used to create indices of chromospheric activity on these stars. The dispersion in these indices will then be compared to model dispersions which take into account stellar inclination, position of active regions, and filling factor. Only a few observations have been made to date. These show that a high signal to noise is achievable over reasonable exposure times. Modelling has only just begun.

1. INTRODUCTION

It has been known for some time that there is a general decrease, wth age, in the level of a stars' chromospheric activity. Wilson (1963) first suggested that the strength of the Ca II H and K reversal is higher in young stars than it is in older stars. Kraft (1967), and many other have subsequently shown that this difference can be interpreted as a general trend toward lower activity in older stars. Taken with the known decrease in rotation rate of stars with age this supports theories for the spinning down of stars by magnetic breaking. Dynamo production of magnetic fields is then linked to the rotational rate of the star, both declining with age.

This simple picture, though instructive, is not the whole story. Dispersion in the above relation, many times the allowable errors in the observations, is found in the majority of these relations. Barry et al. (1981) discovered a bimodal distribution of activity for stars in the old cluster M67. Duncan (1981), in a sample of stars whose ages were determined by Li abundance, finds up to a factor of four difference in the levels of activity for some of the youngest objects in his sample. Finally, Varsik (1984) finds a similar effect in the relation of activity to photometrically determined ages in field stars. He sees both young stars with low levels of activity, and some old stars with high levels of activity. Clearly some additional factor

B. R. Durney and S. Sofia (eds.), The Internal Solar Angular Velocity, 63–67.

is going to have to be considered to account for the large dispersion in the age / rotation / chromospheric activity relationship.

A number of factors could easily be accused of having some effect on activity levels. Differing masses, compositions and stellar activity cycles could all contribute to stars with the same mean activity levels being observed with a range of values. On the other hand, stars of the same age may not all have the same rates of rotation, thus giving them different mean levels of activity. The presence of stellar activity cycles has been well documented in the survey of Wilson (1978). However, the amplitude of these cycles has also been well documented. Cycles have not been observed larger than about 24 percent of the mean, and are normally only about 10 percent. This is not enough to account for the 100's of percent differences that have been observed. The effect of mass on chromospheric activity levels can be estimated through its effect on the convection zone. Noyes (1983) and Noyes et al. (1984) have plotted activity against the Rossby number (a number that measures, in effect, the relative size of the convection zone.) They find that by choosing an appropriate mixing length, they can get a better correlation between age and activity. Their best fit, though, still has a dispersion of around 30 percent of the mean. Stellar composition affects chromospheric activity levels because it determines the underlying spectrum from which the indices are calculated. A number of methods have been used to remove this underlying photospheric contribution. Most have been unsatisfactory because they entail averages over either different masses, wavelengths, or compositions.

The object of this thesis work will be to attempt to account for the effects of as many factors as can be physically accounted for, and then to see what role, if any, remaining factors could play. By observing select groups of stars in the Pleiades, the Hyades and NGC 752, I will be able to create groups of stars with similar ages, masses and compositions. This will minimize the effects of these factors within a given group. Additionally, by modelling the contributions to observed activity of the orientation of stars in the sky, and the position of areas of activity on a star, the effect of these factors can also be reduced. Any unexplained dispersion that then remains can be interpreted in the light of possible differences in rotational rate.

2. MATERIALS AND METHODS

Observations are being made with the University of Hawaii 2.24 meter telescope using the Faint Object Spectrograph with a TI 500X500 3 phase CCD. The spectral coverage is about 350 Angstroms, and the resolution, with a 160 micron slit, is about 1. Angstrom.

Due to the extreme drop off toward the blue of both the sensitivity of the instrument, and the output of the flat lamp, flat frames are first normalized by an average along the slit, fit to a polynomial. All spectral frames are then dark subtracted and flat divided, and finally compressed into a one dimensional spectrum. For each spectrum, a wavelength scale is determined and a new spectrum is generated by interpolation onto uniform wavelength intervals. These spectra are then used to create Ca HK indices.

The indices used on this first data set are patterned after those of Labonte (1986), which in turn are designed to match as closely as possible, with spectral data, the Mount Wilson S index (Vaughan and Preston 1980.) The spectra are first straightened through division by a straight line fit to the intensities at 3912.6 Angstroms and 3988.0 Angstroms.

The two following indices are then created:

HK narrow: The sum of the integrals over 1 Angstrom bands centered on Ca II H and K (3933.69 Angstroms and 3968.49 Angstroms) is divided by the mean of a 5 Angstrom band centered at (3913.0 Angstroms and a 4 Angstrom band centered at 3989.5 Angstroms.

HK broad: The integral over a 56 Angstrom region, centered on Ca H and K, is divided by the mean of the same two regions as above. This index was chosen to experiment with the conclusion by Labonte (1986) that a limit in magnitude of 2 could be gained by integrating over a much broader region in H and K.

Modelling of contributions of stellar inclination and active region position to the Ca index have been started using a simplified model of a stellar surface divided into 1302 equal elements. Further modelling will add surface coverage of active areas, and stellar activity cycles as possible contributors to dispersion.

3. RESULTS

Samples of stars of approximately the same mass were chosen from each of the three clusters NGC 752, the Hyades, and the Pleiades. The specifics are as follows:

Pleiades: 50 stars, .47 <= B-V <= .73
Hyades: 98 stars, .43 <= B-V <= .9
NGC 752: 23 stars, .44 <= B-V <= .73

Note: The larger range in B-V for the Hyades sample is due to the fact that two mass ranges have been included for observation in this nearby cluster.

A set of 23 stars was also chosen from the list of stars determined to be chromospheric activity standards by Wilson (1978). These are to be used in calibrating my activity indices to his system.

Out of the above list, 53 stars have been observed to date. These stars include 11 from NGC 752, 32 from the Pleiades, and 10 of the standards. The averages of all of the spectra in each of these groups can be seen in Fig. 1. The S/N ratio for these spectra is very good, approaching 100 in the core of H and K for a number of stars. However, the activity indices determined for the standards do not reflect this accuracy. I believe this to be primarily due to the way the indices are determined. This is borne out by the 1 to 2 percent accuracy I get when deriving indices developed by a co-worker for the same region, observed with the same instrument. I should be able to minimize this effect by choosing regions in the wing which are not affected by

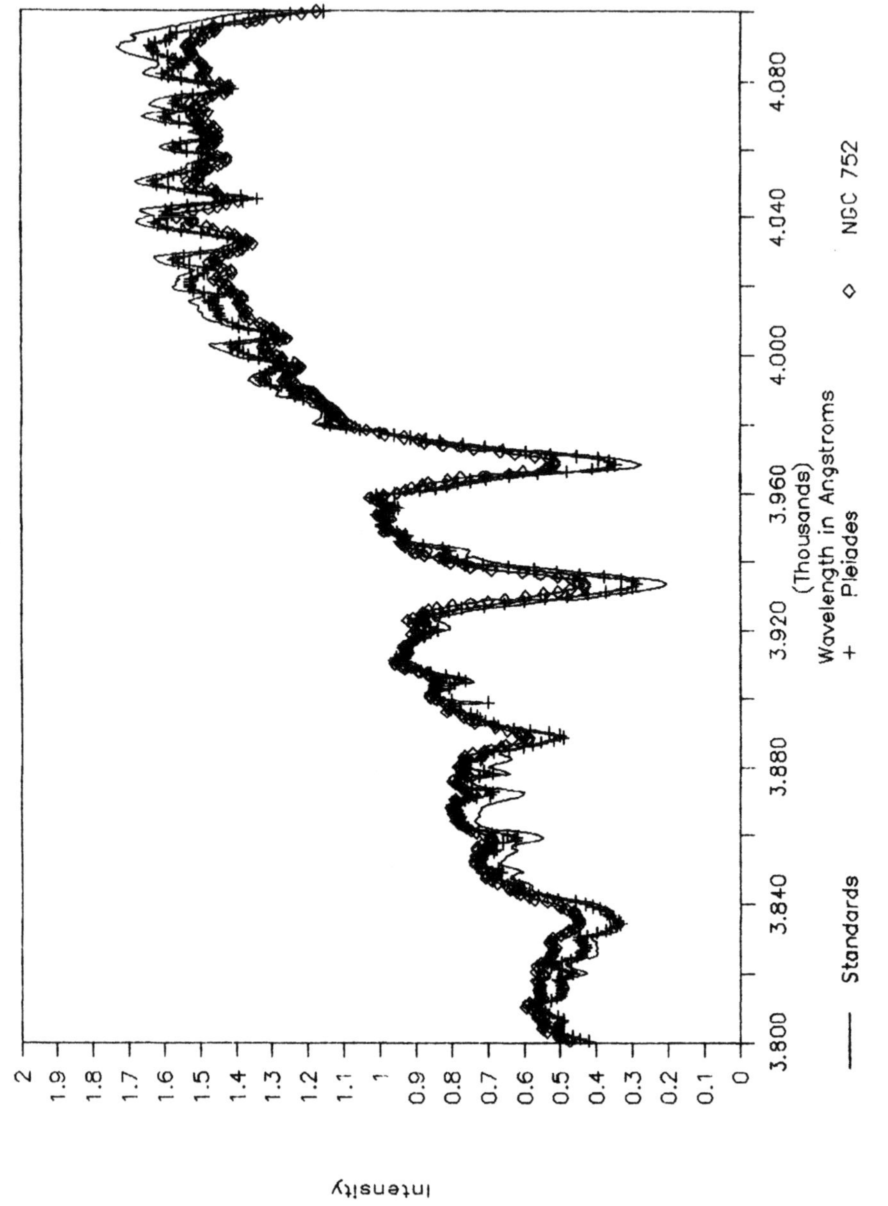

Fig. 1: Average spectrum for each group
(normalized to average over wavelength)

changes in line width, and by correlating my index to some appropriate line width indicator.

4. DISCUSSION

Varsik (1986) has observed that the Ca HK index is most sensitive to changes in activity level for stars of spectral type around middle G. Further observations on my part will concentrate on this range. The S/N ratio for my spectra is sufficient for my needs. However, a better method for creating activity indices will have to be developed. This should include some estimate of the underlying photospheric contamination of the spectra, if possible in a spectral form.

5. CONCLUSION

The remainder of the sample of stars will need to be observed. The underlying data are good, but in order to get more accurate indices a better way of generating them needs to be developed. One important aspect will be the removal of underlying photospheric spectrum, perhaps through the use of model radiative photospheres. The modelling of the contributions from different regions on the surface of a star will need to be expanded to include multiple regions. The addition of better estimates of the center to limb variation of the cores of H and K will also be helpful in this regard.

6. ACKNOWLEDGEMENTS

This work was supported by NASA grant NSG 7536. Dr. Barry Labonte was the source of numerous helpful ideas throughout. Dr. Alan Stockton is to be thanked for explaining and demonstrating to me the workings of the Faint Object Spectrograph.

7. REFERENCES

Barry, D.C., Cromwell, R.H., Hege, K. 1981, Ap. J., 247, 210.
Duncan, D.K. 9181, Ap. J., 248, 651.
Kraft, R.W. 1967, Ap. J., 160, 551.
Labonte, B.J. 1986, Ap. J. Supp., 62, 229.
Noyes, R.P. 1983, in IAU Symposium 102, Solar and Stellar Magnetic Fields: Origins and Coronal Effects, ed. J. O. Stenflo (Dordrecht: Reidel), p. 133.
Noyes, R.W., Hartmann, L.W., Baliunas, S.L., Duncan, D.K., Vaughan, A.H. 1984, Ap. J., 279, 763.
Varsik, J. 1984, in press.
Varsik, J. 1986, private communication.
Vaughan, A.H., and Preston, G.W. 1980, Pub. A.S.P., 92, 385.
Wilson, O.C. 1963, Ap. J., 138, 832.
Wilson, O.C. 1978, Ap. J., 226, 339.

FULL-DISK SOLAR DOPPLERGRAMS OBSERVED WITH A ONE-MEGAPIXEL CCD CAMERA AND A SODIUM MAGNETO-OPTICAL FILTER

Edward J. Rhodes, Jr.
University of Southern California
Los Angeles, California 90089-1342
Jet Propulsion Laboratory
California Institute of Technology
Pasadena, California 91109

Alessandro Cacciani
University of Rome, Rome, Italy

Steven Tomczyk
University of California at Los Angeles
Los Angeles, California 90024

ABSTRACT. We present here the first two full-disk solar Dopplergrams obtained with the new 1024 x 1024-pixel CCD camera which has recently been installed at the 60-Foot Tower Telescope of the Mt. Wilson Observatory. These Dopplergrams have a spatial resolution of 2.2 arcseconds and were obtained in a total of one minute of time. The Dopplergrams were obtained with a magneto-optical filter which was designed to obtain images in the two Na D lines. The filter and the camera were operated together as part of the development of a Solar Oscillations Imager (SOI) esperiment which is currently being designed at JPL for the joint NASA/ESA Solar and Heliospheric Observatory (SOHO) mission. We also include here two difference images obtained by subtracting two pairs of the Dopplergrams from the initial time series.

1. DESCRIPTION OF EQUIPMENT

As part of the development of the SOI experiment at JPL, a 1024 x 1024-pixel CCD camera was fabricated there and then installed at the Mt. Wilson Observatory. This camera is a modification of one orignally designed for the Solar Optical Telescope (SOT) program. It is based upon a Texas Instruments virtual-phase, front-side illuminated CCD array. This array is comprised of 1,048,576 pixels each 18.3 microns on a side. Hence, it has an active area of 18.7 mm square.

Each pixel of the array has a full-well depth of 200,000 electrons, while the complete camera noise floor is about 50 electrons. Hence, a maximum signal-to-noise ratio of 4000 to 1 is possible in a single exposure. To take advantage of this SNR, the A-to-D converter digitizes the analog video signal to 12 bits. The camera electronics has been designed to allow a readout rate of 800,000 pixels per second.

B. R. Durney and S. Sofia (eds.), The Internal Solar Angular Velocity, 69–74.

Hence, a single frame can be read out in 1.3 seconds. For the images employed to create the Dopplergrams shown here, an exposure time of about 0.64 seconds was required. Therefore, the combined exposure and readout time for each filtergram was about 2 seconds. Since two filtergrams were needed for each Dopplergram, the minimum time needed to acquire one Dopplergram was only 4 seconds. However, in these initial tests we employed a shutter cycle of 15 seconds and hence the two Dopplergrams shown here were actually separated by 30 seconds.

Figure 1 contains a close-up photograph of the T.I. CCD chip itself, while the completed camera is shown as it is now installed at Mt. Wilson in Figure 2. The sunlight is passed through the two magneto-optical filter (MOF) cells located to the left of the camera in Figure 2 before it forms an image on the chip near the center of the Figure. A schematic diagram of the data acquisition and processing system which is now in operation at the 60-Foot Tower Telescope is shown in Figure 3. A key part of this system is the high-speed, floating point array processor from CSPI. This processor contains a custom-designed interface port which allows us to store each CCD image directly in a 2-megabyte array during the 2-second exposure and readout interval of the subsequent frame. The array processor can then numerically integrate or subtract the subsequent frames from the first. The data are transferred from disk to tape daily. The camera and the data acquisition system were described in more detail in Rhodes, et al. (1986). Figures 4 and 5 show the first two solar Dopplergrams ever obtained with the MOF and camera combination, while Figure 6 contains an enlargement of one quadrant of Figure 4. Figure 7 shows the image obtained when we subtracted the first and second Dopplergrams, while Figure 8 shows the result of differcing the first and third Dopplergrams. Figure 9 shows an enlargement of one quadrant of Figure 8.

2. REFERENCES

Rhodes, E. J., Jr., Bursch, T. K., Ulrich, R. K., and Tomczyk, S. 'A one megapixel image acquisition and processing system for Solar Oscillation Studies,' to appear in Proceedings of Instrumentation in Astronomy VI, SPIE, in press, 1986.

Figure 1. Close-up photograph of Texas Instruments' 1024 x 1024-pixel virtual phase CCD array. The array is shown in its 43-pin socket. The array is illuminated from the top of the figure.

Figure 2. Installation of CCD camera at Mt. Wilson 60-foot solar telescope. The CCD array is contained within the round aluminum enclosure near the center of the picture, while the signal chain, the timing circuitry, and the control electronics are located within the rectangular box at the right. The magneto-optical filter includes the 6000-Gauss and 2000-Gauss magnet assemblies at the left.

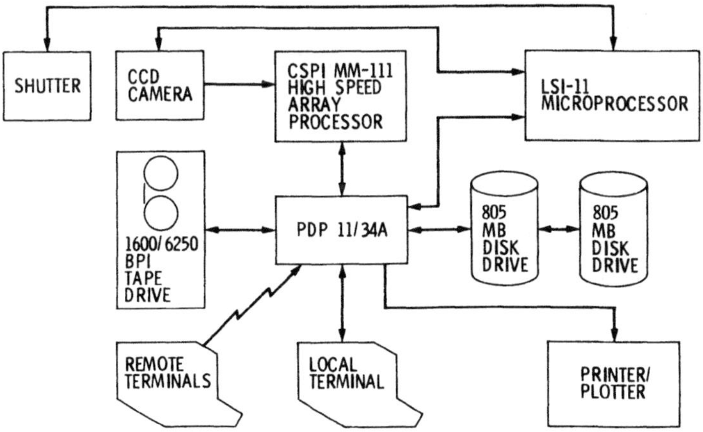

Figure 3. Schematic diagram of the entire data acquisition and processing system.

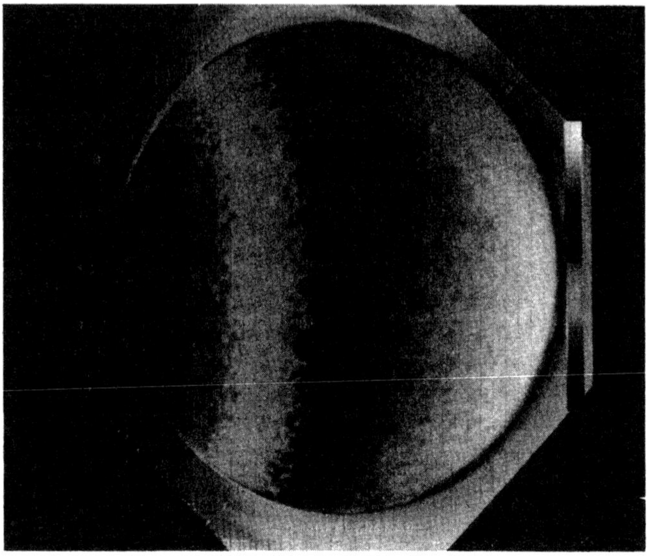

Figure 4. First of five full-disk Dopplergrams obtained with the CCD camera and the Na magneto-optical filter on April 10, 1986. This Dopplergram was obtained by differencing two filtergrams which were taken 15 seconds apart. The solar disk covers about 900 x 900 pixels and has been rotated in order to place the equator in the horizontal direction. Solar rotation is visible as the systematic change in gray levels across the disk.

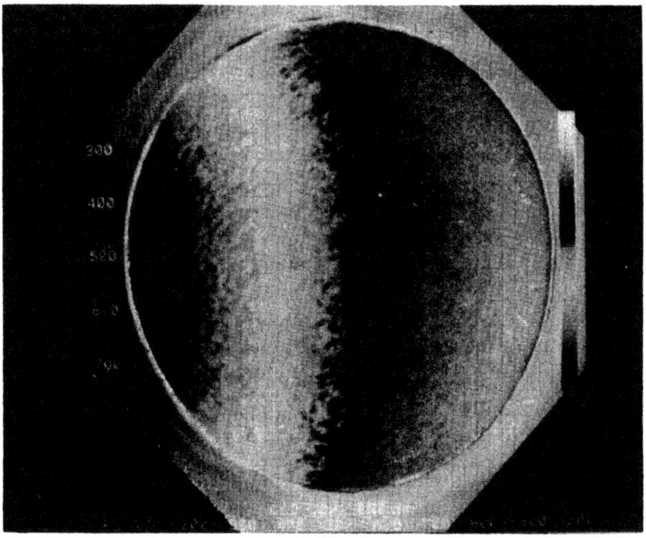

Figure 5. Second Dopplergram which was obtained 60 seconds after that of Figure 4. The diagonal streaks are due to partially-blocked columns in the CCD array itself. Our processing software did not correct for these bad columns.

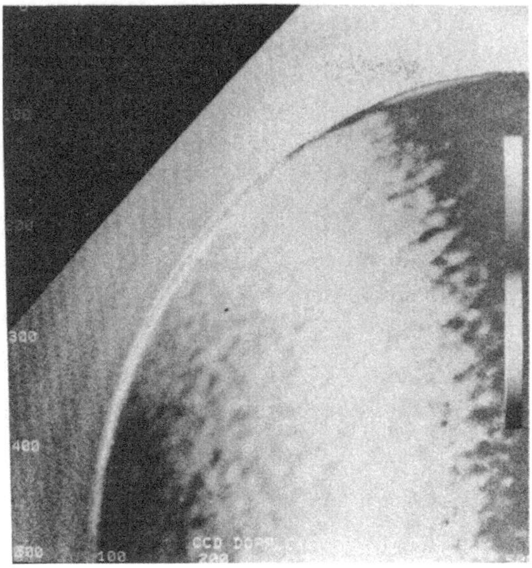

Figure 6. Enlargement of one quadrant of Dopplergram shown in Figure 4. Here every pixel within 512 x 512 pixel area is displayed to accentuate the detail of small-scale features.

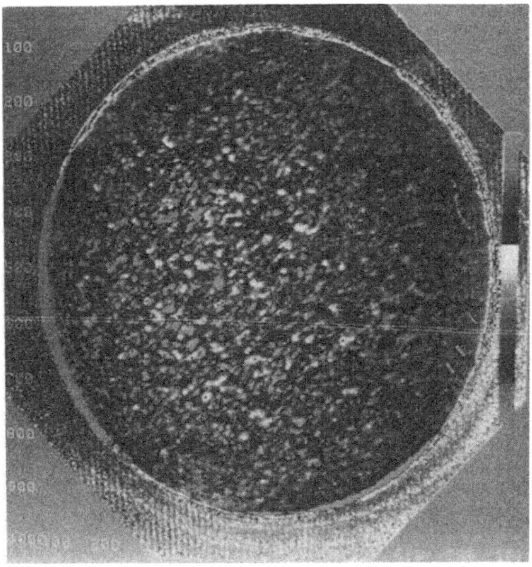

Figure 7. Difference of first and second Dopplergrams. This type of display suppresses solar rotation and allows the oscillating elements to become visible. Also visible is the loss of spatial resolution which becomes evident due to foreshortening near the solar limbs.

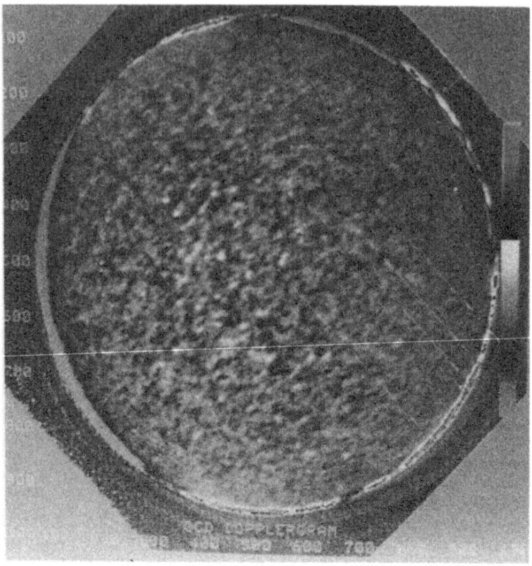

Figure 8. Difference of first and third Dopplergrams. This display accentuates those velocity areas which changed appreciably during the 120 seconds separating Dopplergrams 1 and 3.

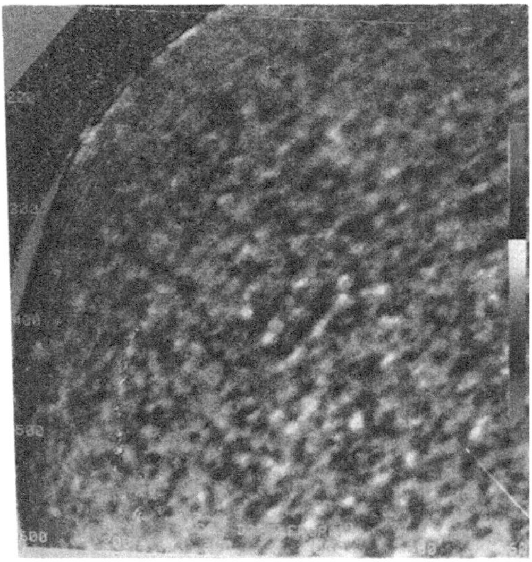

Figure 9. Close-up of one quadrant of difference image shown in Figure 8.

ESTIMATES OF THE SOLAR INTERNAL ANGULAR VELOCITY OBTAINED WITH THE MT. WILSON 60-FOOT SOLAR TOWER

Edward J. Rhodes, Jr.
University of Southern California
Los Angeles, California 90089-1342 and
Jet Propulsion Laboratory
California Institute of Technology
Pasadena, California 91109
Alessandro Cacciani
University of Rome
Rome, Italy
Martin Woodard
Jet Propulsion Laboratory
California Institute of Technology
Pasadena, California 91109
Steven Tomczyk
Sylvain Korzennik
Roger K. Ulrich
University of California at Los Angeles
Los Angeles, California 90024

ABSTRACT. We have obtained estimates of the solar internal angular velocity from measurements of the frequency splittings of p-mode oscillations. Specifically, we have analyzed a 16-day time series of full-disk Dopplergrams obtained during July and August 1984 at the 60-Foot Tower Telescope of the Mt. Wilson Observatory. From this time series we computed power spectra for all of the zonal, tesseral, and sectoral p-modes from $\ell = 0$ to 89 and for all the sectoral p-modes from $\ell = 90$ to 200. We next calculated a mean power spectrum for each degree up to 89 and then computed the frequency differences of all of the different non-zonal modes from these mean power spectra. The resulting frequency-difference curves for $\ell = 1$ to 89 were then expanded as a series of Legendre polynomials. For the odd-order terms in these expressions our results agreed reasonably well with the earlier results of Duvall, Harvey, and Pomerantz[1]. Next we combined the odd-order expansion coefficients into estimates of the solar internal equatorial rotational frequency splitting. We checked these estimates by similarly combining the published coefficients of Duvall, Harvey, and Pomerantz. We also compared our estimates with the earlier results of Hill and his colleagues at SCLERA. Our comparisons showed reasonably close agreement with the Duvall, Harvey, and Pomerantz results but sharp disagreement with SCLERA results.

B. R. Durney and S. Sofia (eds.), The Internal Solar Angular Velocity, 75–82.

1. Introduction and Discussion

Several recent studies have obtained estimates of the solar internal equatorial rotation rate from measurements of the frequency splittings of oscillations[1-6]. All but one of these studies obtained estimates of the rotation rate which were either near or slightly below the surface rotation rate of magnetic features. The one discordant study claimed to have evidence for a much higher rotation rate throughout most of the solar interior[2]. In order to provide an additional independent estimate of the internal rotational velocity, we have developed a helioseismology observing station at the Mount Wilson Observatory. We recently presented the first preliminary results from a small portion of the data we obtained with this system during the summer of 1984[7]. We report here on the results of additional analysis of that data.

During the summer of 1984 we obtained solar full-disk Dopplergrams every 40 seconds for up to 11 hours per day on 90 different days. These Dopplergrams were obtained with a magneto-optical filter feeding a CID camera. We have analyzed a subset of these images which was obtained on 14 out of 16 consecutive days in July and August in order to obtain our own estimates of the internal equatorial rotation rate. Spherical harmonic filtering yielded time series of sectoral

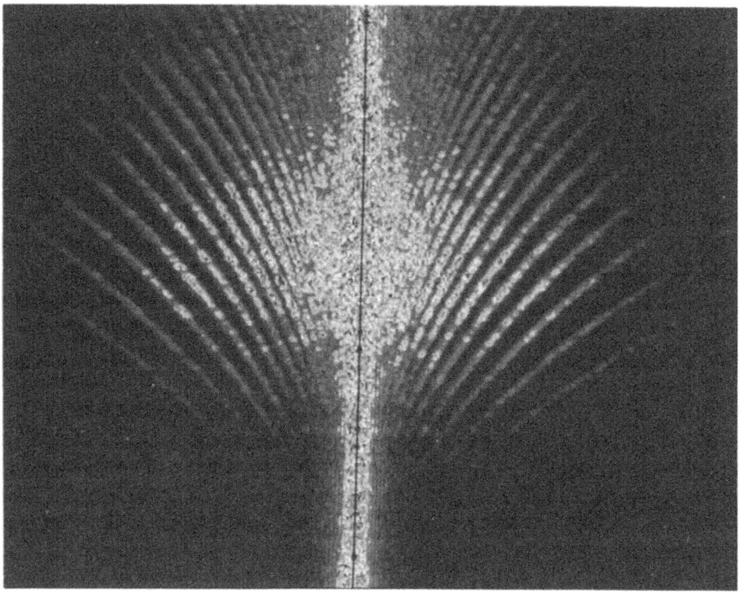

Figure 1. Two-dimensional, ℓ-ν, power spectrum computed from 9.5-day subset of time series of sectoral harmonic amplitudes. The vertical black line in the center is the temporal frequency axis, while the x-axis ranges from ℓ = −200 at the left to ℓ = 200 at the right. The sharp ridges of observed power are due to the solar p-modes. At least 23 ridges can be distinguished in each half of the spectrum.

harmonics which were Fourier transformed to yield the power spectrum of Figure 1. Figure 2 shows that this spectrum clearly contained frequency shifts induced by solar rotation.

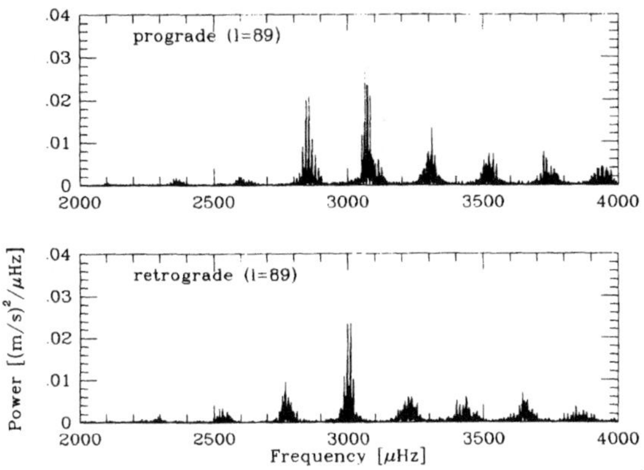

Figure 2. Two vertical slices through the power spectrum shown in Figure 1. Here the prograde and retrograde sectoral harmonic peaks corresponding to a degree of 89 are displayed as a function of the absolute temporal frequency, ν, in mHz. The prograde peaks can be clearly seen to have higher frequencies than do the corresponding retrograde peaks in the lower panel. Since the total prograde-minus-retrograde frequency splitting is equal to 2 x 89 x $\Delta\nu$, where $\Delta\nu$ is the rotational frequency, it is clear that $\Delta\nu$ can be computed easily by dividing the measured splitting by 188. The sidelobes present in each peak are due to both the temporal window function and to spatial leakage from adjacent degrees.

2. Analysis and Conclusions

Figure 3 shows three examples of the m-ν power spectra which resulted when we spatially filtered our Dopplergram time series for every different spherical harmonic between ℓ = 0 and ℓ = 89. Just as Figure 2 showed that the frequency splitting induced by the solar equatorial rotational velocity is clearly visible in the sectoral-harmonic power spectra at ℓ = 89, Figure 3 extends this result down to ℓ = 30 and also illustrates that measureable latitudinal differential rotation (which yields the s-shape of the power spectra) exists at ℓ = 60, 45, and 30. We have made additional m-ν plots for ℓ = 15, 5, and 3 and all of them show a similar slope. On the other hand, a rotational profile similar to that described by H. Hill would have resulted in m-ν spectra having much steeper slopes than we saw in any of our plots. Once we had obtained a preliminary estimate of the solar rotational frequency splitting from these m-ν spectra, we proceded to analyze the spectra in the manner described by Duvall, Harvey, and Pomerantz. Specifically, we first removed to first order the frequency shifts introduced by solar rotation in all of the 2ℓ +1 different spectra at a given ℓ -value and we next collapsed the 2ℓ +1 shifted spectra into a single averaged power spectrum for that degree. We then computed

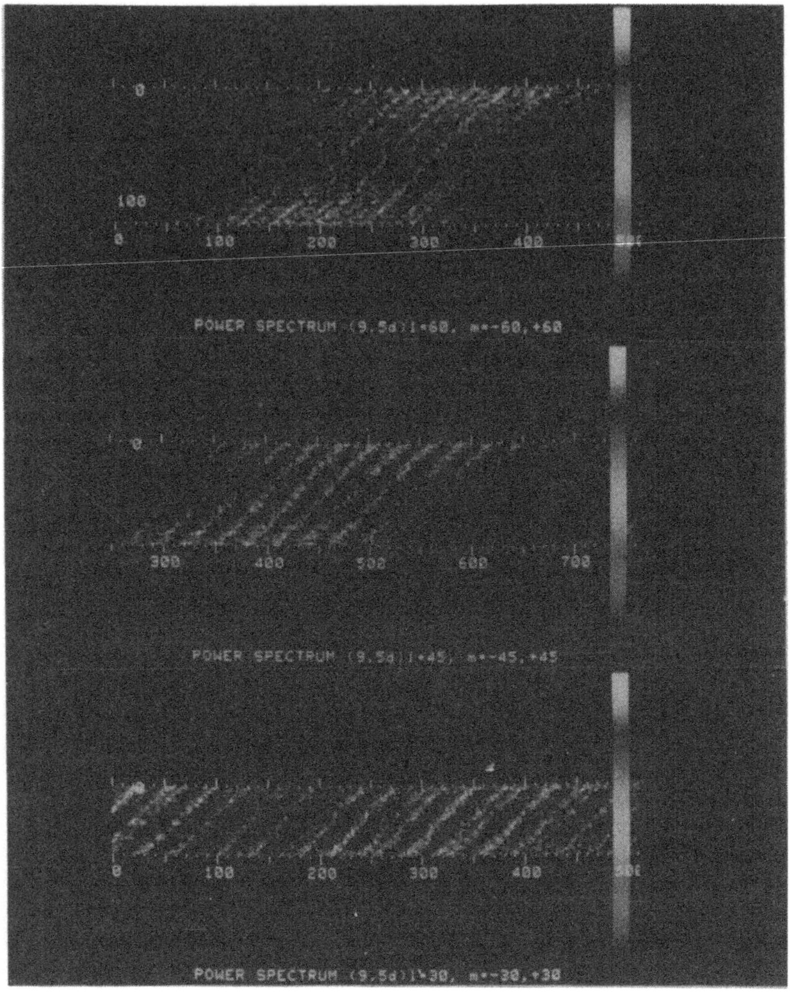

Figure 3. Three m-ν power spectra corresponding to degrees 60 (top), 45 (middle), and 30 (bottom). For all three spectra the observed power is plotted as a set of grey levels as a function of temporal frequency, ν (increasing horizontally), and azimuthal order, m (increasing vertically). Also visible in all three cases are the various spatial and temporal sidelobes which surround each of the true modal peaks. The s-shaped nature of each frequency curve is what is represented by the Legendre polynomial expansion described in the text. The overall slope of each curve is a measure of the equatorial rotational velocity. The much higher rotational velocity described by the SCLERA group would result in a much larger tilt to the curves in all three parts of this figure than is observed.

all of the frequency differences between the individual spectra and the mean spectrum for the chosen degree with a cross-correlation technique. We then iterated this entire procedure two more times until the individual frequency differences which resulted did not change between iterations.

We next represented the frequency variation with m for a given ℓ -value as the series:

$$\nu(n,\ell,m) - \bar{\nu}(n,\ell) = L \sum_{i=0}^{i=N} a_i P_i(-m/L) \qquad (1)$$

where $\bar{\nu}$ is the frequency of the nth p-mode averaged over m, the P_i are the Legendre polynomials of degree i, and $L = \sqrt{\ell(\ell+1)}$. This expansion is exactly that employed by Duvall, Harvey and Pomerantz[1] and was carried out with software which was adapted directly from their codes. These authors also noted that, expressed in terms of equation (1), the surface differential rotation as measured by the motion of magnetic features[8] yields even-indexed coefficients equal to zero and odd-indexed coefficients (nHz sidereal) a_1 = 442.8, a_3 = 21.7, and a_5 = -2.5.

The results of applying this analysis to the power spectra which resulted from our 16-day observing run are illustrated in Figure 4, where we have plotted the one standard deviation error brackets for the three odd expansion coefficients for $3 \leq \ell \leq 89$.

The top panel of Figure 4 shows the variation of the a_1 coefficient with degree. Also shown as the horizontal dashed line is the surface magnetic value of a_1, converted to a synodic rate. The most striking feature of a_1 is its systematic increase with increasing degree. At the highest degrees shown a_1 lies above the surface magnetic value while below ℓ = 10 a_1 lies below that rate. The overall range of a_1 between ℓ = 20 and 89 is similar to that found by Duvall, Harvey and Pomerantz[1] ; however, the variation with ℓ shown here was not seen by those authors. (In synodic terms their results ranged from 414.4 nHz for ℓ = 20-29, to 410.6 for ℓ = 50-59, and to 414.3 for ℓ = 80-89.)

The middle panel of Figure 4 gives the variation of the a_3 coefficient with degree. Again the surface magnetic feature value for a_3 is shown as the horizontal dashed line. Once more for a_3, our results are similar to those of Duvall, Harvey, and Pomerantz.[1]. In both cases neither set of a_3 values shows any ℓ dependence except possibly between ℓ = 10 and 30 in our results. Between ℓ = 20 and 89 the Duvall et al results for a_3 averaged 20.9 nHz, whereas our average value of a_3 over the same range in ℓ was 17.8 nHz. This difference may be significant but it is still too early to say so with much confidence.

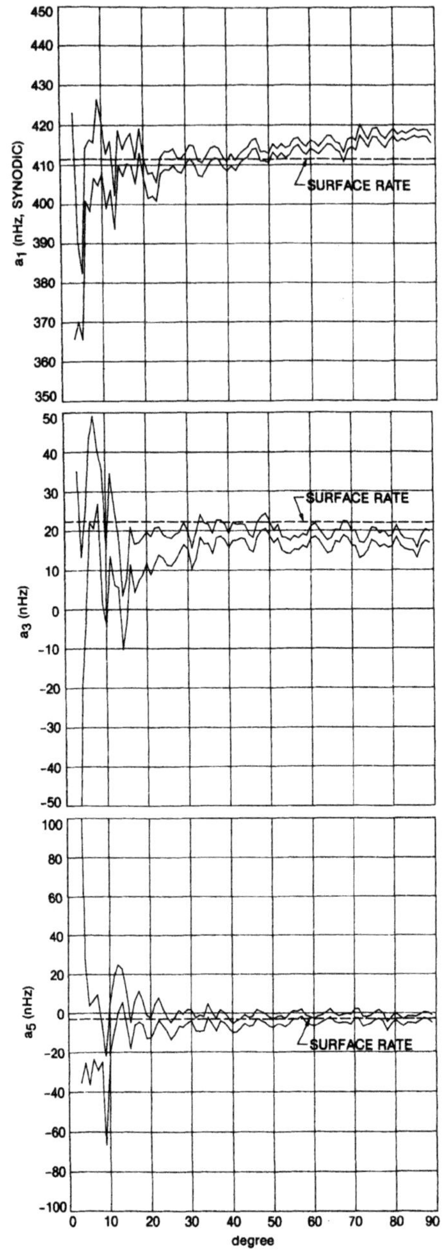

Figure 4. Odd expansion coefficients from equation (1). The two curves shown in each of the three panesl are the $\pm 1\ \sigma$ error bars for each of the individual coefficients. The degree, $\overline{\ell}$, ranged from 3 to 89. The values of a_1 (top), a_3 (middle), and a_5 (bottom) that would have resulted had the entire solar interior been rotating with surface rotation rate for magnetic features[8] are indicated by the dashed horizonatal lines in the three panels.

In the bottom panel of Figure 4 are the values of a_5. To within the scatter of the 1 σ error bars our results appear to be identical to the surface magnetic feature rate of -2.5 nHz. Specifically, over the range ℓ = 20 to 89 our average value of a_5 = -2.8 nHz, while the Duvall, Harvey and Pomerantz results averaged -4.4 nHz. Thus, in summary our a_1 results show more systematic variation with ℓ than do the Duvall et al results; our a_3 values are smaller than either their results or the surface rate; and for a_5 our results are closer to the surface value than are the Duvall et al results.

As soon as we saw the systematic variation of a_1 with ℓ (and the weaker variation of a_3 at a low ℓ) we wanted to determine whether or not these variations in the individual Legendre expansion coefficients were also evident in a plot of the equatorial internal angular velocity, which is simply given by summing a_1, a_3, and a_5. The results of computing this simple sum for all of the degrees between 3 and 89 are illustrated here in Figure 5. In this figure the values of $a_1 + a_3 + a_5$ are given in sidereal, rather than synodic, units and the values themselves are plotted rather than the $\pm$ 1 σ error bars. In this figure the surface magnetic feature frequency splitting

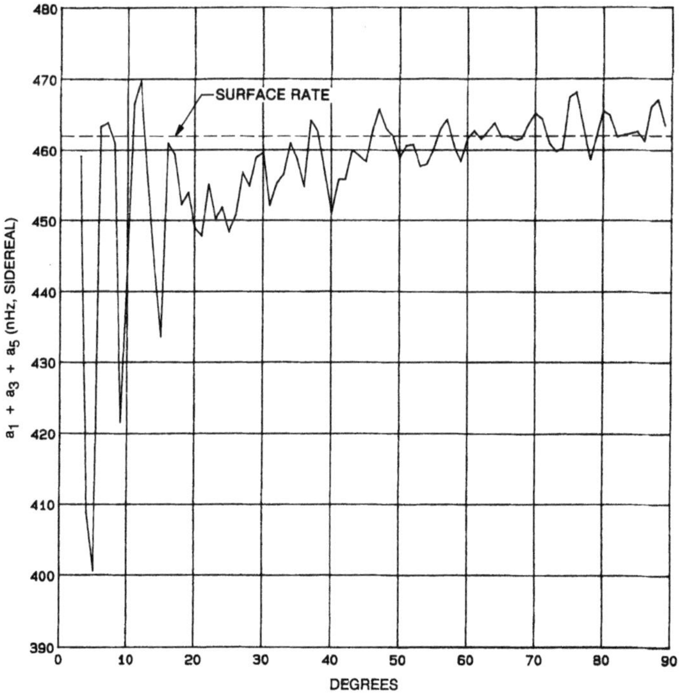

Figure 5. Simple sum of a_1, a_3, and a_5 at each of the individual degrees between 3 and 89. This sum reflects the equatorial rotation rate. The horizontal dashed line at 462 nHz shows the value of the sum that would be expected if the sun rotated with the surface magnetic rotation rate8 at all depths.

is given by the horizontal dashed line at 462 nHz. With the exception of the oscillatory nature of the curve between $\ell = 3$ and 15 Figure 5 shows evidence for a systematic decrease in $a_1 + a_3 + a_5$ with decreasing degree from $\ell = 89$ downward. A preliminary analysis shows that, after the sums shown in Figure 5 have been binned in 10-degree wide bins, the oscillation at low- ℓ disappears and the equatorial internal angular velocity does have a smooth decrease with decreasing degree. The analysis of these binned coefficients is now in progress and will soon be published.

3. Acknowledgements

We wish to acknowledge support from NASA Grant NAGW-13 to USC, NASA Grant, NAGW-472 to UCLA, NSF Grant INT 8400213 to USC. The research conducted at the Jet Propulsion Laboratory of the California Institute of Technology was carried out under contract to the National Aeronautics and Space Administration. Some of this work was carried on while M. Woodard held a NRC/JPL Resident Research Associateship.

4. References

1. Duvall, T. L., Jr. Harvey, J. W., and Pomerantz, M. A., Nature, 321, 500-501 (1986).
2. Hill, H.A., Bos, R.J., and Goode, P.R., Phys. Rev. Lett., 49, 1794-1797 (1982).
3. Duvall, T. L., Jr. & Harvey, J. W., Nature, 310, 19-22 (1984).
4. Brown, T. M., Nature, 317, 591-594 (1985).
5. Brown, T. M. in Seismology of the Sun and the Distant Stars (ed. Gough, D. O.) 199-214.
6. Libbrecht, K. G., Nature, 319, 753-755 (1986).
7. Rhodes, E.J., Jr. Cacciani, A., Woodard, M., Tomczyk, S., Korzennik, S., and Ulrich, R.K., to appear in Proceedings IAU Symposium #123, Christensen-Dalsgaard and Frandsen eds., Reidel, in press, 1987.
8. Snodgrass, H. B., Astrophys. J., 270, 288-299 (1983).

SOLAR ROTATION VARIATIONS FROM SUNSPOT GROUP STATISTICS

I.Tuominen and H. Virtanen
Observatory, University of Helsinki
Tähtitorninmäki,
SF-001300 Helsinki, Finland

ABSTRACT: Variations of solar rotation rate and meridional motion derived using proper motions of sunspot goups from 1874 to 1976 are presented. While the deceleration of the rotation of the groups suggests an inward increase of the rotation near the surface of the solar convection zone, it may also give a possibility to trace the radial dependence of other properties of the mean flow. Rotation varies along the solar cycle in such a way that a minimum occurs near the equator and a maximum in higher sunspot latitudes at the minimum of the solar activity cycle. The rotation was also smaller in the present than in the previous century, correlating with the amplitude of solar activity. The difference of rotation between the N- and S-hemispheres also correlates with the difference of the amplitude of solar activity.

1. INTRODUCTION

The sunspot group positions in Greenwich Photoheliographic Results give a long data base for the studies of solar rotation and other mean properties of the motions in the solar convection zone, and has been used for these by many authors for a long time. In this report we present some examples of our results and some ways to extend them. When seeking the variations of rotation and meridional motion, it may be helpful to keep in mind the following theoretical interpretation. If the motions are interpreted as a feed-back of the dynamo magnetic fields, the equations of motion give the solutions in the form (Rüdiger et al., 1986)

$$\tilde{\Omega} = K^2 \Omega_{\odot} \sum_{1,3,..} (\omega_n + \omega_n^s \sin 2\Omega_m t + \omega_n^c \cos 2\Omega_m t)\, P_n^1 / \sin\theta \qquad (1)$$

for the magnetically produced change of angular velocity and

$$\tilde{u}_\theta = K^2 \sum_{2,4,..} (u_n + u_n^s \sin 2\Omega_m t + u_n^c \cos 2\Omega_m t)\, P_n^1 \sin\theta \qquad (2)$$

for the magnetically produced meridional motion. K is a parameter defining the amplitude of the magnetic field, Ω_m is the frequency of the magnetic cycle, and t time. The expressions (1) and (2) give the following effects: The first term under the sum represents the effect of the magnetic field on the rotation or meridional motion not depending on phase (mean values over the cycle), while the two last ones define the phase dependence. Both effects are proportional to the square of the amplitude of the magnetic field, and the period is always half of the magnetic cycle, i.e. 11 years.

B. R. Durney and S. Sofia (eds.), The Internal Solar Angular Velocity, 83–88.

2. DECELERATION OF THE ROTATION OF SUNSPOT GROUPS

This phenomenon has been studied earlier by Godoli and Mazzucconi (1979), Tuominen and Kyröläinen (1981) and Balthasar et al. (1982). In Fig.1 we give the mean deceleration of all groups (recurrent and non-recurrent) for the two cycles between 1956 and 1976 during eight days after the first observation (Fig.1a) and deceleration of recurrent groups after their first, second, third and more rotations (Fig.1b). We can see that the rotation approaches the photospheric value, given by Doppler measurements (e.g. Howard et al., 1984). The deceleration is about 3 percent, and if interpreted as a faster rotation in the depth where the newly formed spots are anchored, it can be compared to the inward increase of angular velocity near the surface obtained by solar oscillation observations. F.Hill (this workshop) gives an increase of 5 percent, while H.Hill (1984) gives 30 percent increase near the surface.

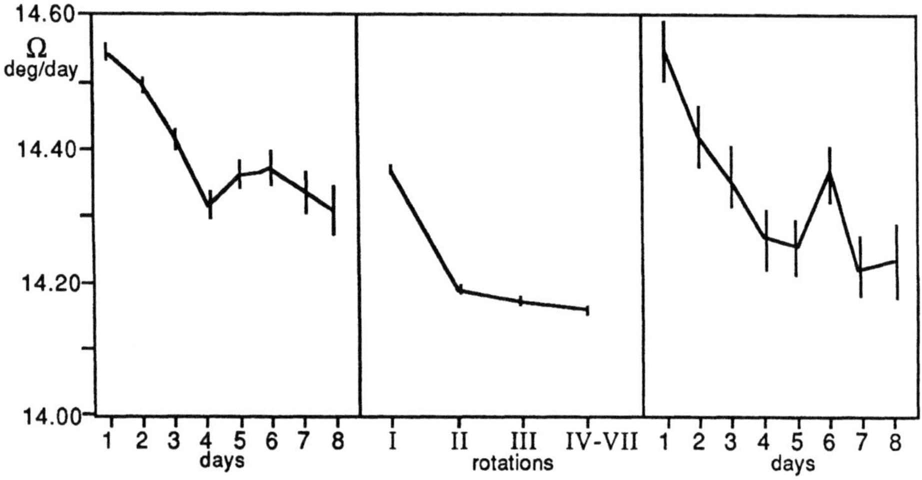

Figure 1. Deceleration of the rotation of sunspot groups, a) newly formed groups during 8 first days of their life, b) recurrent groups in there successive rotations, and c) newly formed groups, which live at least 8 days, during 8 first days of their life. Error bars are 1σ.

It may be argued that the curve in Fig.1a just represents the faster rotation of small spots compared to the large ones, which was found by Howard et al. (1984) using the Mt. Wilson sunspot data. That this may not be the case, can be seen from Fig.1c. It gives the pure deceleration: only those groups are included which live at least eight days. Furthermore, the mean size of the groups of different age in this sample do not correlate with their rotational velocity.

In addition to the information concerning the radial gradient of angular velcity, the investigation of the newly formed groups may give a method to study the other properties of the motions, e.g. the variations of rotation, meridional motions and Reynolds stresses beneath the surface.

3. CORRELATION BETWEEN ROTATION AND THE AMPLITUDE OF THE ACTIVITY

Fig.2 shows the behaviour of the solar angular velocity from 1874. The dotted line gives the number of sunspot groups in these years, and it represent the solar activity. The ordinate is the mean velocity excess for the latitudes 0-35° (deg/day), the mean differential rotation over the whole period being eliminated. Minimum and maximum years are also marked (broken- and continuous vertical lines, respectively).

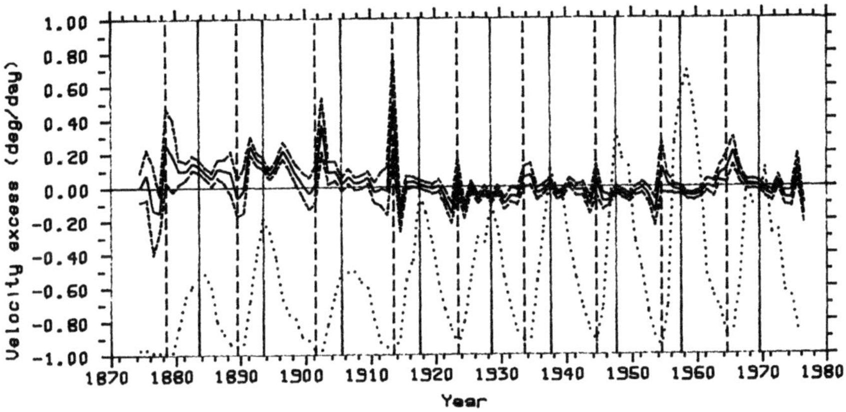

Figure 2. The continuous line represents the mean angular velocity over the sunspot latitudes. The mean differential rotation over the whole data has been subtracted (units are degrees/day).The broken lines give $\pm 1\sigma$ from the mean. The dotted line gives the number of observations, describing the variation of the activity. The broken vertical lines give the minimum years and the solid lines the maximum ones.

As well known the amplitude of the activity is rising from the last century to the present one. We also see that the mean velocity of rotation is at the same time decreasing (the change of the rotational velocity in the beginning of this century was found also by Tuominen and and Kyröläinen, 1981, and by Balthasar et al., 1986). Whether the change of rotation and amplitude are physically related, needs to be confirmed.

To study it further, we have calculated the mean amplitude and the velocity for seven solar cycles, separating northern and southern latitudes (Fig.3). The mean amplitude (Fig.3, the lowest panel) is simply the total number of observations in each cycle.

The three upper panels of Fig.3 show the same velocity excess as Fig.2, but the mean is taken also over each cycle. This is done separately for all groups, for non-recurrent (short living), and for recurrent (living more than one solar rotation), respectively. The panel for all groups clearly shows a negative correlation between the amplitude and the velocity: when the activity is smaller at N- than at S-latitudes, the velocity excess is larger and vice versa. This correlation is even more clear for non-recurrent groups, which on the average are younger. For recurrent groups (only groups on their second rotation are included) the correlation disappears or may even be positive. This means a difference of behaviour between younger and older groups (less decelerated and more decelerated). We notice the strong amplitude dependence of the magnetically produced part of the rotation, given by Eq.(1). This must be, however, further investigated.

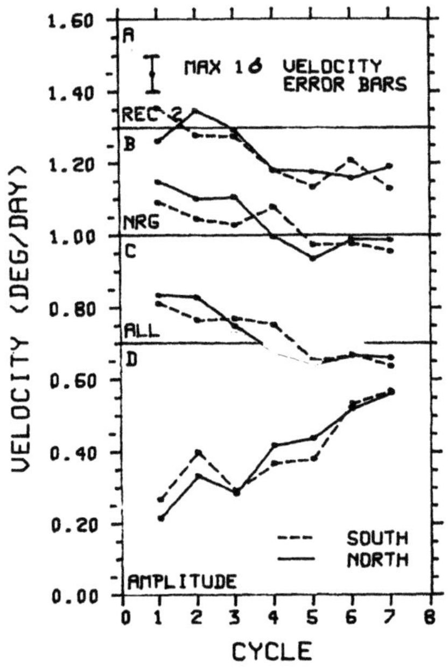

Figure 3. Depedence of the rotation on the amplitude of the magnetic cycle. (A) The mean velocity of recurrent groups in their second rotation for different solar cycles. (B) The velocities of non-recurrent groups, (C) The velocities for all observations, (D) The mean amplitude of the activity. The continuous line is for northern latitudes and the broken line for the southern ones.

4. PHASE-DEPENDENCE OF THE ROTATION

Another question is the dependence of the rotational velocity on phase in the activity cycle. We can see from Fig.2 that there is usually maximum velocity at solar minimum. Another smaller maximum velocity at the maximum activity can be seen more clearly from Fig.4a , where all cycles are combined. This double-peak phenomenon was found by Gilman and Howard (1984) from Mt. Wilson sunspot observations. Our expression (1) does not predict that kind of phenomenon. Fig.4b and c give low and high sunspot latitudes separated. We see that the peak at the minimum occurs at higher latitudes, and that at the equator there is only a minimum. So the apparent double peak reflects only the variable profile of the differential rotation, and the Greenwich data give only one maximum for the equatorial velocity, and this occurs in the activity maximum.

5. MERIDIONAL MOTION

Mean meridional motion from Greenwich data was first found by J. Tuominen (1941), giving poleward flow at higher sunspot latitudes and equatorward flow near the equator. Later it has been found to be phase-dependent (for further references see Rüdiger et al. 1986). In Fig.5 we give, as an extension for the results given by Tuominen and Virtanen (1984), the meridional velocity in the combined butterfly diagram for the 5 cycles between the years 1874 and 1933, N- and S-latitudes separated. The tendency is clear: at the pole-side of the most active belt (centre of the butterfly diagram) the motion is poleward and at the equator-side it is equatorward. For the northern and southern latitudes the motion is quite symmetrical, giving no evidence for that kind antisymmetric behaviour as obtained by Howard and Gilman (1986) from Mt. Wilson data.

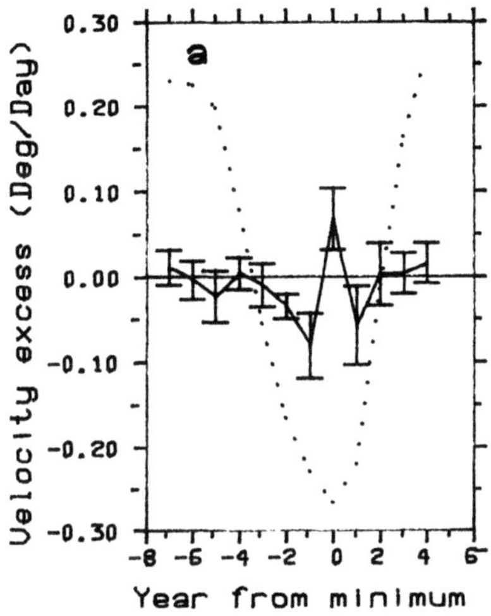

Figure 4a. Dependence of the rotation on the phase of the cycle. The lines represent the mean angular velocity over the sunspot latitudes subtracting the mean differential rotation and combining different cycles. The error bars are 1σ. The dotted line is the number of observations, representing the amplitude, in arbitrary units.

Figure 4b. Same for lower (0° ±15°) sunspot latitudes.

Figure 4c. Same for higher (±15° ±35°) sunspot latitudes.

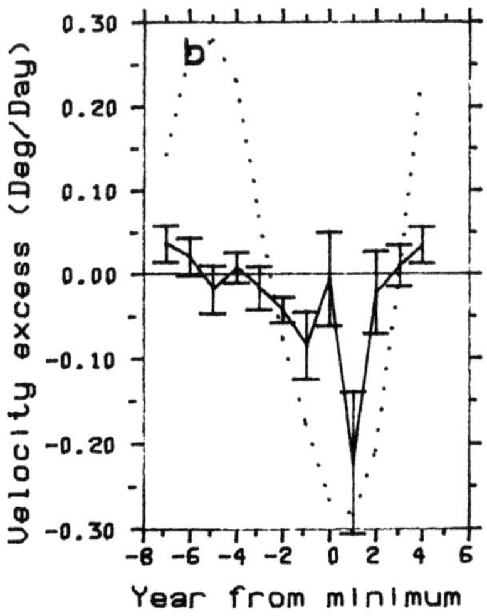

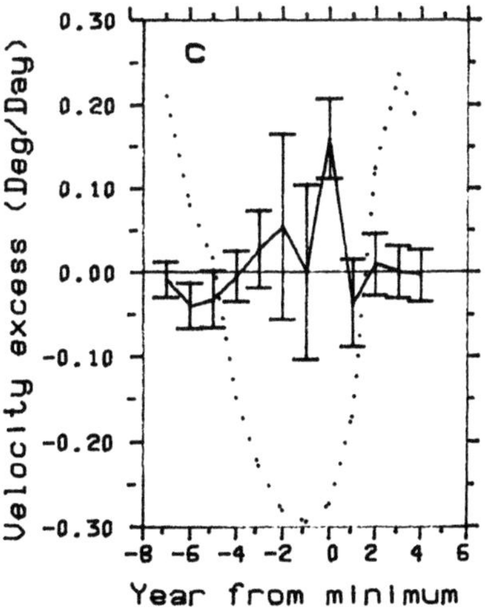

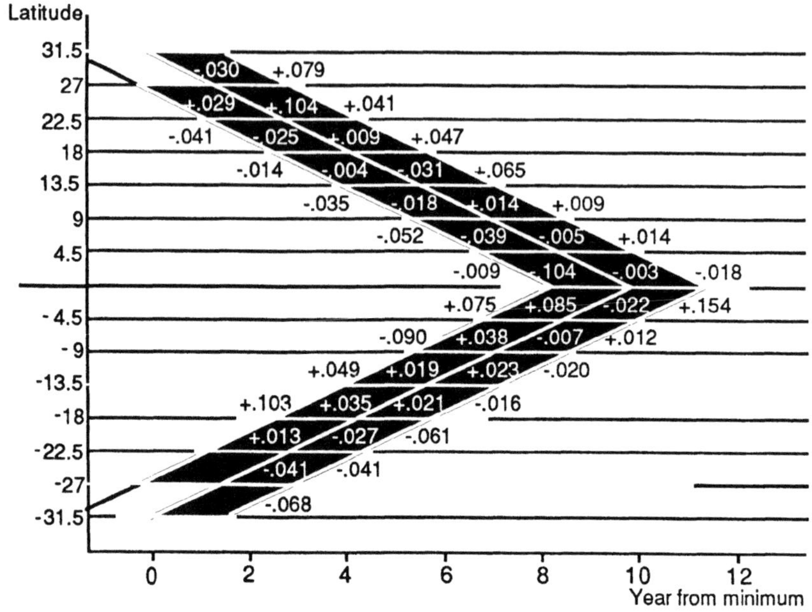

Figure 5. Mean meridional motion for the 5 cycles during the years 1874-1933 in the combined butterfly diagram. Positive values mean poleward motion and negative values equatorward motion at northern latitudes and vice versa at southern latitudes.

The data in Fig.5 are for the recurrent groups. Ribes et al. (1985) found for the newly formed spots a similar type of meridional motion but with faster velocity. Our first inspection concerning the motion of newly formed groups also give larger velocity and also the pattern of the circulation may be different than for recurrent groups, and may in this way represent the motion in deeper layers, as indicated by the deceleration curve in Fig.1.

REFERENCES

Balthasar, H.,Schüssler, M., Wöhl, H.: 1982, *Solar Phys.* **76**, 21

Balthasar, H., Vázques, M., Wöhl, H.: 1986, *Astron. Astrophys.* **155**, 87

Gilman, P.A., Howard, R.: 1984, *Astrophys.J.* **283**, 385

Godoli, G., Mazzucconi, F.: 1979, *Solar Phys.* **64**, 247

Hill, H.A., Yakowitz, D.S., Rosenwald, R.D., Campbell, W.: 1984, *Proc. of 4th European Meeting of Solar Phys.* (ESA SP- 220), 187

Howard, R., Gilman, P.A., Gilman, P.I.: 1984, *Astrophys. J.* **283**,373

Howard, R., Gilman, P.A.:1986, *Astrophys. J.* **307**, 389

Ribes, E., Mein, P., Mangeney, A.: 1985, *Nature* **318**, 170

Rüdiger G., Tuominen I., Krause F., Virtanen H.: 1986, *Astron. Astrophys.* **166**, 306

Tuominen, J.: 1941, *Z. Astrophys.* **21**, 96.

Tuominen, J., Kyröläinen, J.: 1981, *Solar Phys.* **74**, 153

Tuominen, I., Virtanen, H.: 1984, *Astron. Nachr.* **305**, 225

THE DETECTION OF GLOBAL CONVECTION ON THE SUN BY AN ANALYSIS OF LINE SHIFT DATA OF THE JOHN M. WILCOX SOLAR OBSERVATORY AT STANFORD UNIVERSITY

Hirokazu Yoshimura*
Space Science Laboratory
George C. Marshall Space Flight Center
National Aeronautics and Space Administration
Marshall Space Flight Center, Alabama 35812
U.S.A.
and
Department of Astronomy
University of Tokyo
Tokyo 113
Japan

ABSTRACT. The first significant signatures of the real existence of the global convection in the Sun were found in the absorption line shift data of the John M. Wilcox Solar Observatory at Stanford University. The signatures consist of persistent periodic time variations in the east-west component of the velocity fields defined by fitting a slope to the line shift data in a certain longitude window at a specified latitude and longitude by a least square method. The variations show that the amplitude of the velocity fields is on the order of 100 m/s. The spacial structure of the east-west flows on the visible hemisphere of the Sun propagates along with the rotation of the Sun at a slightly slower speed than that of the Sun at the equator. The structure extends across the equator as predicted by theory. The periodicity of the time variation indicates that the wave numbers of the structure are 7 and 5 in Septembers of 1979 and 1981 with faster propagation velocity for the larger wave number during these two time intervals, when clear signals of the global convection were found. This implies that several modes of the global convection are coexisting in the solar convection zone. A personal account of the analysis and its interpretation are given.

1. BACKGROUND OF THE ANALYSIS

The global convection, which is the third kind of convection in the solar convection zone besides granulation and supergranulation, has been expected to exist in the Sun since the 1960's. However its existence has been inferred only from various magnetic field indicators observed on the surface of the Sun and by various theories of fluid dynamics of the solar convection zone. These inferences suggest that the spacial scale of the convection is on the order of 300,000 km and that its lifetime is on the order of months to years. The existence of the convection was proposed from two different kinds of observations and analyses. One was by Bumba and Howard in 1965 at Mt. Wilson Observatory. They systematically observed global magnetic field patterns on the surface of

* National Academy of Sciences / National Research Council Senior Resident Research Associate on leave of absence from Department of Astronomy, University of Tokyo.

B. R. Durney and S. Sofia (eds.), The Internal Solar Angular Velocity, 89–95.

the Sun, which were discovered and named Unipolar Magnetic Regions by Babcock earlier at Mt. Wilson. Bumba and Howard noted the similarity of the Unipolar Magnetic Regions and the magnetic pattern caused by supergranules, which were just discovered at that time by Leighton, Noyes, and Simon in 1962 also at Mt. Wilson. They thought that the magnetic pattern of Unipolar Magnetic Regions was also caused by a similar kind of convection but with much larger spacial and time scales, which they called giant cells. Another proposal was made by Ward in 1965 at Massachusetts Institute of Technology. He analysed proper motions of sunspots observed at Greenwich to infer global flows of the Sun. He found negative correlation between east-west and north-south flows, which meant that angular momentum was being transported toward the equator. Since this mechanism is similar to angular momentum transport due to global fluid motions called Rossby waves in the Earth's atmosphere, he inferred that there must be similar Rossby waves in the solar convection zone. Much theoretical work was done along this line since this could explain the latitudinal differential rotation of the Sun called the equatorial acceleration. However, this possibility was eventually dropped since its excitation was found to be extremely difficult in the solar convection zone. Rossby wave flows are mostly horizontal so that horizontal forcing, like a temperature gradient between the poles and the equator, is most efficient in driving these flows. Such a temperature difference was found to be too small to observe in the case of the Sun. A second possible mode of flow in a rotating sphere like the Sun is the convective (or gravity) mode driven by radial forcing like gravity and a radial temperature gradient. The most significant result of studies of convection at this stage was that the convective mode that is most easily excited in a slowly rotating sphere is a sectorial mode whose pattern crosses the equator (often called banana-like or melon-shaped). This result was found either by arguments concerning the efficiency of heat transport, or by directly calculating growth rates of linear modes. It was also found that only this mode can transport angular momentum toward the equator from all the other latitudes. The basic importance of this mode for the global magnetic structure inside the convection zone was that one single cell of this mode can produce two magnetic cells with opposite polarities in the northern and southern hemispheres since the basic general magnetic field lines inside the solar convection zone, on which the sectorial convective mode work, are nearly parallel to the equator with opposite directions. One magnetic cell of Unipolar Magnetic Regions does not correspond to one convective cell according to this concept. This is the basic difference between the concept of giant cells of Bumba and Howard and the sectorial convective mode. Furthermore, the more fundamental role of this kind of convection for the basic structure and generation of the magnetic fields of the Sun in the form of the solar cycle changing its polarity in every eleven years was proposed by myself in 1972 and studied in detail afterwards. That is, this kind of convection can work as a dynamo, generating the magnetic fields of the Sun and driving the solar cycle oscillations of the fields. Feeling that this kind of convection is ubiquitous in general astrophysical contexts including planets, I began to call it the global convection, including other tesseral modes of the convection in rotating fluid systems besides the sectorial modes of the convection. Although a variety of indirect evidence shows that the global convection should in fact be playing a basic role not only in the global dynamics of the magnetic fields of the Sun but also in more local dynamics of the magnetic fields of the active regions, direct evidence for the existence of the global convection in terms of velocity fields has been felt to be of urgent necessity and importance to understand the basic mechanism of the solar activity and its solar cycle. This is the background of my series of analyses of the velocity fields of the Sun, first at Mt. Wilson with Dr. Robert F. Howard and later at Stanford University with the late Dr. John M. Wilcox and Drs. Philip H. Scherrer, J. Todd Hoeksema, and Richard S. Bogart.

2. METHOD OF ANALYSIS AND RESULTS

The basic feature of the present analysis was invented when I visited Mt. Wilson Observatory from August 1974 to September 1975. At that time, the latitudinal differential rotation profile was determined by fitting a polynomial formula for the differential rotation to the line shift data of the whole visible hemisphere, taking into account the limb shift and other effects as described by Howard and Harvey in 1970 in *Solar Physics*. Differences in the line shift data from this mean differental rotation were called residual velocities and studied for the purpose of detection of the so-called giant cell convection. However, no systematic features were found. Since the measurements of the line shift data was done in a boustrophedonic manner from north or south pole, taking about one hour and a half, an extremely stable wavelength measuring instrument was required. By visually inspecting the residual velocity pattern, I found the Mt. Wilson magnetograph of the 150 foot telescope at those times was not stable enough for the purpose of detecting the global convection if we used such a method. The time drift of the zero wave length reference (absolute wavelength measurements) was so severe that artificial features were clearly evident in the residual velocity pattern. I thought that the situation would be better if we analysed the data for each latitude strip since it took much less time to measure the line shift across one latitude scan line. The zero shift and other important requirements of stable wavelength measurements (relative wavelength measurements) would be smaller for shorter time intervals. If we draw line shift values for each latitude scan in the line shift - central meridian distance plane, the differential rotation would appear as a straight line with a slope if there is no limb shift and no residual velocity. The value of the line shift at the central meridian could be taken as the zero reference point for this latitudinal scan. The limb shift makes the line warped, but since the shift is large only near the edge of the Sun, we could take the data avoiding the extreme limb area. Dr. Howard devised a method to lessen the effect of the limb shift by folding the eastern and western parts of the data to determine the slope. Another important factor in this kind of analysis is noise caused by small-scale velocity fields of the Sun especially by five-minute oscillations and supergranulation. In order to reduce the effects of such noise, we took spacial averages of the data. To achieve this in the Mt. Wilson data, we averaged many latitudinal scan lines over five or ten degree intervals. Then we could determine a detailed latitudinal differential rotation profile. The rotation of each latitude being determined independently from the data of other latitudes. This eventually led to the discovery of the torsional oscillations of the Sun by Howard and LaBonte in 1981. By examining the difference between the line shift data and the mean slope, we thought we could detect the velocity fields of the global convection. However, we could not find any definitive evidence. Wavelength calibration fluctuations from day to day were found to be too serious for the purpose of detecting the velocity fields of the global convection. I found that some reference lines must be installed in the magnetograph to achieve the required stability of the instrument. So, absolute wavelength reference lines using neon gas in the laboratory was installed at Mt. Wilson and began to operate in 1982. It was found that this improved the quality of the data to an impressive degree as reported by Snodgrass and Howard in 1984 in the *Astrophysical Journal*. The results of the analysis of 1974-1975 was reported in the *International Astronomical Union Symposium* No. 71 very briefly. The full implication of these results was never published. One important result of the analysis, which was not mentioned before in the literature, but only among my close colleagues, is that the differential rotation profile was remarkably steady, although its absolute value changed in time, most likely due to the calibration fluctuations. This is in sharp constrast to the earlier results by Howard and Harvey in 1970, which showed large fluctuations in the differential rotation profiles. This steadiness implied that the angular momentum transport due to the global eonvection might not be resposible for maintenance of the differential rotation of the Sun. Fluctuations of the global convection due to its very convective nature would produce fairly large fluctuations in the differential rotation of the Sun, unless the flows of the global convection are remarkably steady due to some nonlinear phase-locking mechanism. However this latter possiblity is found to be unlikely by the present data analysis. The implication of the results of the analysis in 1974-1975 was that some other mechanism, possibly due to anisotropy of the small-scale convection, might be responsible for the driving and maintenance of the differential rotation.

The fact that I did not publish this result in the literature caused some confusion in understanding the solar cycle. In my paper on the solar cycle in the *Astophysical Journal* in 1972, I proposed that the global convection could be the driving force of both the solar cycle and the differential rotation. In accordance with this concept, models of the solar cycle incorporating the dynamics of the differential rotation driven by the global convection were built and studied in detail by my colleagues. However, none of these models have succeeded in simulating the observed differential rotation and the solar cycle self-consistently. On the other hand, my present approach to this problem, which assigns the profile of the differential rotation and the structure of the global convection independently, was criticized by my colleagues on the basis that I could not assign the two kinds of fluid motions independently because they should be dynamically interrelated. However, if the differential rotation is determined by another mechanism, then a third dynamical process, possibly due to small-scale convection, should come into the models. The structures of the differential rotation and of the global convection should become less dependent on each other. Also, the structure of the global convection that I assumed in my models of the solar cycle is, in some sense, consistent with the structure of the differential rotation since the structure of the convection satisfied linearized Navier-Stokes equations with a given differential rotation profile. At any rate, my models have succeeded in simulating the solar cycle to a remarkable degree. This approach has also answered the basic question of the dynamo problem positively, that is, there exist fluid motions that can amplify an infinitesimal seed magnetic field to a finite amplitude and can maintain it. This is possible only when the effect of rotation on the fluid motions is strong enough to give rise to sufficiently strong helicity in the motions to act as a dynamo. The basic reason that I did not publish this key concept was that I thought this was the result of cooperation between Dr. Howard and myself. I hesitated to publish the key concept by myself since it seemed to me that we had different views concerning the basic mechanisms of the solar global dynamics, especially after the discovery of the torsional oscillations. However, I think now I should have published this key concept. I relied on this concept that I could feel safe that I could assume the structures of the differential rotation and of the global convection independently. When I solved the dynamo equation for the first time at National Center for Atmospheric Research in 1973-1974, I took this approach, thinking that this was a numerical experiment to understand the basic dynamo and the solar cycle processes. The justification of the approach came later by the analysis of the Mt. Wilson data. Similarly, I believe now that we have found important observational characteristics of the global convection, besides the evidence of its existence, which we can use to understand the basic mechanisms of the global dynamics of the Sun by building theoretical models. This is why I am now publishing my view of the results of the Stanford data analysis independently from my colleagues in Stanford despite the fact that we have cooperated on this subject but we have different interpretations of the results at present.

Stanford Solar Observatory was established by the late Dr. John M. Wilcox and Dr. Philip H. Scherrer. The telescope of the observatory measures the same quantities as those of Mt. Wilson, i.e., magnetic, line shift, and brightness fields, but with a much larger aperture of the slot feeding sunlight to the spectrograph. From the beginning, this instrument has been suitable for studying global dynamics of the Sun. The noise due to small-scale velocity fields of the Sun such as of five-minute oscillations and supergranulation can be reduced substantially even at the time of observation due to the large aperture of the slot. Spacial averaging is made at the time of measurements over a fairly large area of the Sun. After an initial preliminary stage of observations, the telescope began to accumulate data from 1976. The quality of the data improved from 1979 after the problem of contamination by scattered light in the telescope was found and removed substantially. In November of 1983, I started to work on data from the Stanford Solar Observatory in collaboration with Drs. Scherrer and Hoeksema, and later with Dr. Bogart. Dr. Wilcox was deceased just before I visited Stanford at that time. The first task of this project was to examine the continuity and quality of the data. Summer to early autumn is the best time for the observations. However, in the prime time of August, the telescope was often devoted to measuring the solar oscillations. This is why the two time intervals in which we eventually found significant signatures of the global convection both happened to be in September. To test the stability of the telescope, I chose to take a half hemipshere, either eastern or western half of the visible hemisphere, to apply the slope method. We then can compare the results of the two hemispheres to estimate of the reality of the results. We inspected time variations of the values of the slope and continuity of the data. A power spectrum analysis was employed to see whether there was any periodic variation in the slope, corresponding to an east-west component of velocity fields. If global convection with sectorial modes exists on the surface of the Sun, it should travel across the disk so that the slope should undergo periodic variations. I searched for the time intervals in which this power spectrum method could give any hint of such periodic variations for the entire set of data from 1976 to 1983. I found two time intervals which showed single peaks in the power spectrum. One is in September of 1979 and the other is in September of 1981.

Next step of the analysis was to study the spacial structure of the global convection and to prove that these power spectrum signals in fact meant that the global convection appeared on the surface of the Sun during those time intervals. This was done during my two short stays in Stanford in February-March and July-August of 1985. In order to determine the spacial structure of the east-west flows, we tried at first to map the residual velocity in the same way as done at Mt. Wilson. However, the resulting maps were so noisy that even the large aperture of the Stanford telescope proved to be inadequate in sufficiently reducing the noise caused by small-scale velocity fields to allow detecting the global convection using such a simple method. Then, feeling the need to average the data over larger spacial areas, I chose to use the same slope analysis as in the case of the Mt. Wilson data analysis but with a longitudinal window, which is smaller than that of half hemisphere as was used in the power spectrum analysis and which is large enough to reduce the noise. By shifting the position of the longitudinal window, we could map the spacial structure of the east-west flows in longitude-latitude space, the longitude in this case being defined as the value at the center of the longitudinal window. This is a sort of running mean of the velocity fields over longitude. We constructed three kinds of diagrams from this analysis. The first was made in longitude-time space, abscissa being longitude and ordinate being time. The data of the flows at a fixed latitude was used in one diagram. These diagrams showed remarkably regular periodic wave patterns translating across the disc of the Sun along with the rotation of the Sun. This confirmed that the signals were of solar origin. Combining the information derived by the power spectrum analysis with the wave patterns, I could determine the wave numbers for the two time intervals. The period was 27.0 days (28.9 days) and wave number was 7 (5) in September of 1979 (1981). This tendency for a shorter period for larger wave number conforms to theory for the global convection. The second kind of diagram was daily maps made in longitude-latitude space for one observation. These diagrams showed that the spacial structure of the global flow pattern was very similar to the sectorial mode of the global convection deformed by the effect of rotation as the linear theory predicted by myself in 1974. The pattern is not exactly sectorial. The polar parts are tailing after the equatorial parts. The third kind of diagram was synoptic maps made in time-latitude space and was made by putting together latitudinal strips of the daily maps for a month at a fixed central meridian distance, mainly at the central meridian itself. Notice that, because we used a finite longitudinal window to define the slope, we could determine the east-west horizontal flows even at the central meridian.

Comparing the synoptic maps with the magnetic maps we noticed a strong correlation between the magnetic active regions and the global convective pattern, although the latitudinal width of the magnetic regions was much smaller than that of the velocity convective pattern. There often were strong magnetic active regions at the convergent areas of the flow pattern. At this point, my interpretation of the correlation and that of Drs. Scherrer and Bogart became clearly different. Their interpretation of the correlation was that this indicated that most of the east-west flow signal came from downdrafts in magnetic field regions which affected the slope analysis. They took the extreme view that any region with weak general magnetic fields should have some downdrafts since these weak fields should consist of small flux tubes with strong magnetic fields and the downdrafts should accompany the strong fields no matter whether the downdrafts are really velocity fields or due to line profile asymmetry caused by the presence of the magnetic fields. So, I examined the correlation between the two kinds of synoptic maps closely. There was in fact strong correlation but not one to one correspondence. If the east-west flow signals came from the presence of the magnetic fields in general, there must be one to one correspondence. Also, if small scale flux tubes affect this analysis in a fundamental way, the daily and synoptic maps should show almost exactly the same patterns because the effects of the small scale tubes should always be present on the Sun due to their statistical nature. Since magnetic field daily and synoptic maps almost always show organized long-lived patterns, the velocity field maps should also show organized pattern according to this concept. But this was not found in the observed maps. To examine whether or not downdrafts which occasionally happen in active regions affect the analysis, we took out any areas with magnetic fields above some value and reanalysed the data. Although the pattern of the daily maps as well as synoptic maps became slightly different, especially because of the smaller number of data in the reanalysis, the main features that led to the initial conclusions did not change. From these arguments, I concluded that the analysis is in fact showing us that the global convection appeared on the surface of the Sun at least during the two time intervals mentioned above. The different wave numbers and propagational velocities of the patterns in the two intervals show that different modes of the convection appeared in those times, which indicates that several modes of the convection can occur in the Sun. This means that if several convective modes appear on the surface simultaneously, this simple method of detection could not show strong evidence for the existence of the global convection. Since different modes have different propagational velocities, the resulting flow pattern would show complex time and spacial structures. More sophisticated methods would be necessary to study these structures. Another interesting fact of the present analysis which supports the present conclusion that the signals that we found were in fact of the global convection, is that the levels of signals of the slope analysis were also high in quiet times when there were no large active regions although the signals were not so well organized as in the two time intervals mentioned above. This could mean that the velocity signals could be present also in magnetically quiet times but was not so well organized because of presence of multiple modes. At any rate, the results obtained by this simple method strongly recommend continuous observation of the global velocity as well as magnetic and brightness fields. The correlation also indicates that strong magnetic active regions often exist at the converging areas of the convection. According to my interpretation, the correlation is of a physical nature and shows a basic clue for understanding formation and evolution of magnetic active regions. Although a continous sequence of the synoptic maps of the east-west flows cannot be made because of the scarcity of data, incomplete maps before and after the two time intervals show development of the flow pattern followed by the appearance of the active regions. There appears to be a time lag between the two developments.

I present this view notwithstanding the disagreement between my colleagues in Stanford and myself because of the utmost importance of the role of the global convection, not only for the global dynamics of the Sun but also for the local dynamics of magnetic active regions. If the above interpretation is in fact true, the global convection should strongly act on magnetic field lines of the active regions and thus strongly affect their various activities, including solar flares. My colleagues in Stanford agreed that I present this view. I would like to thank them for their cooperation, discussions, and support. I also would like to thank Dr. Howard and Dr. Horace W. Babcock for making the present series of data analyses possible by inviting me to Mt. Wilson Observatory in 1974-1975 as a Carnegie Fellow. Dr. Howard invited me also in the Summer of 1977 to Mt. Wilson Observatory. I would like to thank him for many enlightening discussions. I am deeply grateful to the late Dr. Wilcox for making this cooperation with Stanford colleagues possible and for recognizing my research work from its infancy. The first announcement of the present detection was made in the abstract for a poster presentation for the Annual Meeting of the Solar Physics Division of American Astronomical Society held in May 13-15, 1985 (*Bull. Amer. Astron. Soc.*, 17, 639, 1985). However, results of a completely different analysis were presented by my Stanford colleagues. At about the same time, I presented the results of the present analysis at the Spring Annual Meeting of Astronomical Society of Japan on May 22, 1985. The newspapers "*Yomiuri Shimbun*" and "*Nihon Keizai Shimbun*" in Japan reported the results in May 19 and 21, 1985 issues, respectively. A science magazine "*Newton*" in Japan also reported the results in its March, 1986 issue (pp.33-43). I reported the present results also in the First Solar Cycle Workshop / Minimum Phase, held in August 17-20, 1986 at Big Bear Solar Observatory. Dr. Scherrer reported the results with his own views at the NATO Advanced Research Workshop on Seismology of the Sun and the Distant Stars held in June 17-21, 1985 at Cambridge University. The readers could find difference of approach between his and my analyses, which had already existed at that time, in the proceedings of the workshop. The interested readers will find some relevant references in *Ap. J.* 178, 865, 1972; 201, 740, 1975; 220, 692, 1978; 226, 706, 1978; 227, 1047, 1979; 247, 1102, 1981; 280, 865, 1984; 283, 870, 1984; 285, 325, 1984; *Ap. J. Suppl.* 29, 467, 1975; 52, 363, 1983; *Solar Phys.* 18, 417, 1971; 22, 20, 1972; 33, 131, 1973; 47, 581, 1976; 50, 3, 1976; 52, 41, 1977; 54, 229, 1977; 87, 251, 1983; *Publ. Astron. Soc. Japan*, 23, 57, 1971; 26, 9, 1974; 37, 171, 1985; *I. A. U. Symp.* 71, 19, 1976; 71, 137, 1976; 71, 409, 1976. The readers should notice that the literatures cited here and the references mentioned in the above literatures are not complete by any means.

ADIABATIC NONRADIAL OSCILLATIONS OF A DIFFERENTIALLY ROTATING STAR

W. Unno
Research Institute for Science and Technology, Kinki University
Kowakae, Higashi-Osaka-shi, Osaka 577, Japan

and

H. Saio
Department f Astronomy, University of Tokyo
Bunkyo-ku, Tokyo 113, Japan

ABSTRACT. Adiabatic nonradial oscillations in a differentially rotating star are formulated in a general form by use of the general variational principle, the centrifugal distortion of the unperturbed configuration, however, being neglected. Further calculations are performed for the case in which the rotation law $\Omega\ (r)$ is a function of the radius r from the center.

The corotation resonance and the Lindblad resonances can appear, if the core rotates sufficiently fast or if the frequency of an oscillation is sufficiently low. Some of the tesseral (azimuthal quantum number, $m \neq 0$) g-modes in a superadiabatic core (convectively stabilized by rotation) can be unstable even without the resonance. The co-rotation resonance seems to have a destabilizing influence to the oscillation. However, for all the resonance cases, terms of all orders in Ω become important in the characteristic equation, and the stability criterion requires numerical calculations even for slow rotaters.

1. INTRODUCTION

If the internal differential rotation turns out to be responsible for the oscillations of β Cep and the related stars, not only the origin of the oscillations could be satisfactorily understood but also the behavior of the internal differential rotation should be known by the asteroseismolgical means. To make the problem simple, we restrict ourselves to study the adiabatic oscillation, although the nonadiabatic effects can also be formulated without much difficulty. Nevertheless, the problem requires self-revealing mathematical formalism, since there are so many modes characterized by different harmonic numbers l and m and the radial order n for each (acoustic, gravity, or inertia) type of oscillations. In the absence of the usual variational principle for differentially rotating stars, the most suitable formalism appears to be the general variational principle (c.f. Unno, 1968). In fact, the standard procedure to derive the characteristic equation for the stability problem leads us to show the possible presence of the corotation resonance and the Lindblad resonances in a straightforward way. The purpose of the present paper is to provide an unambiguous formalism for numerical studies and to give some qualitative insight on nonradial oscillations of a differentially

B. R. Durney and S. Sofia (eds.), The Internal Solar Angular Velocity, 97–106.

rotating star.

2. BASIC EQUATION

Basic equation of the adiabatic oscillation is the equation of motion which can be expressed in the general variational principle (Prigogine and Glansdorff 1964 and 1965),

$$\delta\Psi = \int dM_r \delta v \cdot \left(\frac{dv}{dt} + \frac{1}{\rho}\nabla P + \nabla\Phi\right) = 0, \tag{1}$$

supplemented by the adiabatic relation and the continuity in linearized form,

$$P_1/P_0 = \gamma\rho_1/\rho_0 = -\gamma\nabla\cdot\xi, \tag{2}$$

where the subscript 0 (to be omitted hereafter) and 1 represent the unperturbed state and the Lagrange perturbation, respectively. In the above equations, δv denotes a variation in the velocity v measured from the fixed coordinate, d/dt the time derivative for a mass element dM_r , Φ the gravitational potential, ξ the displacement, γ the adiabatic exponent, and P and ρ the pressure and the density, respectively. Introducing the Lagrange variations of velocity and acceleration such that

$$u = v_1 = d\xi/dt \text{ and } a = (dv/dt)_1 = du/dt, \tag{3}$$

we reform the general variational equation (1) in linearized equation (Unno 1968),

$$\delta\Psi = \int dM_r \delta u^* \cdot \Big[ac^2\nabla(\nabla\cdot\xi)+(\gamma-1)g(\nabla\cdot\xi)\nabla r+\nabla(g\xi_r)+$$
$$+\frac{1}{2}(\xi\cdot\nabla\Omega^2)\nabla(r\sin\theta)^2+(\Omega\times\xi)\times\Omega+\nabla\Phi'\Big] = 0, \tag{4}$$

where the asterisk represents the complex conjugate,

$$c^2 = \gamma P/\rho \text{ and } g = \partial\Phi/\partial r. \tag{5}$$

The baroclinic term, $\rho^{-2}\xi\times(\nabla P\times\nabla\rho)$, that is important for the circulation has been neglected in the above expression (4). The unperturbed equilibrium configuration is assumed to be spherically symmetric. From equations (3), we have

$$u = \left(\frac{\partial}{\partial t}+\Omega\frac{\partial}{\partial\phi}\right)\xi = i(\sigma+m\Omega)\xi + \Omega\times\xi, \tag{6}$$

and

$$a = \left(\frac{\partial}{\partial t}+\Omega\frac{\partial}{\partial \phi}\right)u = i(\sigma+m\Omega)u + \Omega\times u, \tag{7}$$

if the perturbation is assumed to be proportional to $e^{i\sigma t}e^{im\phi}$. The general variational formulation (4), however, requires the expression of ξ in terms of u instead of equation (6) for more compact and more convenient treatment. (The statistical mechanical reasoning is beyond the scope of the present paper). Solving equation (6) for ξ , we obtain

$$\xi = \frac{1}{i\sigma_0}u + \frac{1}{\sigma_0^{\Omega}2-\Omega^2}(\Omega\times u) - \frac{1}{i\sigma_0(\sigma_0^2-\Omega^2)}\Omega\times(\Omega\times u), \tag{8}$$

where

$$\sigma_0 = \sigma + m\Omega. \tag{9}$$

Equations (4), (7), and (8) suffice to determine the adiabatic oscillation of a differentially rotating star, since the remaining variable Φ' is given in terms of ξ by

$$\Phi'(r) = G\int\frac{\mathrm{div}_{r'}(\rho\xi)_{r'}}{|r-r'|}dr'. \tag{10}$$

Equation (8) is a purely kinematical relation. But, when it is inserted in equation (4), the presence of the resonance on the surfaces:

$$\sigma_0 = \sigma + m\Omega(r,\theta) = 0, \tag{11}$$

and

$$\sigma_0 \pm \Omega = \sigma + (m\pm 1)\Omega(r,\theta) = 0, \tag{12}$$

is well revealed. The condition (11) represents the corotation resonance, while the condition (12) corresponds to the Lindblad resonance for the Keplerian rotation. For given m and σ, there are three resonance surfaces where $\Omega(r,\theta)=-\sigma/(m-1)$, $-\sigma/m$, and $-\sigma/(m+1)$. The nature of the corotation resonance, however, is not the same as the other two. The expansion of the eigen-function u in the spherical surface harmonics,

$$Y_l^m(\theta,\phi)=\left[\frac{(2l+1)(l-|m|)!}{(l+m)!}\right]^{1/2}P_l^{|m|}(\cos\theta)e^{im\phi}, \quad \left(\int|Y_l^m|^2\frac{d\omega}{4\pi}=1\right), \tag{13}$$

where P_l^m denotes the associated Legendre polynomial, seems to be adequate to define modes to which the nature of the oscillation depends, although each mode cannot be represented by a single Y_l^m harmonic. For simplicity, the rest of the formulation will be restricted to the case of $\Omega = \Omega\ (r)$.

3. DERIVATION OF THE CHARACTERISTIC EQUATION

Since there are so many eigen-modes belonging to a single m value, a systematic complete representation of eigen-functions is dispensable. We express the vector $\boldsymbol{u}$ to be

$$\boldsymbol{u}=e^{i\sigma t}\sum_{l=|m|}^{\infty}\left[u_R^{(l)}(r)Y_l^m(\theta,\phi)\nabla r+u_P^{(l)}(r)\frac{r}{\Lambda}\nabla Y_l^m(\theta,\phi)+u_T^{(l)}(r)\frac{r}{\Lambda}\nabla r\times\nabla Y_l^m(\theta,\phi)\right], \tag{14}$$

where

$$\Lambda^2 = l(l+1). \tag{15}$$

If the equilibrium model is a slow rotater with a simple structure, the eigen-frequency σ is characterized by m, the representative l value for which the radial function ($u_R^{(l)}$ for the p-mode, $u_P^{(l)}$ for the g-mode, and $u_T^{(l)}$ for the inertia mode) shows the maximum amplitude, and the nodal number n of the representative radial function ($u_R^{(l)}$,$u_P^{(l)}$, or $u_T^{(l)}$) of the mode. For brevity, we rewrite hereafter the expression (14) in a simplified form,

$$\boldsymbol{u} = u_R Y\nabla r + u_P\frac{r}{\Lambda}\nabla Y + u_T\frac{r}{\Lambda}\nabla r\times\nabla Y, \tag{14'}$$

omitting the time factor, the summation symbol and the quantum numbers l and m. The first term in equation (14') represents the radial component, and the second and the third term represent tangential curvilinear components on the tesseral surface defined by Y_l^m. The factor r/Λ is a normalization factor to measure u_R, u_P, and u_T in the same dimension and scale. Any vector can be expanded in the form (14) because of the orthogonality and the completeness of Y_l^m . Conversely, we have

$$u_R = \int Y^*\nabla r\cdot\boldsymbol{u}\frac{d\omega}{4\pi}, \quad u_P = \int\frac{r}{\Lambda}\nabla Y^*\cdot\boldsymbol{u}\frac{d\omega}{4\pi}, \tag{16a,b}$$

and

$$u_T = \int\frac{r}{\Lambda}(\nabla r\times\nabla Y^*)\cdot\boldsymbol{u}\frac{d\omega}{4\pi}. \tag{16c}$$

Applying these rules to equation (8), we obtain, for instance,

$$i\sigma_0\xi_R=\left\{\left[1+\frac{\Omega^2}{\sigma_0^2-\Omega^2}(1-J_l^2-J_{l+1}^2)\right]\delta_{l'l}-\frac{\Omega^2}{\sigma_0^2-\Omega^2}(J_{l-1}J_l\delta_{l'l-2}+J_{l+1}J_{l+2}\delta_{l'l+2})\right\}u_R{}'+$$

$$+\left\{\left[\frac{\sigma_0\Omega}{\sigma_0^2-\Omega^2}\frac{m}{\Lambda}+\frac{\Omega^2}{\sigma_0^2-\Omega^2}\left(\frac{l+1}{\Lambda}J_l^2+\frac{l}{\Lambda}J_{l+1}^2\right)\right]\delta_{l'l}+\right.$$

$$+\frac{\Omega^2}{\sigma_0^2-\Omega^2}\left(\frac{l-2}{\Lambda'}J_{l-1}J_l\delta_{l'l-2}-\frac{l+3}{\Lambda'}J_{l+1}J_{l+2}\delta_{l'l+2}\right)\Bigg\}u_P'$$

$$-i\frac{\Omega}{\sigma_0^2-\Omega^2}\left\{\left(\frac{l-1}{\Lambda'}\sigma_0+\frac{m}{\Lambda'}\Omega\right)J_l\delta_{l'l-1}+\left(\frac{l+2}{\Lambda'}\sigma_0+\frac{m}{\Lambda'}\Omega\right)J_{l+1}\delta_{l'l+1}\right\}u_T', \tag{17}$$

where

$$J_l=\left(\frac{l^2-m^2}{4l^2-1}\right)^{1/2}, \tag{18}$$

the prime denotes the quantity for $l=l'$, and the summation convention over l' should be understood. Similarly, the integration in angular variables can be performed for the general variation (4). After some algebra, we obtain

$$\delta\Psi=\int dM_r\delta u_R{}^*\left\{\frac{N^2-\sigma_0^2}{i\sigma_0}u_R+\frac{N^2}{g}\Phi_R+\Omega_1\left(1+\frac{N^2}{\sigma_0^2}\right)(\nabla z\times\boldsymbol{u})_R\right.$$

$$-\frac{\Omega\Omega_1}{i\sigma_0}\left(1+\frac{N^2}{\sigma_0^2}\right)[\nabla z\times(\nabla z\times\boldsymbol{u})]_R-\kappa\Omega^2[\xi_r\sin\theta\nabla(r\sin\theta)]_R$$

$$-\left(\frac{d}{dr}-\frac{N^2}{g}\right)\frac{r}{\Lambda}\left[i\sigma_0u_P+\Omega_1(\nabla z\times\boldsymbol{u})_P-\frac{\Omega\Omega_1}{i\sigma_0}[\nabla z\times(\nabla z\times\boldsymbol{u})]_P\right.$$

$$\left.\left.-\kappa\Omega^2[\xi_r\sin\theta\nabla(r\sin\theta)]_P\right]\right\}+$$

$$+\delta u_P{}^*\left\{\frac{L_l^2-\sigma_0^2}{i\sigma_0}u_P+\Omega_1\left(1+\frac{L_l^2}{\sigma_0^2}\right)(\nabla z\times\boldsymbol{u})_P\right.$$

$$\left.-\kappa\Omega^2[\xi_r\sin\theta\nabla(r\sin\theta)]_P-\frac{\Lambda}{r}\left[c^2\frac{d}{r^2dr}(r^2\xi_R)-g\xi_R-\Phi_R\right]\right\}+$$

$$+\delta u_T{}^*\left\{i\sigma_0u_T+\Omega_1(\nabla z\times\boldsymbol{u})_T-\frac{\Omega\Omega_1}{i\sigma_0}[\nabla z\times(\nabla z\times\boldsymbol{u})]_T\right.$$

$$-\kappa\Omega^2[\xi_r \sin\theta \nabla(r\sin\theta)]_T\Big\} = 0, \tag{19}$$

where the subscripts R, P, and T have the same meaning as in equations (16a, b, c),

$$\xi_r = \xi_R Y, \tag{20}$$

$$\Phi_R = G\int \frac{dM_{r'}}{2l+1}\frac{[\min(r,r')]^l}{[\max(r,r')]^{l+1}}\left[\frac{d}{\rho r^2 dr}(\rho r^2\xi_R) - \frac{\Lambda}{r}\xi_P\right]_{r=r'}, \tag{21}$$

$$N^2 = g\left(\frac{1}{\gamma}\frac{d\ln P}{dr} - \frac{d\ln\rho}{dr}\right) \quad : \text{Brunt-Vaisala frequency}, \tag{22}$$

$$L_l^2 = \Lambda^2 c^2/r^2 \quad : \text{Lamb frequency}, \tag{23}$$

$$\kappa = -d\ln\Omega^2/d\ln r, \quad \text{and} \quad \Omega_1 = \sigma 2_0^2\Omega/(\sigma_0^2 - \Omega^2). \tag{24,25}$$

Since δu_R^*, δu_P^*, and δu_T^* are arbitrary variations, the formulae in curly brackets must vanish, forming a set of linearized equations to determine u_R, u_P, and u_T. If $\Omega = 0$, these equations reduce to a familiar set of equations (cf. Unno et al. 1979, eqs. [13.1,2]) for u_R and u_P (or $u_{P'}$) belonging to a single l value. If $\Omega \neq 0$, however, the equation to determine $u_R^{(l)}$ contains $u_{R,P}^{(l\pm 2)}$ and $u_{R,P}^{(l\pm 4)}$ as well as $u_{R,P}^{(l)}$ through ξ_r-term or other so that infinite number of simultaneous equations ($l=|m|, |m|+2, \ldots$; or $l=|m|+1, |m|+3, \cdots$) have to be solved, in principle, to determine a single mode of oscillation.

A standard way to derive the characteristic equation in the general variational formulation is to take the amplitude of the eigen-function as the variational parameter. We set $u_R = C_R^{(l)} u_R^{(l)}$ and $\delta u_R^* = \delta C_R^{(l)*} u_R^{(l)*}$, and similarly for u_P and u_T. Then, for each $\delta C_{R,P,T}^{(l)*}$ which is arbitrary, the variational equation (19) gives a linear equation in $C_{R,P,T}^{(l)}, C_{R,P,T}^{(l\pm 2)}$, etc. The decoupling between $l = |m| +$(even number) and $l = |m| +$(odd number) is due to the assumption of $\Omega = \Omega(r)$. The condition that the set of linear homogeneous equations in $C_{R,P,T}^{(l)}$ etc. should have nontrivial solutions is the determinant of the coefficient matrix to vanish, which is the characteristic equation to determine the eigenvalue σ as a function of m, l_M, and n, l_M being the l of the maximum amplitude eigen-function. It is to be noted that the terms with the Ω^2 factor cannot be neglected even for slow rotaters if the resonance exists (large Ω_1, large ξ_R). Therefore, the characteristic equation in that case has the determinant with a large number of elements, even if the truncation in l for large $|l-l_M|$ can be made in practice. Numerical study is necessary to see the stability of the resonant case.

4. THE Ω^1-APPROXIMATION

In spite of what has been concluded above, we will go to the Ω^1-approximation, essentially neglecting the centrifugal distortion and taking account of only the Coriolis effect. The resonance factor $(\sigma_0^2-\Omega^2)$, however, will be retained with the hope that at least a part of the resonance effects can be studied theoretically. The advantage of the Ω^1 -approximation is in the mathematical simplicity that the eigen-function $\boldsymbol{u}$ [equation (14)] can be expressed by only one Y_l^m harmonic similarly as in the nonrotating case. As seen from the last bracket expression in equation (19), the value of u_T is of the first order in Ω (which is not true, however, in the resonance cases). Also, in ξ_R and $(\nabla z\times u)_{R,P}$, only $u_{R,P}^{(l)}$ appears in the Ω^1-approximation, and $u_T^{(l\pm1)}$ appears together with another Ω factor and is negelected. Therefore, the u_T component which is characteristic in the helical motion decouples with the p- and g-mode oscillations in the Ω^1-approximation. We will neglect also Φ_R (the Cowling appproximation) which is small in the nonradial oscillation.

Introducing new variables $u_{R'}$ and $u_{P'}$ such that

$$u_R = u_{R'} - \omega u_{P'}, \qquad u_P = u_{P'} + \omega\left(u_{R'}+\frac{1}{\Lambda}u_{P'}\right), \tag{26a,b}$$

where

$$\omega = m\Omega_1/\Lambda\sigma_0 = m\sigma_0\Omega/\Lambda(\sigma_0^2-\Omega^2), \tag{27}$$

we rewrite equation (19) in the Ω^1-approximation in the following form,

$$i\delta\Psi=\int dM_r\left\{\frac{\sigma_0\delta u_R{}^*}{\rho^{1/2}r^2}\left[\frac{N^2-\sigma_0^2}{\sigma_0^2}(\rho^{1/2}r^2u_{R'})+\frac{1}{\Lambda}\left(r\frac{d}{dr}-\frac{N^2r}{2g}+\frac{gr}{2c^2}+2\Lambda\omega\right)(\rho^{1/2}r^2u_{P'})\right]+\right.$$
$$+\frac{L_l^2\delta u_P{}^*}{\sigma_0\rho^{1/2}r^2}\left[\left(1+\frac{2\omega}{\Lambda}-\frac{\sigma_0^2}{L_l^2}\right)(\rho^{1/2}r^2u_{P'})-\frac{1}{\Lambda}\left(r\frac{d}{dr}+\frac{N^2r}{2g}-\frac{gr}{2c^2}-2\Lambda\omega\right)\times\right.$$
$$\left.\left.\times(\rho^{1/2}r^2u_{R'})\right]\right\}=0, \tag{28}$$

where either $(\rho^{1/2}r^2u_{R'})$ or $(\rho^{1/2}r^2u_{P'})$ are the dependent variables to match the standard WKB formulation so that $r(d/dr)$ can be approximated by ik in the propagation region.

To obtain the characteristic equation, we write u_R and u_P as Au_l and Bv_l and take the amplitude variation so that $\delta u_R{}^*=u_l{}^*\delta A$ and $\delta u_P{}^*=v_l{}^*\delta B$. Then, equation (28) gives two simultaneous linear homogeneous equations of A and B for arbitrary δA and δB variations. Nontrivial roots exist if the following characteristic equation is satisfied.

$$\int\left(\frac{N^2-\sigma_0^2}{\sigma_0}+\frac{\omega V_1}{\Lambda}\sigma_0-ik\frac{\omega\sigma_0}{\Lambda}\right)|u_R|^2dM_r\times$$

$$\times \int \left[\left(1-\frac{\omega}{\Lambda}\right)(L_l^2\sigma_0^2)+\frac{\omega L_l^2}{\Lambda}(2+V_1)-ik\frac{\omega L_l^2}{\Lambda}\right]\frac{|u_p|^2}{\sigma_0}dM_r$$

$$-\int \left[\frac{\sigma_0}{\Lambda}\left(1-\frac{\omega}{\Lambda}\right)V_1+\omega\frac{N^2-\sigma_0^2}{\sigma_0}+ik\frac{\sigma_0}{\Lambda}\left(1-\frac{\omega}{\Lambda}\right)\right](u_R{}^*u_P)dM_r\times$$

$$\times \int \left[\frac{V_1}{\Lambda}L_l^2-\omega(L_l^2-\sigma_0^2)-ik\frac{L_l^2}{\Lambda}\right]\frac{(u_R u_P{}^*)}{\sigma_0}dM_r = 0, \tag{29}$$

where

$$V_1 = \frac{r}{2}\left(\frac{g}{c^2}-\frac{N^2}{g}\right)+2\Lambda\omega = -\frac{d\ln P}{d\ln r}\left(\frac{1}{\gamma}-\frac{d\ln\rho}{2d\ln P}\right)+2\Lambda\omega. \tag{30}$$

The characteristic frequencies L_l and N as well as Ω are functions of r . Therefore, the non-dimensional wave number k is also a function of r which, however, is assumed here to be constant in the core and in the envelope.

Following Osaki (1974), we consider an f-mode which oscillates in comparable energy in the core and in the envelope. For early type stars slightly evolved off the main sequence, the f-mode behaves like p-mode in the envelope and like g-mode in the rapidly spinning core. These characteristics of f-modes may be qualitatively represented by

$$v_l \approx \pm iu_l, \quad |u_l| \sim |v_l| \sim \text{const.}, \tag{31}$$

$|A|>>|B|$ in the envelope and $|A|<<|B|$ in the core. For the quantitative analysis, the eigen-functions should be represented in analytic or numerical (finite elements) schemes in which parameters are determined variationally by use of equation (19). The approximation (31) is only to see the qualitative property of the characteristic equation (29). In the Ω^1-approximation, we then have

$$\left[\frac{N^2-\sigma_0^2}{\sigma_0}+\frac{\omega V_1}{\Lambda}\sigma_0\right]_{C+E}\left[\left(1-\frac{\omega}{\Lambda}\right)(L_l^2-\sigma_0^2)+\frac{\omega L_l^2}{\Lambda}(2+V_1)\right]_{C+E}$$

$$-\left[\frac{\sigma_0}{\Lambda}\left(1-\frac{\omega}{\Lambda}\right)V_1+\omega\frac{N^2-\sigma^2}{\sigma_0}\right]_{C+E}\left[\frac{V_1}{\Lambda}L_l^2-\omega(L_l^2-\sigma_0^2)\right]_{C+E}$$

$$-k^2\left[\left(1-\frac{\omega}{\Lambda}\right)\frac{L_l^2}{\Lambda^2}\sigma_0\right]_{C+E} = 0, \tag{32}$$

where each square-bracket factor constitutes of the sum of the core contribution and the envelope contribution. Also, the resonance contributions are to be added in each square-bracket factor, if the conditions of the corotation resonance ($\sigma_0=0$) and of the Lindblad resonance ($\sigma_0 = \pm\Omega$) are fulfilled somewhere in the region of the core-envelope transition. The rotation law near the corotation point r_0 is approximated by

$$\Omega = -\frac{\sigma_r}{m}\left(1 - \frac{r-r_0}{H_{rot}}\right), \quad H_{rot} = -\left(\frac{d\ln\Omega}{dr}\right)^{-1}. \tag{33}$$

Then, the corotation resonance gives additional terms through the factor σ_0^{-1} in equation (32) by the replacement,

$$\begin{aligned}\sigma_0^{-1} &\rightarrow -i\pi\ \mathrm{sign}(\sigma_i/\sigma_r)\sigma_r^{-1}H_{rot}\delta(r-r_0)\\ &\rightarrow -i\pi\ \mathrm{sign}(\sigma_i/\sigma_r)\sigma_r^{-1}(H_{rot}/r_0),\end{aligned} \tag{34}$$

where σ_r, and σ_i are the real and imaginary parts of σ. Similarly, the Lindblad resonance gives contributions through ω,

$$\omega \rightarrow \pm i\ \mathrm{sign}(\sigma_i/\sigma_r)\frac{m\pi}{2\Lambda(-m\pm 1)}(H_{rot}/r_\pm), \quad (m \neq \pm 1). \tag{35}$$

These Lindblad resonance contributions generally differ in sign and mostly cancel each other, unless one of the resonance points, $r_\pm$ does not take place within the star.

5. THE OVERSTABLE OSCILLATION IN THE CORE

Neglecting the envelope contribution in equation (32), we will first study the oscillation of the superadiabatic ($N^2 < 0$) spinning core. Without the resonance, the rotation efficiency factor ω is assumed to be small and neglected except in the first square-bracket factor in equation (32) representing the convection stability. Then, we have

$$\left[N^2 + \left(\frac{\omega V_1}{\Lambda} - 1\right)\sigma_0^2\right](L_l^2 - \sigma_0^2) - (V_1^2 + k^2)\frac{L_l^2\sigma_0^2}{\Lambda^2} = 0. \tag{36}$$

If $\omega < \Lambda V_1^{-1}$, the convection is unstable monotonically. If $\omega > \Lambda V_1^{-1}$, the stability requires that

$$\left|\left(\frac{\omega V_1}{\Lambda} - 1\right)^{1/2} L_l - |N^2|^{1/2}\right| > (V_1^2 + k^2)^{1/2}\frac{L_l}{\Lambda}, \tag{37}$$

a condition which is hardly satisfied even for small k.. There is such a k value for which the overstable slow mode has nearly the same frequency as the frequency of an envelope oscillation.

The effect of the corotation resonance may be studied in the Ω^1-approximation by use of equations (32) and (34). We have

$$\left[N^2-\sigma_0^2\pm i\pi\frac{\sigma_0}{\sigma_r}\left(\frac{H_{rot}}{r}N^2\right)_{r_0}\right](L_l^2-\sigma_0^2) - (V_1^2+k^2)\frac{L_l^2\sigma_0^2}{\Lambda^2} = 0, \tag{38}$$

where N^2, L_l^2, σ_0^2 etc. represent the effective values for the whole star while $(H_{rot}N^2/r)_{r_0}$ represents the value at the resonance point r_0. The perturbation treatment results in

$$\Delta(\sigma_0^2) = \pm i\pi\frac{(\sigma_0/\sigma_r)[(H_{rot}/r)N^2]_{r_0}(L_l^2\sigma_0^2)}{L_l^2+N^2+(V_1^2+k^2)(L_l^2/\Lambda^2)2\sigma_0^2} \,. \tag{39}$$

Since σ^2 becomes complex, there should be always one mode which is destabilized by the resonance effect. The rotation couples with the oscillation at the corotation point. Whether a stable mode ($\sigma_i/\sigma_r>0$) is destabilized or an overstable mode ($\sigma_i/\sigma_r<0$) is further destabilized depends, however, on whether the corotation point is situated in the superadiabatic core ($N^2<0$) or in the subadiabatic envelope ($N^2>0$). Also, a quantitative study of the resonance effect is not possible in the Ω^1-approximation and should await for the precise numerical calculation without neglecting higher orders in Ω.

REFERENCES

Osaki, Y. 1974, *Astrophys. J.*, **189**, 469.

Prigogine, I. and Glansdorff, P. 1965, *Physica*, **31**, 1242, see also 1964, *Physica*,**30**, 351.

Unno, W. 1968, *Publ. Astron. Soc. Japan*, **20**, 356.

Unno, W., Osaki, Y., Ando, H., and Shibahashi, H. 1979, *Nonradial Oscillations of Stars*, (University of Tokyo Press, Tokyo).

INERTIAL OSCILLATIONS AND THE ROTATION RATE PROFILE OF THE SUN

D. B. Guenther
Center for Solar and Space Research
Yale University
P. O. Box 6666
New Haven, CT 06511
U.S.A.

ABSTRACT. We have calculated the stream functions of axisymmetric inertial oscillations in a detailed model of the Sun's convective envelope, using a finite-difference form of the linearized fluid flow equations in a rotating spherical shell. Inertial oscillations are long period oscillations (of the order of one rotation period) for which the Sun's Coriolis force provides the principal restoring force. The frequencies and low latitude wave numbers of these oscillations are sensitive to both the rotation rate profile and the depth of the convective envelope. We show the effect different rotation rate curves have on the modes by using both frequency versus wave number diagrams and two-dimensional contour plots of the stream functions. We anticipate that if these modes are observed they will provide a useful and needed check of the rotation curves inferred from p-mode oscillations.

1. INTRODUCTION

If a fluid element is displaced (adiabatically) from its equilibrium position in a rotating adiabatic fluid then the Coriolis force will act to bend the fluid's path back to its original position. An oscillation will result about the equilibrium position. These oscillations are called inertial oscillations and are distinct from R-modes (or toroidal modes) waves in that inertial oscillations exhibit large radial motions.

Curious as to the nature of these modes in an adiabatically stratified spherical shell i.e., a model of the Sun's convective envelope, we began a series of numerical experiments investigating the behavior of the i-modes in response to changes in the shell's depth and the rotation rate (Guenther and Gilman 1985, Gilman and Guenther 1985). We discovered that for those i-modes whose amplitudes are strongest near the equator the frequencies and wavenumbers of the modes are sensitive independently to both the shell depth and the rotation rate profile. These numerical experiments have been repeated using the convective envelope of the standard solar model generated by the Prather-Demarque stellar evolution program. The inertial oscillation computer program was rewritten so that we could include the effects of the superadiabatic layer immediately below the Sun's photosphere.

In the next section we describe the numerical model used to describe inertial oscillations in the Sun's convective envelope. In §3 we describe some of our results and in §4 we briefly state some conclusions.

B. R. Durney and S. Sofia (eds.), The Internal Solar Angular Velocity, 107–114.

2. MODELING INERTIAL OSCILLATIONS

2.1. The Physics

Beginning with the general nonlinear fluid flow equations (Gilman and Glatzmier 1981, eqs. [1]-[7]) we use the anelastic approximation to remove the effects of acoustic modes. Acoustic modes have periods less than thirty minutes and can be ignored when compared to the much longer two week to more than one year periods of the inertial modes. To simplify the problem we only solve for axisymmetric modes and ignore buoyancy terms. In addition, we linearize the equations (retaining terms of first order in the perturbation term) by expressing the physical variables as the sum of a nonrotating term, a rotating term, and a small perturbation term. The nonrotating terms solve the equations of motion in a nonrotating inertial reference frame. The rotating term, which is small because of the Sun's slow rotation rate, is the correction to the nonrotating term to account for rotation. And the perturbation term describes the motion of the inertial oscillation with respect to the rotating reference frame. For the models presented here, further simplification was obtained by taking the velocities of the oscillations to be zero at the base and surface of the convection zone.

2.2. The Math

The linearized equations were cast into finite difference form on a two-dimensional staggered grid in radius and latitude. The algebraic equations were then mapped into a form characteristic of a modified eigenvalue problem:

$$i\,\Omega\,\mathbf{U} = \mathbf{A}\;\mathbf{S}, \tag{2.1}$$

$$i\,\Omega\,\mathbf{P} = \mathbf{B}\;\mathbf{S}, \tag{2.2}$$

$$i\,\Omega\,\mathbf{D}\;\mathbf{S} = \mathbf{C}_1\,\mathbf{U} + \mathbf{C}_2\,\mathbf{P}, \tag{2.3}$$

where Ω is the eigenfrequency of the inertial oscillation, **U** represents the east-west velocity, **P** represents the entropy perturbation (to model the superadiabatic layer), **S** represents the stream function for the velocities in the north-south-radius grid plane, and **A** **B**, $\mathbf{C}_1$, $\mathbf{C}_2$, and **D** are constant arrays which depend on the structure of the solar convection zone which in turn depends on the rotation profile.

The modified eigenvalue problem is solved in the following manner: *Step 1.* The two linear systems,

$$\mathbf{D}\;\mathbf{X}_1 = \mathbf{C}_1, \tag{2.4}$$

$$\mathbf{D}\;\mathbf{X}_2 = \mathbf{C}_2, \tag{2.5}$$

are solved for $\mathbf{X}_1$ and $\mathbf{X}_2$. *Step 2.* The matrix **Y** is calculated from

$$\mathbf{Y} = -\,(\mathbf{X}_1\;\mathbf{A} + \mathbf{X}_2\;\mathbf{B}). \tag{2.6}$$

Step 3. The eigenvalue problem

$$\Omega^2 \mathbf{S} = \mathbf{Y}\,\mathbf{S} \qquad (2.7)$$

is solved for Ω^2 and **S**. *Step 4.* **U** is calculated from

$$\mathbf{U} = -\,i\,\Omega^{-1}\,\mathbf{A}\,\mathbf{S}. \qquad (2.8)$$

Numerically, for a 30 in latitude by 18 in radius grid, this amounts to solving a 540 by 540 linear eigenvalue problem. The solution from a computer run contains the entire spectrum, to the resolution of the grid, of inertial oscillations.

3. THE RESULTS

3.1. Contour Plots of the Stream Functions

Figure 1(a) and 1(b) show respectively the stream functions for an i-mode with a period of two weeks and and an i-mode with a period of one year. For solid body rotation inertial oscillations must have periods greater than 1/2 the rotation period (Greenspan 1968). As the period of the oscillation increases, the motion becomes more strongly aligned with the rotation axis. This is expected since rotation does not affect a fluid element of the mode when its velocity component is parallel to the rotation axis i.e., the Coriolis force on the fluid element is zero when the fluid element is moving parallel to the rotation axis, hence the frequency decreases for those modes whose fluid element path is primarily parallel to the rotation axis.

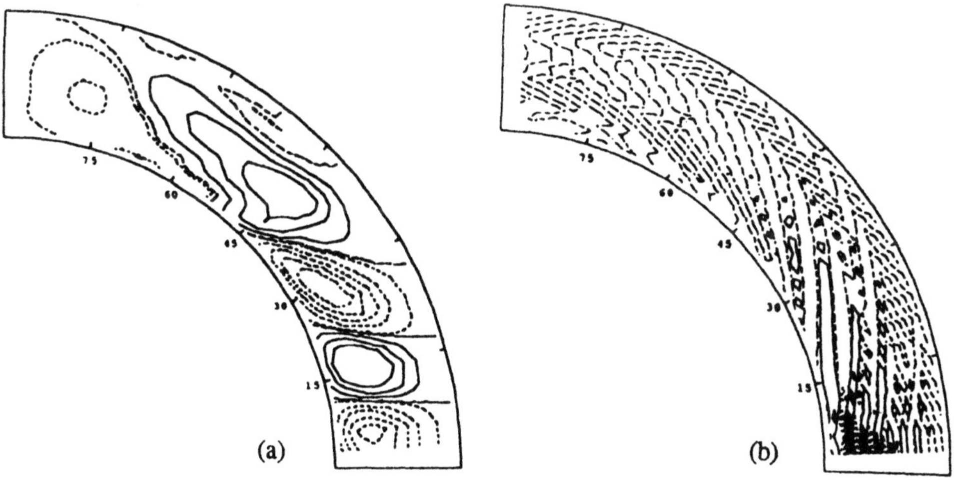

Figure 1(a) and (b). Stream function of two inertial oscillation modes confined to the solar convective envelope. The i-mode in (a) has a period of approximately two weeks and the i-mode in (b) has a period of approximately one year.

Figure 2(a), (b), and (c) show $n = 0$, 1, and 2, respectively, i-modes. We define n as the radial node number between the surface and the base of the convection zone. We find for many of the i-modes that the largest velocity perturbation amplitudes of the modes are found in the equatorial regions typified by the i-mode in Figure 2(c). In fact, very few modes have large velocity amplitudes at all latitudes similar to those shown in Figure 1(a), 2(a), and 2(b); the modes appear to be confined to either the equatorial region, the polar region, or the mid-latitudes.

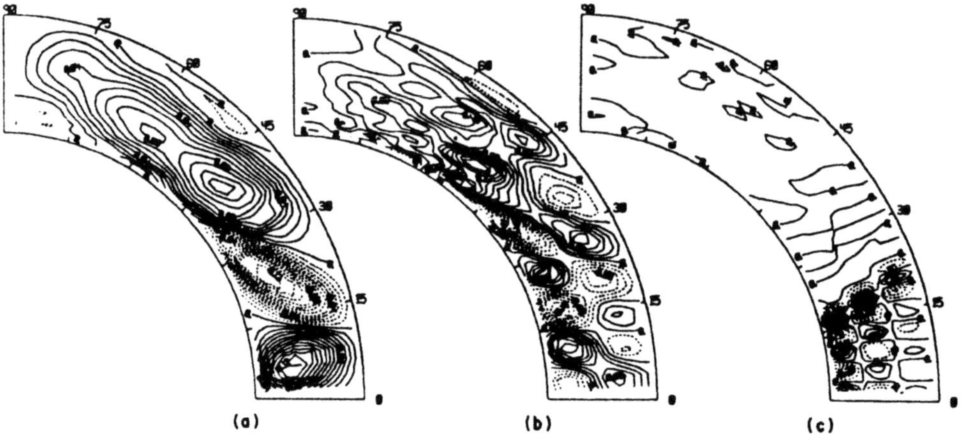

Figure 2(a), (b), and (c). Stream function of inertial oscillation modes with with radial order $n = 0, 1, 2$ (from left to right).

Figure 3(a),...,(n) shows a sequence of similar modes for different rotation rate profiles. Rotation is taken to be constant along cylinders and is defined by:

$$U_0(r, \phi) = r \cos(\phi) \left[1 + b \left[1 - (r / r_0)^2 \cos^2(\phi) \right] \right] U_{equ}, \tag{3.1}$$

where r is the radial position, ϕ is the latitudinal position, r_0 is the star's radius, U_0 is the unperturbed fluid velocity in the east-west direction, and U_{equ} is the equatorial surface rotation velocity. For b negative the rotation rate profile decreases with depth, for b zero the rotation is solid body, and for b positive the rotation rate profile increases with depth. We see that as the rotation rate goes from b = -0.20 in Figure 3(a), through solid body rotation, Figure 3(i), to b = 0.20 in Figure 3(n), the wavelength in the latitudinal direction of the cell at the equator decreases. Notice, also, that the number of distinct cells at low latitudes goes from two to three as a large cell at the mid-latitudes is pinched in half by another cell. We find that for some rotation profiles a given mode may disappear or split into several similar modes.

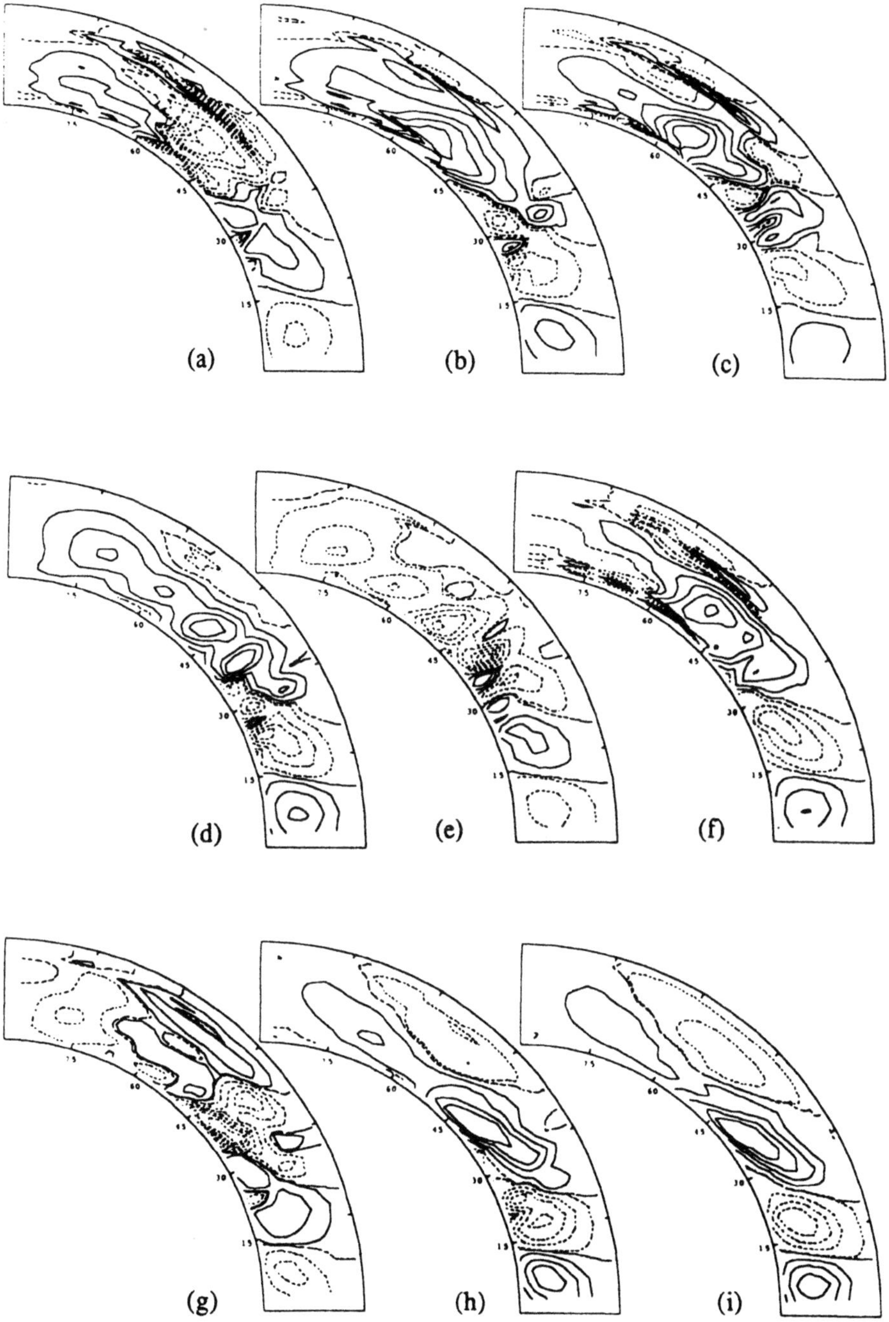

Figure 3(a),...,(n) Stream function of the same inertial mode as the rotation rate parameter b is increased from b = -0.20 in (a) to b = 0.20 in (n). (a) through (n) are, respectively, b = -0.20, -0.20, -0.17, -0.17, -0.13, -0.13, -0.07, -0.03, 0.00, 0.03, 0.07, 0.10, 0.13, 0.17.

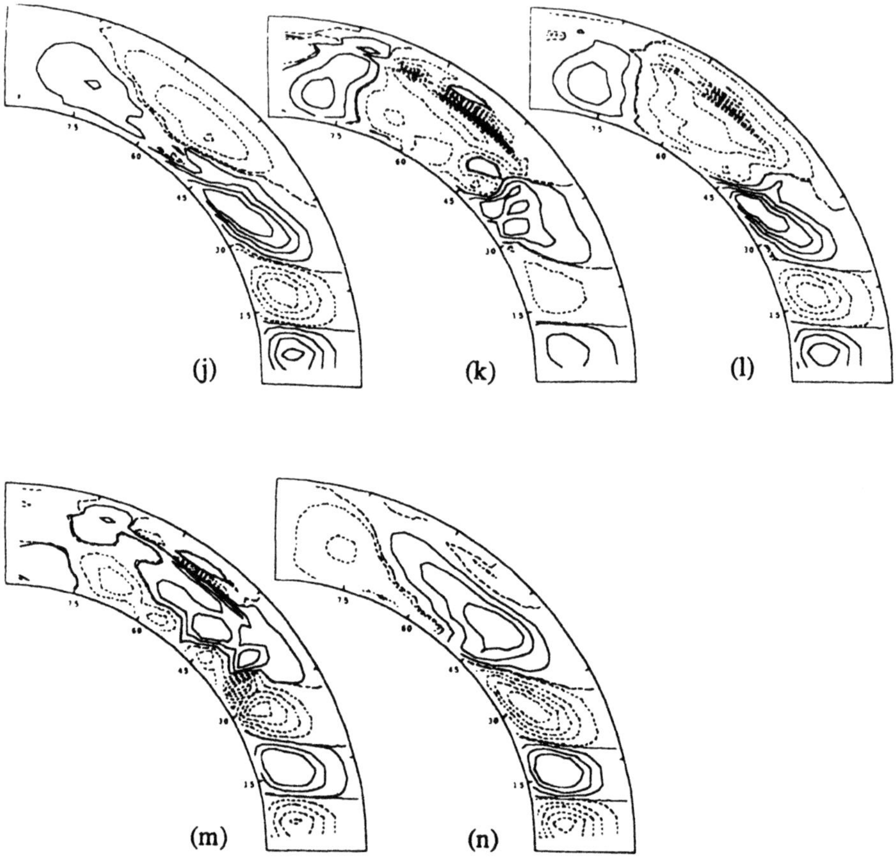

Figure 3 *continued*.

3.2. Frequency Versus Wavenumber Diagrams

It is very difficult to gather any useful information from the 540 stream functions generated from each computer run. Noticing that many of the modes are confined to the equatorial regions we define a low-latitude wave number as the average wavenumber in the north-south direction of the stream function cells between the equator and 20° latitude. Plotting the wavenumber of the i-modes versus their frequency we obtain graphs like those represented in Figure 4(a), (b), and (c).

The modes fall along ridges in frequency versus low-latitude wavenumber diagrams. The ridges correspond to modes which have the same radial order, with the highest frequency ridge having $n = 0$, the second highest frequency ridge $n = 1$, etc. The ridge spacing decreases with increasing convective envelope depth. The ridges shift

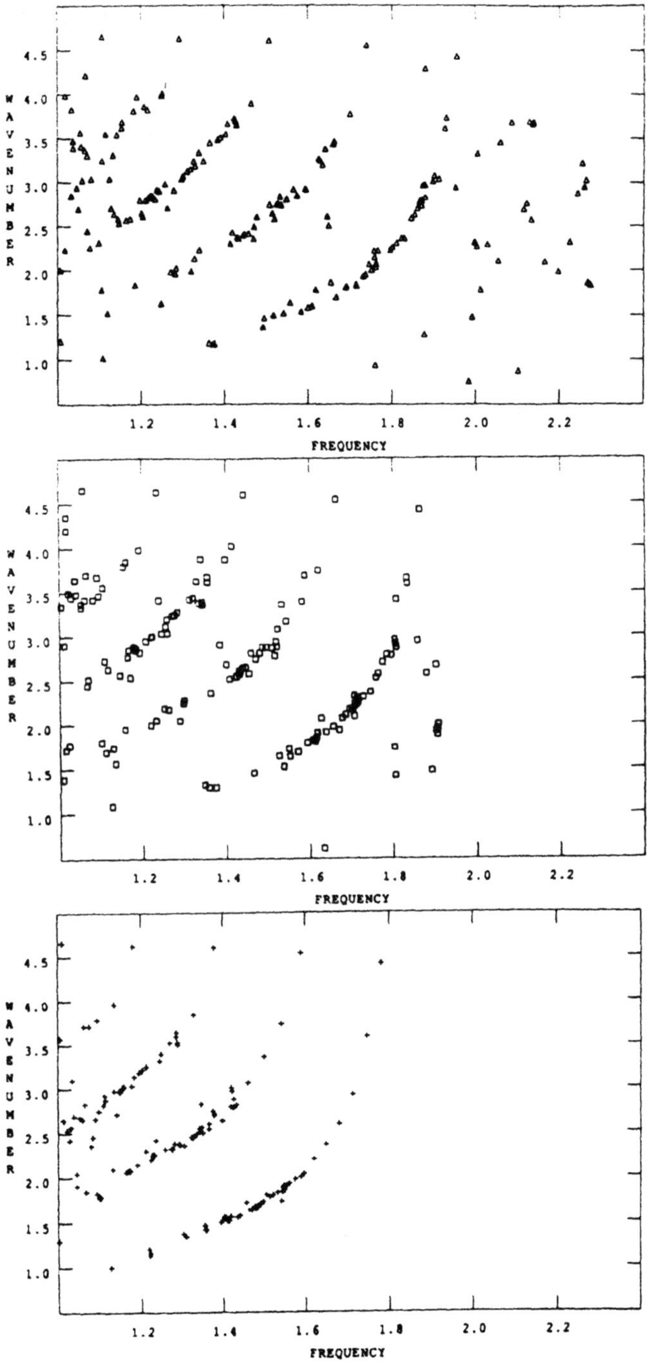

(a) b = 0.20

(b) b = 0.00

(c) b = -0.20

Figure 4(a), (b), and (c). Frequency versus wavenumber for i-modes at b = 0.20, b = 0.0, and b = -0.20.

towards lower frequencies when b increases from b = -0.20 to b = 0.20. This can be seen by comparing Figure 4a where b = -0.20 with Figure 4b and c where b = 0.0 and b = 0.20, respectively.

The points which appear scattered over the region to the right of the $n = 0$ ridge in Figure 4b and 4c are due to modes which have their maximum velocity amplitudes near the poles and are not correctly interpreted by a low-latitude wavenumber. Curiously, these modes do not appear for rotation rate profiles which decrease with depth.

4. CONCLUSIONS

We believe that if inertial oscillations can be observed at the surface of the Sun they will provide a very useful check on other methods used to determine the depth of the convection and the rotation rate profile. Clearly though, our model of inertial oscillations is very simple and cannot tell us, for example, what is the expected velocity amplitude of i-modes at the photosphere, or whether the i-modes well be swamped out by the myriad of other oscillations appearing at the surface. If the amplitudes are of the order of 1 m/s then with respect to spatial and temporal resolution we calculate that one year of data from the Fourier tachometer (Tim Brown, private communication) should be able to resolve the ridge structure in the frequency versus low-latitude wavenumber diagrams.

We have not included the superadiabatic term in the computer runs because of the increased grid resolution required to realistically model the narrow superadiabatic region. I would not surprise us, though, if this region significantly alters the frequency versus wavenumber diagrams (Gilman, private communication).

There exists a set of high latitude i-modes which produce the scattered points on the right of Figure 4b and 4c which have not been discussed here because of the impracticalities involved in observing high solar latitudes. If these modes could be observed (from a highly inclined heliocentric orbiting satellite, for example), they would provide latitudinal dependent information about the rotation profile of the Sun's convective envelope.

The author wishes to thank Peter Gilman for introducing the subject of inertial oscillations to him. The author is grateful for support provided by Control Data Corporation through their award of a PACER Postdoctoral Fellowship.

5. REFERENCES

Peter A. Gilman and D. B. Guenther 1985, *Inertial Oscillations in the Solar Convective Zone. II. A Cylindrical Model for Equatorial Regions, Ap. J.*, **296**, 685-695.

Greenspan, H. P. 1968, *The Theory of Rotating Fluids* (Cambridge: Cambridge University Press).

D. B. Guenther and Peter A. Gilman 1985, *Inertial Oscillations in the Solar Convective Zone. I. Spherical Shell Model* , *Ap. J.*, **295**, 195-212.

A SPHERICAL HARMONIC DECOMPOSITION TECHNIQUE FOR ANALYSING STEADY PHOTOSPHERIC FLOWS

David H. Hathaway
Mail Code ES52
Space Science Laboratory
NASA Marshall Space Flight Center
Huntsville, Alabama 35812
USA

ABSTRACT. Steady flows in the photosphere, including differential rotation, meridional circulation and convection, can be represented by a spectrum of modes. A technique is described in which the spherical harmonic functions are used to determine the characteristics of this spectrum and the nature of these flows. Some information about the spectrum is lost because only one hemisphere is seen, only the line of sight velocity is measured and the measurements contain noise. This produces an apparent mixing between spectral modes. By analysing synthetic data it is found that, in spite of this mixing, differential rotation can be accurately measured, meridional circulations with small amplitudes can be extracted from the data and giant cells might be separated from supergranules.

1. INTRODUCTION

The solar internal rotation can be influenced by a variety of mechanisms. It can be forced by the action of the Coriolis force on a thermally driven meridional circulation (Gierasch, 1974; Belvedere and Paterno, 1976; Bogart and Gierasch, 1979; Tassoul and Tassoul, 1984). It can also be driven more indirectly through the transport of momentum by nonaxisymmetric waves or convective flows. In theoretical models of the solar differential rotation this indirect forcing can be represented by an anisotropic viscosity (Biermann, 1951; Kippenhahn, 1963; Cocke, 1967; Kohler, 1970; Schmidt, 1982), a stress tensor (Durney and Spruit, 1979; Rudiger, 1980; Hathaway, 1984) or by explicitly calculating the convective flows themselves (Busse, 1970; Durney, 1970; Yoshimura and Kato, 1971; Gilman, 1977; Glatzmaier, 1984). The rotation profile may also be influenced by the Lorentz force on large scale magnetic fields (Yoshimura, 1981; Hinata, 1982). In order to understand the solar internal rotation we must understand the different forcing mechanisms and determine the nature of these other flows in the solar interior.

Analyses of the solar p-mode oscillations promise to provide us with the best information about the actual internal rotation of the Sun (Gough, 1985). However, in order to understand the nature of this rotation profile we must also measure the meridional circulation and the convective flows as well as the rotation profile of the surface layers for which the p-mode oscillations provide little information. Here a technique is described for analysing Doppler velocity measurements of photospheric flows. While the technique uses data like that required for the p-mode analyses the details of the analysis are somewhat different.

B. R. Durney and S. Sofia (eds.), The Internal Solar Angular Velocity, 115–120.

The spherical harmonics are used as the natural representation of flows in the photosphere. Solar rotation, differential rotation and meridional circulation are represented by the axisymmetric or zonal harmonics. Supergranules and global convection are represented by a spectrum of nonaxisymmetric spherical harmonic components. The decomposition technique uses the properties of the spherical harmonics to separate the different flow types and accurately determine their characteristics. An overview of the technique is presented here. Further details are given by Hathaway (1987).

2. DATA REQUIREMENTS

The analysis requires data in the form of spatially resolved, full disk velocity measurements of the line of sight velocity component for the "steady" flows. The word steady is qualified to include velocity components like supergranules and giant cells which evolve slowly over many hours or more. It is meant to exclude granules and the p-mode oscillations. Adequate spatial resolution is required to resolve the supergranules (8 arcsec resolution or a 256 by 256 pixel array should be sufficient). The temporal resolution is limited by the presence of the 5 minute oscillations. Measurements must be taken about once a minute and then averaged over about an hour to remove this oscillatory signal from that of the steady flows. Data in this form should be produced by instruments which are designed for studying the p-modes. This would include the Solar Oscillations Instrument (SOI) proposed for SOHO and the Global Oscillations Network Group (GONG) instruments.

Another useful requirement for the data is to have the solar rotation axis aligned with one dimension of the data array. Although the data array can be rotated in software to fulfill this requirement, the process introduces noise that could be avoided if the alignment is made at the time of the data acquisition. The analysis technique employs a spherical polar coordinate system whose axis is in the plane of the sky. The observed spectrum is transformed afterward to a coordinate system whose axis is coincident with the Sun's rotation axis.

In this paper, synthetic data representing the actual observations is constructed in order to test the analysis technique. This synthetic data includes the limb shift, solar rotation and differential rotation, a meridional circulation, supergranules, giant cells and a component of random noise. This data is constructed to represent full disk velocity observations on a 256 by 256 pixel array with characteristics similar to those previously determined for the Sun.

The limb shift is assumed to be independent of latitude and to vary from a 300 m/s blue shift at disk center to a 100 m/s red shift at the limb. The noise in the data is represented by random contributions at each pixel with some predetermined amplitude limit (10 m/s and 100 m/s). The other velocity components are calculated from a spherical harmonic representation of the full vector velocities on the surface of a sphere (e.g. Chandrasekhar, 1961). The steady flows of interest are assumed to have no radial component to their flows. The horizontal velocity components in the polar and azimuthal directions are given by

$$U_\theta(\theta,\phi) = \sum_\ell \sum_m \left[S_\ell^m \frac{\partial Y_\ell^m}{\partial\theta} + T_\ell^m \frac{1}{\sin\theta} \frac{\partial Y_\ell^m}{\partial\phi} \right] \tag{1}$$

$$U_\phi(\theta,\phi) = \sum_\ell \sum_m \left[S_\ell^m \frac{1}{\sin\theta} \frac{\partial Y_\ell^m}{\partial\phi} - T_\ell^m \frac{\partial Y_\ell^m}{\partial\theta} \right] \tag{2}$$

where the S's and T's are the spectral amplitudes of the poloidal and toroidal modes respectively, and the Y's are the spherical harmonics. These spherical harmonics are given by

$$Y_\ell^m(\theta,\phi) = P_\ell^m(\theta)\, e^{im\phi} \tag{3}$$

where the P's are the normalized, associated Legendre polynomials.

The spectral components are chosen so that the rotation profile matches that given by Howard and Harvey (1970) with an equatorial velocity of 2000 m/s. The meridional circulation is in the form of a single cell per hemisphere with a poleward flow of 20 m/s in the midlatitudes. The supergranules are constructed from a spectrum of modes with l=100 and amplitudes such that the maximum veolcities are on the order of 500 m/s. The giant cells are constructed from a spectrum of modes with l=24 and amplitudes such that the maximum velocities are on the order of 40 m/s.

The horizontal velocities are calculated at each pixel using these different flow components. The line of sight velocity is then calculated from

$$V(\theta,\phi) = U_\theta(\theta,\phi)\cos\theta\sin\phi + U_\phi(\theta,\phi)\cos\phi \tag{4}$$

It should be noted that information about the spectrum is lost in constructing this synthetic data. Only the line of sight velocity is used in the analysis, only one hemisphere is seen and random noise is added. An image of the synthetic data is shown in Hathaway (1987). The only obvious flows are the solar rotation and the supergranulation. Some evidence of differential rotation and limb shift can be seen but the other flow components, being much weaker, are well hidden by these stronger components.

3. THE ANALYSIS TECHNIQUE

In principle the analysis of the different velocity field components can proceed without regard to the order in which they are taken. In practice the analysis is cleaner if it is done in three sequential steps. First the limb shift is calculated and removed. Next the axisymmetric flows, rotation and meridional circulation, are calculated and a correction is made to the limb shift. The final step is to find the nonaxisymmetric convective flows, the supergranules and giant cells. This final step can proceed much faster if the original data is reformatted onto a grid with data points equally spaced in azimuth (longitude) for each latitude band. Since this process introduces additional noise, the axisymmetric flows, which do not require this step, are calculated before the nonaxisymmetric ones.

The limb shift is first calculated in the manner described by Snodgrass (1984). The velocity averaged around the disk is found for a series of different radial positions from disk center to the limb. This limb shift profile is then represented by a set of orthonormal polynomials. Since this averaging procedure will include parts of the meridional circulation signal, a correction to the limb shift is added after the meridional circulation is reconstructed.

Although the differential rotation and the meridional circulation are in fact axisymmetric flows, the observed line of sight component of these flows appears as wavenumber m=1 patterns. Rotation produces a pattern in which the eastern hemisphere is moving toward the observer and the western hemisphere is moving away. A meridional circulation produces a pattern in which the flow is away from the observer at a given latitude over the visible hemisphere but would be toward the observer at that latitude on the backside of the Sun. The spectral components for the rotation profile are given by

$$T_\ell^0 = [\ell(\ell+1)]^{-1/2} \frac{2}{\pi} \int_0^\pi \int_0^\pi V(\theta,\phi)\, P_\ell^1(\theta) \cos\phi \sin\theta \, d\theta \, d\phi \tag{5}$$

and the spectral components for the meridional circulation are given by

$$S_\ell^0 = [\ell(\ell+1)]^{-1/2} \left[\frac{(2\ell+1)(2\ell-1)}{(\ell+1)(\ell-1)}\right]^{1/2} \frac{2}{\pi} \int_0^\pi \int_0^\pi V(\theta,\phi)\, P_\ell^1(\theta) \sin\phi \sin\theta \cos\theta \, d\theta \, d\phi - \frac{(\ell-2)}{(\ell+1)} \left[\frac{(2\ell+1)}{(2\ell-3)}\right]^{1/2} S_{\ell-2}^0 \tag{6}$$

These integrals can be transformed by using the data in its original form with pixels equally spaced across the disk of the Sun rather than equally spaced in longitude. An integration by parts gives the rotation components in terms of an average slope in the line of sight velocity across the disk. The meridional circulation components are given by a weighted average of the line of sight velocity at each latitude. This is very similar to the method used for calculating the limb shift and illustrates how the signals for these two components tend to be folded together. These signals can be unfolded by determining the response of the analysis technique to the different meridional circulation components and then correcting for the effects of this response.

The results of this analysis for the synthetic data indicates the power of the technique. The recovered limb shift and differential rotation profiles agree with the input profiles to within about 3 m/s even when the noise level is artificially high (100 m/s). The meridional circulation profile, which has an amplitude two orders of magnitude smaller, is recovered without too much distortion. The results for the meridional circulation are shown in figure 1.

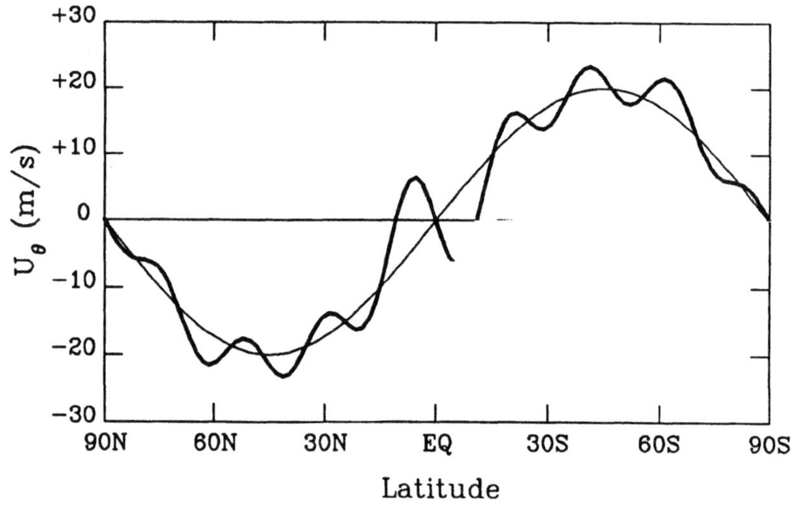

Figure 1. The meridional circulation profile extracted from the synthetic data (thick line). The input profile is plotted as a thin line.

The variations in the measured profile are mostly due to crosstalk with the signal from the supergranules. These large scale flows are relatively insensitive to random noise because they are measured by averaging over many pixels.

It should be noted that one benefit of this analysis is that the total angular momentum of the surface layer is wholly contained in the lowest order rotational mode. Any exchange of angular momentum with other layers is readily seen in variations in the amplitude of this mode with time. Variations in the other rotation modes, the differential rotation components, reflect a redistribution of angular momentum that does not involve changing the net angular momentum of the surface.

The last step in the analysis is to determine the spectrum of the nonaxisymmetric flows. This is done by finding the amplitudes

$$V_\ell^m = \frac{2}{\pi} \int_0^\pi \int_0^\pi V(\theta,\phi)\, Y_\ell^m(\theta,\phi) \sin\theta \, d\theta \, d\phi \tag{7}$$

The spectral amplitudes S, of the poloidal modes and T, of the toroidal modes cannot be determined from a single observation because of the mixing between the various components produced by the lack of information about the transverse velocity and the velocity field in the unseen hemisphere. In spite of this apparent drawback, the spherical harmonic transform still serves as an effective spatial filter. Much can be determined from the observed amplitudes, V. The spectrum of these amplitudes can be broken up to produce a low pass filter for giant cells and a high pass filter for supergranules. When this is done for the synthetic data the giant cell pattern that was included in the data is accurately reconstructed from

$$V(\theta,\phi) = \sum_{\ell=1}^{32} \sum_{m=1}^{\ell} V_\ell^m \, Y_\ell^m(\theta,\phi) \tag{8}$$

The reconstructed image of the giant cells is substantially altered only when the noise level is artifically high (100 m/s).

4. CONCLUSIONS

This spherical harmonic decomposition technique is found to be a powerful, accurate and useful means of investigating steady flows in the solar photosphere. By using fairly realistic synthetic data it is found that the solar rotation profile is accurately measured and cast in a form that is useful for determining secular variations in the solar rotation. Low amplitude meridional circulation patterns can be accurately extracted from the data and separated from the limb shift signal. Large convective cells with small amplitudes can be separated from supergranules with much larger amplitudes.

The analysis technique uses hourly averages of full disk Doppler velocity measurements like those needed for the p-mode analyses. However, the results are greatly enhanced if the observations are continued for more extended periods. Accumulating and averaging the spectral amplitudes for the rotation and meridional circulation will improve their accuracy. Retaining these amplitudes over the course of many years will indicate the nature of any solar cycle variations in these flows. The existence and nature of any large scale convective flows would be better determined by comparing the low pass filtered images over the course of a solar rotation. If persistent patterns are seen then observations from different days can be combined to provide stereoscopic views of the velocity field that can be used to determine the total vector velocity.

An accurate determination of these various velocity fields is important for understanding the internal differential rotation of the Sun. The technique described here should enhance our ability to separate and measure the different flow components and determine the role they play in driving the differential rotation.

REFERENCES

Belvedere, G., and Paterno, L. 1976, *Solar Phys.* **47**, 525.
Biermann, L. 1951, *Zs. Ap.* **28**, 304.
Bogart, R. S., and Gierasch, P. J. 1979, *Ap. J.* **228**, 624.
Busse, F. H. 1970, *Ap. J.* **159**, 629.
Chandrasekhar, S. 1961, *Hydrodynamic and Hydromagnetic Stability* (Oxford: Clarendon Press).
Cocke, W. J. 1967, *Ap. J.* **150**, 1041.
Durney, B. R. 1970, *Ap. J.* **161**, 1115.
Durney, B. R., and Spruit, H. C. 1979, *Ap. J.* **234**, 1067.
Gierasch, P. J. 1974, *Ap. J.* **190**, 199.
Gilman, P. A. 1977, *Geophys. Ap. Fluid Dyn.* **8**, 93.
Glatzmaier, G. A. 1984, *J. Comput. Phys.* **55**, 461.
Gough, D. 1985, *Solar Phys.* **100**, 65.
Hathaway, D. H. 1984, *Ap. J.* **276**, 316.
Hathaway, D. H. 1987, *Solar Phys.* (in press).
Hinata, S. 1982, *Ap. J.* **256**, L23.
Howard, R. F., and Harvey, J. 1970, *Solar Phys.* **12**, 23.
Kippenhahn, R. 1963, *Ap. J.* **137**, 664.
Kohler, H. 1970, *Solar Phys.* **13**, 3.
Rudiger, G. 1980, *Geophys. Ap. Fluid Dyn.* **16**, 239.
Schmidt, W. 1982, *Geophys. Ap. Fluid Dyn.* **21**, 27.
Snodgrass, H. B. 1984, *Solar Phys.* **94**, 13.
Tassoul, M., and Tassoul, J. L. 1984, *Ap. J.* **286**, 350.
Yoshimura, H. 1981, *Ap. J.* **247**, 1102.
Yoshimura, H., and Kato, S. 1971, *Publ. Astr. Soc. Japan* **23**, 57.

THE SOLAR-STELLAR CONNECTION: INTERNAL ROTATION IN LOW-MASS STARS

Pierre Demarque
Center for Solar and Space Research
Yale University
P.O. Box 6666
New Haven, CT 06511

1. INTRODUCTION

Since the theme of this workshop is the internal rotation of the Sun, I will discuss primarily the problem of the internal rotation of stars with masses and internal evolution which are similar to that of the Sun. Observations of solar-type stars in different stages of evolution throw light on the past and future history of the Sun. At the same time, the ability to study stars with different rates of surface rotation, different masses and chemical compositions, provides an opportunity to test more thoroughly and convincingly the physical principles which underlie our models of the solar interior and its state of internal rotation.

The role of rotation in stellar evolution has many diverse facets, and can be observed in a variety of ways. How angular momentum is distributed and transferred within the star, and removed from its surface, affects the formation process in a gas cloud and the internal evolution throughout the pre-main-sequence phase and the early post-main-sequence phase, where we now find the Sun. The interaction of rotation with convection is responsible for the generation of the magnetic fields associated with chromospheric activity and the activity cycles. Rotation induces internal mixing of chemical elements through meridional circulation and through the instabilities which it causes, and in this way can modify surface abundances. Perhaps less often noted is the fact that rotationally induced mixing can also change the rate of evolution and thus affect the luminosity function of star clusters. Finally, rotation should in principle be detectable by seismological means through the splitting of non-radial oscillation modes. The distribution of low degree p- and g-modes also tests sensitively the central mass concentration of stars.

I will review briefly each of these topics. Each one is, or promises soon to be, a large and active area of research. In the process, I will discuss in more detail a seemingly rather technical, but crucial, point about the need to construct models of stellar convection zones which are consistent with full interior models. The significance of this point is best illustrated in the context of two important astrophysical problems, i.e. the calculation of the Rossby

B. R. Durney and S. Sofia (eds.), The Internal Solar Angular Velocity, 121–130.

number in stellar dynamos and the evaluation of surface lithium abundances in solar-type stars. In both cases, the answer depends sensitively on the detailed structure of the convection zone, i.e. its geometrical depth, run of temperature and density, and convective velocities. A frequently adopted procedure in this kind of study is to construct envelope models integrated inward without applying realistic constraints at the lower boundary. Often this procedure yields quite incorrect or inconsistent results.

2. STAR FORMATION AND HISTORY OF SURFACE ROTATION RATES DURING PRE- AND EARLY POST-MAIN-SEQUENCE EVOLUTION

The study of star formation, pre-main-sequence and early post-main-sequence evolution provides information on surface braking mechanisms and on the transport of angular momentum inside stars. As our understanding of the place of magnetic fields in early evolution grows, we can also hope to learn more about the strength of initial magnetic fields and how they are affected by rotation (Fox and Bernstein 1987). Motions in star forming regions in interstellar clouds suggest the presence of much more specific angular momentum in proto-stars than is actually observed in T Tauri stars. The problem is then one of finding mechanisms which expel angular momentum from the forming star. Larson (1984) has suggested gravitational torques in non-axisymmetric configurations as such a mechanism and has applied it to the case of FU Orionis. Magnetic torques have been invoked by Pudritz and Norman (1983). From the T Tauri stage to the main sequence stage, stellar winds are believed to provide an efficient braking mechanism (Schatzman 1962). Indeed, the models of pre-main-sequence evolution for the Sun proposed by Endal and Sofia (1981), which include the braking effect at the surface of a wind parameterized according to the models of Belcher and McGregor (1976), are in quite satisfactory agreement with recent observations of young star clusters.

Early work on rotational velocities of F and G dwarfs in the Hyades and Pleiades suggested to Kraft (1967) a correlation between age and rotational velocity for these spectral types. Kraft (1965) had noted that the observed slowdown in rotational velocity coincides in surface temperature with the point where the depth of the convection zone increases rapidly. Wilson (1966) also noted the coincidence between the appearance of magnetic activity and surface convection zone, a point which was then discussed in the light of stellar evolutionary tracks by Demarque and Roeder (1967). The Schatzman torque due to the coupling of the stellar wind (which is associated with the convection zone) to the interstellar magnetic field, provided a physical explanation for these suggestive coincidences. The "Skumanich law", which relates the surface rotational velocity to the inverse square root of time, was derived from observations of the Pleiades, Hyades and the Sun (Skumanich 1972). More recent work has tended to confirm "Skumanich's law" reasonably well for stars older than the Pleiades, although age determinations based solely on the magnetic activity index are quite uncertain for old disk stars in the field (Soderblom 1985). And as discussed in Section 4, one must be cautious with "lithium ages" since, for some masses and compositions, the surface lithium abundance is not correlated with age in a simple way, first decreasing, then increasing, and finally decreasing again while the star is near the main sequence.

In recent years, observations of rotational velocities of pre-main-sequence stars in the T Tauri phase (Hartmann et al 1986), and in young star clusters and associations (Stauffer et al 1985), have shed important new light on the time dependence of stellar surface rotation rates. Here, the "Skumanich law" does not hold. Although still preliminary, a clear picture already emerges, based on the pioneer work of Endal and Sofia (1981)(see also Sofia 1987), and work currently in progress at Yale. The early phases of pre-main-sequence evolution are characterized by an increase in surface rotation rate: this is because during this phase little angular momentum is removed from the star as it contracts toward the main sequence. Only when the star approaches the main sequence is most of the angular momentum then rapidly transferred out of the convection zone, due to the braking action of the stellar wind. The evolutionary models predict that this phase of rapid spindown gives rise to a steep gradient of differential rotation between the radiative interior and the convective envelope.

3. CONVECTION ZONES AND MAGNETIC ACTIVITY

It is generally believed that the observed magnetic activity cycle in the Sun is due to a dynamo seated near the bottom of the convection zone, in a region with differential rotation, at the interface between the still rotating interior and the outer convective layers which have been braked down by the solar wind (Parker 1975; Durney and Latour 1978; Gilman 1983; Glatzmaier 1985). The same kind of interaction between convection and rotation must then take place in all late-type stars, as indeed the discovery of activity cycles in late-type dwarfs by Wilson (1978) demonstrates. The study of these objects thus provides an opportunity to test our theoretical ideas about the generation of magnetic activity cycles in stars with a mass slightly different from that of the Sun, and with different ages and chemical compositions.

Dimensional arguments have been advanced which relate the strength of the dynamo to the Rossby number, the ratio of the surface rotation period to the convective overturn timescale in the region where the dynamo operates (Durney and Latour 1978; Parker 1979). Durney and Latour (1978), constructed convection zone models from which they evaluated Rossby numbers using the overturn timescale of the pressure scale height at the bottom of the convection zone. In a subsequent study, Durney and Robinson (1982) introduced a new number, i.e. the product of the Rossby number times the square root of the quantity (r/H), where r denotes the geometrical depth of the convection zone, and H is the pressure scale height at the bottom of the convection zone. They argued that this modified Rossby number might be a more representative measure of the star's activity index. More recently, Noyes et al (1983) have analyzed the sample of Wilson stars studied at the Mt Wilson Observatory during the last few years, and correlated their magnetic activity index with theoretical estimates of the overturn timescale in the convection zone (Gilman 1980). They reached the interesting conclusion that magnetic activity is better correlated with the Rossby number than with the rotation period, as some other investigators had previously suggested. They also found their correlation to be less tight for the modified Rossby number. Finally, since the Gilman models predicted a dependence of overturn time on alpha, the mixing length to scale height ratio adopted in the subphotospheric models, Noyes et al (1983) pointed out that the data favored a particular value of alpha, around 1.9.

There is currently controversy as to whether it is physically meaningful to correlate the Rossby number, or another parameter related to it, with the magnetic activity index. Part of the controversy has to do with the manner in which the magnetic activity index itself is evaluated (Rutten and Schrijver 1986; Zwaan 1986) and is beyond the scope of this paper.

It is also natural to enquire about any possible correlation between activity cycle periods and the Rossby number (Robinson and Durney 1982). There is some evidence from the available data for a linear relationship between cycle periods and Rossby numbers, which favors a dynamo in which the saturation occurs through buoyant loss of magnetic flux (Noyes et al 1984).

Although it is premature at this stage to draw definite conclusions regarding the significance of the Rossby number in stellar activity research, it is evident that any discussion of the dependence of the Rossby number on stellar physical parameters should be based on properly evaluated Rossby numbers. It is a dangerous practice to construct model stellar envelopes, integrated from the surface inward with different values of alpha, which are not subjected to the constraints of a fully consistent stellar model at the lower boundary. Because of the non-linearity of the equations of stellar structure (Schwarzschild 1958)(SES), small departures amplify rapidly in inward integrations, and can yield quite incorrect runs of pressure and temperature, and consequently also incorrect convective velocities at the bottom of the convection zone. This point has also been made recently by Gilliland (1985).

A striking example can be provided with a standard model of the Sun (e.g. Demarque et al 1986). The point can be illustrated quantitatively by using the Milne non-dimensional UV-plane used in SES (see p.109). In Schwarzschild's nomenclature, the specific entropy of the adiabatic convection zone is proportional to the parameter E. The larger E, the deeper the convection zone. However, given the constraints of a standard solar model, there is a unique E which matches the solar radius, which corresponds to a unique value of alpha. In our case, alpha = 1.37, and E = 17, which fits the core integration (it must be the point where convective instability sets in) at about U = 0.45 and V = 7.6. Integrations from the surface for different values of alpha yield E = 5, 15, 18 and 23 respectively for alpha = 0.5, 1.0, 1.5 and 2.0. The mismatch in the UV-plane is substantial (see e.g. Fig.23.2 of SES). In terms of the overturn timescale tau, the four (inconsistent) envelopes yield a sensitive dependence on alpha, similar to that found by Gilman (1980), i.e. tau can vary by an order of magnitude, when alpha varies from 1.0 to 2.0. However, consistent models with the temperature, luminosity and chemical composition of the Sun, but with different assumed values of alpha, must have slightly different masses. To each assumed alpha corresponds a unique mass, for any given age. Our calculations show that the dependence of tau on alpha is very weak in the range of 3.65 < logTeff < 3.78, for 1.0 < alpha < 2.0, and closely follows the same curve as that derived by Noyes et al for alpha about 2.0. This is true whatever the age of the star.

This weak dependence of tau on alpha and age explains why all stars in Noyes et al 's sample of disk field stars fall on the same curve, despite the fact that the sample includes a wide range of ages and rotation rates, in which one might therefore expect a scatter in the effective value of alpha. Indeed, detailed studies of nearby stars do suggest variations in effective alpha from star to star (Guenther and Demarque 1986; Guenther 1987). Similarly, variations in effective alpha for a given star as a function of age, which have usually been ignored for the sake of simplicity, might be expected. A plausible case can be made that the effective alpha, which is after all only meant to provide an approximate measure of convective efficiency, depends on rotation rate and magnetic activity in the convection zone (Tayler 1986; Guenther 1987).

Unfortunately, for the cooler stars considered by Noyes et al (1983), the weak dependence of tau on temperature makes it difficult to demonstrate a preference for dependence on Rossby number rather than rotation period. But for logTeff above about 3.78, tau depends on alpha more sensitively as the convection zone becomes increasingly shallow and eventually disappears. The temperature at which convection disappears also depends on alpha. Preliminary models yield best agreement with Noyes et al's data for alpha in the range 1.3-1.5. We note that this is in agreement with our estimate of 1.37 based on fitting the solar radius precisely. On the other hand, it might be possible to distinguish between the predictions of models for alpha equal to 1.0 and 1.5. It is for stars in this transition region, where the activity index shows a steep dependence on stellar color, that it may be possible to establish a convincing correlation with Rossby number, and to distinguish at the same time between different convection zone models. More detailed modeling in the transition region where the convection zone vanishes, including evolutionary effects, will be needed to determine whether the Rossby number, or some other similar quantity, can be used to parameterize magnetic activity. It would be interesting to find out whether these stars, which are also all relatively young and more rapid rotators, do differ from later-type stars in their dynamo regimes (Durney et al 1981; Knobloch et al 1982).

4. SURFACE CHEMICAL ABUNDANCES

Surface abundances can provide information about the dynamics of stellar interiors, because one of the ways angular momentum may be transferred in stars is through internal circulation and mixing.

Spectroscopic studies in field red giants reveal the existence of objects in which surface CNO abundances show evidence for completion of the first step of the CNO cycle, which converts C12 into N13 through proton capture in the interior of the star and changes the C12/C13 isotopic ratio (Lambert and Ries 1981). Such surface abundance modifications could be due either to very efficient mass loss on the giant branch, or to mixing from deep layers to the surface, or a combination of both processes (Dearborn et al 1978). In the case of the brighter giants in globular cluster stars, Sweigart and Mengel (1979) have made a very plausible case that meridional circulation is the cause of the observed CNO anomalies, although their theory does not explain the appearance of these anomalies in some fainter giants (Carbon et al 1984), and needs some modification.

In solar-type stars near the main-sequence, the light elements Li, Be and He3 are well-known to provide a test of the efficiency of mixing deep in the stellar interior (Schatzman and Maeder 1981). I would like to report briefly on research currently in progress at Yale in which we attempt to account for the observed low lithium abundance in the Sun (where it is depleted by two orders of magnitude) and in dwarf stars of the halo population [in which Spite and Spite (1982) report little lithium depletion], in terms of self-consistent stellar evolutionary models (Deliyannis and Demarque 1987).

The physical processes which affect surface Li abundances are complex and highly sensitive to the temperature and density at the bottom of the convection zone and to the efficiency of mixing mechanisms. Because the relative importance of each relevant process varies markedly within narrow ranges of mass and chemical composition, Li abundance generally is not a meaningful age parameter for halo stars and is at best only a crude indicator in special circumstances.

Four processes must be taken into account:

(1) near the main-sequence, Li depletion through nuclear burning at the bottom of the convection zone.

(2) as the convection zone receeds, Li diffuses below the convection zone, thus draining Li from the convection zone into the radiative layers right below it (Michaud et al 1984).

(3) the dredge-up of diffused, but unburnt Li back to the surface, as the convection zone deepens due to evolutionary turnoff.

(4) Li dilution as the convection zone deepens further and dredges up layers in which Li has been previously depleted.

All of these processes do not necessarily take place in any given star, nor do they operate in the same way in stars with different masses and chemical composition, hence the complexity of the observed behavior. The warnings of the previous section on the need to use fully consistent stellar models, as well as taking into account the evolutionary history of the star, apply even more strongly in this context. The use of incomplete model envelopes can lead to incorrect or strangely inconsistent results. Preliminary models also indicate that pre-main-sequence evolution should also be considered with higher numerical accuracy than has been done in the past in order to interpret the lithium observations, because of the extreme sensitivity of the rate of lithium burning to temperature and density.

For the Sun, the prediction of our standard model is found consistent with the observed surface Li abundance, taking into account the uncertainties of the models. The presence of beryllium in the solar spectrum is also found to be consistent. Our result thus puts stringent constraints on the effects of rotation on lithium burning and mixing in the interior. Interestingly, we note that our solar model does not require any appeal to the process of turbulent diffusion invoked in recent years by several authors to explain the observed solar Li abundance (Schatzman and Maeder 1981). This result is encouraging in view of the difficulties already encountered in reproducing theoretically the observed p-mode spectrum of the Sun for models including the extensive internal mixing predicted by turbulent diffusion (Ulrich and Rhodes 1983).

Turning now to field halo stars, we find that using the same theoretical assumptions as for the Sun, our standard models also yield satisfactory agreement with the lithium observations of Spite and Spite (1982). This is achieved in terms of a sequence of successive phases of Li burning, diffusion, dredge-up and dilution, which in each case depends sensitively on mass and chemical composition. These models suggest that the role of rotation on lithium mixing and depletion is, as in the Sun, small. Full results and discussion will be published elsewhere (Deliyannis and Demarque 1987).

We see that the Li and Be tests thus probe the whole previous history of the star. But, of course, they probe the history of the stellar envelope only. To answer the question whether the inner parts of the Sun are now affected by rapid rotation, or have been in the past, will require tests involving the rate of nuclear burning or seismological studies. We must also keep in mind that transfer of angular momentum does not necessarily induce chemical element mixing, and that magnetic fields may play an important role in transporting angular momentum in solar-type stars (Fox and Bernstein 1987).

5. WHAT CAN WE LEARN FROM THE ADVANCED EVOLUTIONARY PHASES?

The effects of internal rotation near the main-sequence can also be felt in later phases of evolution. We know, for example, that some blue horizontal branch stars have measurable surface rotation rates (Peterson 1983). In the absence of a transfer of angular momentum by interaction with another star, this indicates that angular momentum must have remained hidden in these stars' interiors throughout their evolution, and appears at the surface after most of their envelope has been lost. It is also well-known that internal rotation can affect the development of the helium flash (Deupree 1986), as well as the helium core mass of horizontal-branch stars (Mengel and Gross 1976).

Since luminosity functions mirror the rate of internal evolution, the subgiant and giant branch luminosity functions in globular clusters can be used to test the internal history of low mass stars. The use of PDS microdensitometers and now of CCD detectors make it possible to measure accurate magnitudes to faint levels for large numbers of stars and to construct luminosity functions with high resolution and precision.

I will mention here two fine structure features which have been detected on the giant branch luminosity function of some globular clusters. The first is a peak which is predicted by standard stellar evolution theory (Thomas 1967). The other is a gap (or dip), the interpretation of which is still unclear.

The Thomas peak tells us about the maximum extent of mixing of the convective envelope when it reaches its deepest, just above the base of the giant branch in the HR-diagram. At this point, convection deepens into layers in which some hydrogen has previously been converted into helium during the main sequence phase of evolution, and leaves behind a step in the run of the helium abundance. At a later time, when the hydrogen-burning shell passes through this composition step, the evolutionary rate of the star is increased, resulting in a peak in the luminosity function. A comparison of theory with observed luminosity functions indicates that mixing of the order of a pressure scale height beyond the predicted mass fraction is required to reproduce the position of the Thomas peak (King et al 1985).

The other feature, still unexplained, may also teach us about the interior physics of subgiants and low-luminosity giants of low mass. A gap has been found near the base of the giant branch in all metal-poor globular clusters for which we have statistically significant data (Cannon and Lee 1981; Da Costa and Villumsen 1981; Buonanno et al 1984). On the other hand, 47 Tuc, the only metal-rich cluster for which we have adequate data, shows no gap in its luminosity function.

Armandroff and Demarque (1984) have proposed an explanation based on the postulated presence of a discontinuity in He3 in the interior. Yahil and van den Horn (1985) have noted that the gap is located at a point of discontinuous transition between two classes of solution of the equations of stellar structure. Work in progress (Deliyannis and Demarque 1987) suggests that this gap may be related to the dredge-up to the stellar surface of helium which has previously diffused below the convection zone near the main sequence.

6. WHAT CAN WE HOPE TO LEARN ABOUT ROTATION FROM SEISMOLOGICAL STUDIES?

Stellar seismology already has remarkable achievements to its credit in the case of planetary nuclei, where rotational splitting of g-modes has been observed and used to determine accurate masses for these objects (Kawaler et al 1986).

Solar-type non-radial oscillations have small amplitudes and are difficult to detect in stars. At this point only preliminary results have been reported on a few stars. Much thought and effort is being put on solving instrumental and data reduction problems and there seems little doubt that these obstacles will be overcome in the not too-distant future. As in the case of the Sun, rotation could affect the oscillation spectrum in two possible ways:

(1) the splitting of non-radial modes, p-modes and g-modes. The latter, if observable, would be particularly sensitive to central rotation since they penetrate most deeply in the star.

(2) the effect of rotationally induced mixing on the mass distribution, which in turn modifies the non-radial oscillation spectrum. Low-degree p-modes and g-modes could be sensitive to this effect.

This work was supported in part by grants NAGW-777 and NAGW-778 from the National Aeronautics and Space Administration.

5. REFERENCES

Armandroff, T. E. and Demarque, P. 1984 *Astron. Astrophys.* **139**, 305.
Belcher, J. W. and McGregor, K. B. 1976 *Astrophys. J.* **210**, 498.
Buonanno, R., Corsi, C. E. Fusi-Pecci, F., Alcaino, G. and Liller, W. 1984 *Astron. Astrophys.* **159**, 189.
Cannon, R. D. and Lee, S. W. 1981 in IAU Colloq. 68, *Astrophysical Parameters for Globular Clusters*, ed. A. G. D. Philip and D. S. Hayes (Schenectady: Davis), p. 501.
Carbon, D. F., Langer, G. E., Butler, D., Kraft, R. P., Suntzeff, N. B., Kemper, E. Trefzger, C. F. and Romanishin, W. 1984 *Astrophys. J. Suppl.* **49**, 207.
Da Costa, G. S. and Villumsen, J. V. 1981 in IAU Colloq. 68, *Astrophysical Parameters for Globular Clusters*, ed. A. G. D. Philip and D. S. Hayes (Schenectady: Davis), p. 527.
Dearborn, D. S. P., Blake, J. B., Hainebach, K. L. and Schramm, D. N. 1978 *Astrophys. J.* **223**, 552.
Deliyannis, C. and Demarque, P. 1987, in preparation.
Demarque, P., Guenther, D. B. and van Altena, W. F. 1986 *Astrophys. J.* **300**, 773.
Demarque, P. and Roeder, R. C. 1967 *Astrophys. J.* **147**, 1188.
Deupree, R. G. 1986 *Astrophys. J.* **303**, 649.
Durney, B. R. and Latour, J. 1978 *Geophys. Astrophys. Fl. Dyn.* **9**, 241.
Durney, B. R., Mihalas, D. and Robinson, R. D. 1981 *Pub. Astron. Soc. Pacific* **93**, 537.
Durney, B. R. and Robinson, R. D. 1982 *Astrophys. J.* **253**, 290.
Endal, A. S. and Sofia, S. 1981 *Astrophys. J.* **243**, 625.
Fox, P. A. and Bernstein, I. B. 1987, these Proceedings.
Gilliland, R. 1985 *Astrophys. J.* **299**, 286.
Gilman, P. 1980 in IAU Colloq. 51, *Stellar Turbulence*, eds. D. Gray and J. Linsley (New York: Springer), p. 19.
Gilman, P. 1983 *Astrophys. J. Supp.* **53**, 243.

Glatzmaier, G. 1985 *Astrophys. J.* **291**, 300.
Guenther, D. B. 1987 *Astrophys. J.* **312**, 211.
Guenther, D. B. and Demarque, P. 1986 *Astrophys. J.* **301**, 207.
Hartmann, L. Hewett, R., Stahler, S., Mathieu, R. D. 1986 *Astrophys. J.* **309**, 275.
Kawaler, S. D., Winget, D. E., Hansen, C. J. and Iben, I. Jr, 1986 *Astrophys. J.* **306**, L41.
King, C. R., Da Costa, G. S. and Demarque, P. 1985 *Astrophys. J.* **299**, 674.
Knobloch, E., Rosner, R. and Weiss, N. O. 1982 *Mon. Not. R. Astron. Soc.* **197**, 45 P.
Kraft, R. P. 1965 *Astrophys. J.* **142**, 681.
Kraft, R. P. 1967 *Astrophys. J.* **150**, 551.
Lambert, D. L. and Ries, L. M. 1981 *Astrophys. J.* **248**, 228.
Larson, R. B. 1984 *Mon. Not. R. Astron. Soc.* **206**, 197.
Mengel, J. G. and Gross, P. G. 1976 *Astrophys. Space Sci.*, **41**, 407.
Michaud, G., Fontaine, G. and Beaudet, G. 1984 *Astrophys. J.* **282**, 206.
Noyes, R. W., Hartmann, L. W., Baliunas, S. L. Duncan, D. K., and Vaughan, A. H. 1983, *Astrophys. J.* **279**, 763.
Noyes, R. W., Weiss, N. O., and Vaughan, A. H. 1984, *Astrophys. J.* **287**, 769.
Parker, E. N. 1975 *Astrophys. J.* **198**, 205.
Parker, E. N. 1979 *Cosmical Magnetic Fields* (Oxford: Clarendon Press).
Peterson, R. 1983 *Astrophys. J.* **275**, 737.
Pudritz, R. E. and Norman, C. A. 1983 *Astrophys. J.* **274**, 677.
Robinson, R. D. and Durney, B. R. 1982 *Astron. Astrophys.* **108**, 322.
Rutter, R. G. M. and Schrijver, C. 1986, in *Cool Stars, Stellar Systems and the Sun*, eds. M. Zeilik and D. Gibson (Berlin: Springer), p. 120.
Schwarzschild, M. 1958, *Structure and Evolution of the Stars*, (Princeton: Princeton U. Press).
Schatzman, E. 1962 *Astron. Astrophys.* **25**, 18.
Schatzman, E. and Maeder, A. 1981 *Astron. Astrophys.* **96**, 1.
Skumanich, A. 1972 *Astrophys. J.* **171**, 265.
Soderblom, D. R. 1985 *Astron. J.* **90**, 2103.
Sofia, S. 1987 these Proceedings.
Spite, F. and Spite M. 1982 *Astron. Astrophys.* **115**, 357.
Stauffer, J. R., Hartmann, L., Burnham, N. and Jones, B. 1985 *Astrophys. J.* **289**, 247.
Sweigart, A. V. and Mengel, J. G. 1979 *Astrophys. J.* **229**, 624.
Tayler, R. J. 1986 *Mon. Not. R. Astron. Soc.* **220**, 793.
Thomas, H.-C. 1967 *Zeit. Astrophys.* **67**, 420.
Ulrich, R. K. and Rhodes, E. J. 1983 *Astrophys. J.* **265**, 551.
Wilson, O. C. 1966 *Astrophys. J.* **144**, 695.
Wilson, O. C. 1978 *Astrophys. J.* **226**, 379.
Yahil, A. and van den Horn, L. 1985 *Astrophys. J.* **296**, 554.
Zwaan, C. 1986, in *Cool Stars, Stellar Systems, and the Sun,* eds. M. Zeilik and D. Gibson, (Berlin: Springer), p. 19.

NUMERICAL STUDIES OF SOLAR EVOLUTION WITH ROTATION AND COMPARISON WITH STELLAR ROTATION DATA

A. S. Endal
Applied Research Corporation
8201 Corporate Drive
Landover, MD 20785

ABSTRACT. Models of rotating stars involve a number of assumptions regarding initial rotation, surface angular momentum losses (due to winds), and internal angular momentum redistribution (due to convective flows, fluid instabilities, magnetic torques, etc.). Such models can be used to predict surface rotational velocities, which may be compared to the rotation rates observed for stars at various masses and ages. These comparisons are useful for checking the assumptions in the models and, more indirectly, inferring the physical mechanisms which govern the internal rotation of the Sun and other stars. Recent improvements in the observational data allow more meaningful comparisons to the model predictions. In young clusters, the K dwarfs rotate much more rapidly that the G dwarfs, suggesting that the convective envelopes are spun down more easily than the radiative interiors. This creates a shear layer at the core-envelope interface. Post-main-sequence stars show clear evidence for strong rotational braking at the onset of deep envelope convection. These observations are reviewed in the context of models for rotating stars.

1. INTRODUCTION

During the past decade, I have been involved in modelling the evolution of rotating stars, in collaboration with S. Sofia and D.F. Gray. In this effort, we have developed a stellar evolution code which, starting from a prescribed initial rotation, predicts the rotational state at a later time by computing changes in the internal angular momentum due to stellar winds, internal fluid flows, fluid instabilities, etc. We have used this code to follow the evolution of solar-type and more massive stars in both the pre-main-sequence and post-main-sequence stages.

Given our present limited understanding of stellar fluid dynamics, calculations of this type necessarily involve a number of assumptions which are reasonable, but certainly not justified with a large degree of confidence. This is even more true concerning the interactions of rotational shear flows with magnetic fields. Therefore, we have emphasized the importance of comparing the

B. R. Durney and S. Sofia (eds.), The Internal Solar Angular Velocity, 131–141.

predicted surface rotation rates from the models with the rates inferred from observations of stellar photospheres, in order to check the validity of the model assumptions.

Most of these comparisons (Endal and Sofia 1979; Endal 1982) have involved stars on, or relatively near, the upper main sequence. Such stars rotate rapidly enough (> 10 km/s) that rotational broadening dominates over other factors contributing to the absorption line profiles. For these stars, the model predictions are in excellent agreement with the observations, indicating that the code is able to simulate the behavior of stars with radiative envelopes and with a moderate degree of radial differential rotation. Here, the dominant angular momentum redistribution mechanism is Eddington circulation within the deep radiative envelopes.

We would also like to be able to compare predictions and observations for stars in the lower and right portions of the HR diagram. Unfortunately, the surfaces of these stars rotate so slowly that it is difficult to disentangle rotational broadening of the line profiles from broadening due to other atmospheric mass motions. Inspite of these difficulties, the observers have triumphed in obtaining accurate rotational velocities for both solar-type and more massive stars in the lower and right portions of the HR diagram. This important advance provides us with an opportunity to compare model predictions with observations in a regime where the envelope angular momentum distribution is controlled by convection and the entire star loses angular momentum due to the torque exerted by a magnetized stellar wind.

In the following sections, I will first summarize the techniques and assumptions used in modelling the evolution of rotating stars, and then compare the predictions to recent observations, both for solar-type stars and for more massive stars evolved to the right side of the HR diagram. Finally, I will draw these various comparisons together to see what may be inferred regarding the influence of convection and magnetic fields on stellar (and, in particular, solar) rotation.

2. MODELS OF ROTATING STARS

The techniques used to calculate stellar models and the evolution of the angular momentum distribution have been described in detail by Endal and Sofia (1976, 1978, and 1982), so I will limit the present discussion to those features most relevant to comparisons with observations presented in subsequent sections. The stellar structure is computed with a standard hydrostatic (Henyey) code, modified to account for the effects of rotation. In all of the models discussed here, rotational effects on the structure are very minor. Starting from standard initial conditions (uniform composition either on the zero-age main sequence or on the pre-main-sequence Hayashi track), a model sequence is computed by updating the composition for nuclear burning and convective mixing as the structure evolves in response to the altered composition and under the influence of the thermal relaxation terms of the energy equation.

To compute the rotational velocities, we specify the distribution of angular momentum (or the angular velocities) in the initial model. For the models discussed here, we used uniform (solid-body) rotation as the initial condition. The rotation at later stages in the evolution is controlled by three factors:

- expansion/contraction of various regions of the star alter the angular velocities and create rotation shears, even if angular momentum is locally conserved;
- internal mass motions associated with circulation currents, convection, and instabilities of shear flows change the distribution of the angular momentum within the star; and
- for solar models, the total angular momentum decreases with time due to the magnetically coupled solar wind.

This last factor is the only explicit recognition of the influence of magnetic fields in our models. In all cases, we assume that the angular velocity is constant on equipotential shells, so it is the rate of rotation of these shells which is computed.

Rigid-body rotation is enforced within convective regions, and (in the solar models) the effect of the solar wind is simulated by gradually reducing the angular momentum contained in the convective envelope. Within radiative regions, we use a diffusion equation to model angular momentum redistribution by circulation currents. The circulation currents are created by: Eddington circulation, the Goldreich-Schubert instability, and the secular shear instability. When the evolution leads to a dynamically unstable gradient in angular velocity, the gradient is reduced to the marginal stability limit (as defined by the Solberg-Hoiland criterion) by redistributing the angular momentum within the affected region.

3. SOLAR-TYPE STARS

Figure 1 displays the evolution in the HR diagram of a solar model, starting in the fully convective pre-main-sequence stage and continuing up to the present age of the Sun. The initial model contained about 60 times as much angular momentum as the Sun presently has, if the bulk of the Sun rotates at the same rate as the surface. This value was chosen to be consistent with the velocities measured for more massive stars, which do not show evidence for significant rotational braking by stellar winds. In the subsequent evolution, angular momentum was removed from the convective envelope at the rate predicted by the solar-wind model of Belcher and MacGregor (1976). This model assumes that the strength of the surface magnetic field is directly proportional to the rotation rate.

Figure 2 shows the interior rotation of this model as a function of time. Note that the interior rotation is not strongly affected by the spin down of the convective envelope. As a result, we predict that the Sun presently has a rapidly rotating core. In our models, this core retains too much angular momentum to be consistent with measurements of the solar oblateness. This problem is discussed in some detail by Endal and Sofia (1981).

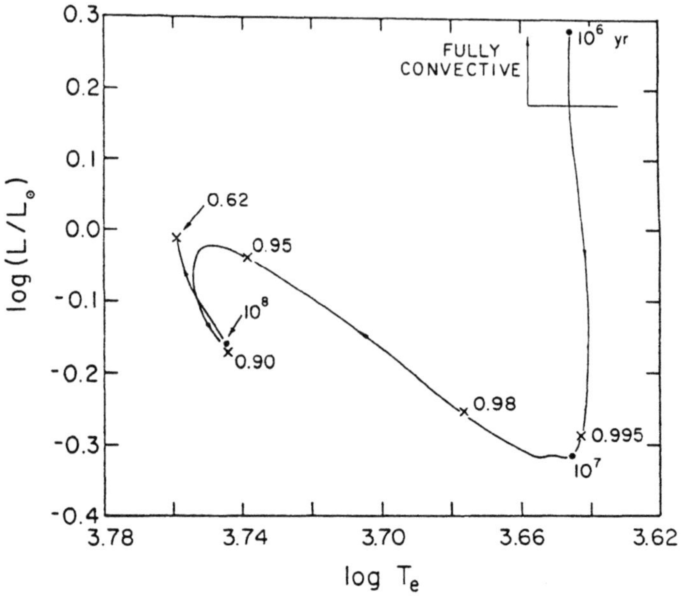

Figure 1. The evolution in the HR diagram of a 1 $M_\odot$ star from the fully convective pre-main-sequence stage to the present age of the Sun. The numbers by the heavy dots give the age, in years; the values near x's given the fraction of the initial angular momentum remaining in the interior. From Endal and Sofia (1981).

3.1 Comparison to Observations of Pre-Main-Sequence Stars

Our pre-main-sequence models predict that the stars initially rotate very slowly, and then spin up as they contract toward the ZAMS (see the M_r=1.0 $M_\odot$ curve in Figure 2). Predicted surface rotation rates are typically in the 10-100 km/s range. The rapid evolution through this stage, combined with the small angular velocities, limit the rotational braking by the wind so that the rotation rates reflect primarily the conditions at the time that hydrostatic equilibrium is attained.

Obviously, it would be of interest to compare the model velocities to vsini measurements for pre-main-sequence stars of approximately 1 $M_\odot$. The difficulty of making meaningful comparisons is illustrated by the data shown in Figure 3. These measurements refer to the G stars in the Orion Ic cluster. Two types of evolutionary tracks are also shown in the figure. Based on conventional hydrostatic evolution, we would assign masses to these stars in the 1.2-2.25 $M_\odot$ range. However, hydrodynamic models with spherical accretion from a protostellar cloud (*e.g.*, Larson 1969; Winkler and Newman 1980) predict very different trajectories in the HR diagram, such that the masses assigned to the stars in Figure 3 would be near and below 1 $M_\odot$. This difference in the assigned masses is critical for interpretation of the data. At present, it is not clear which type of model is preferred.

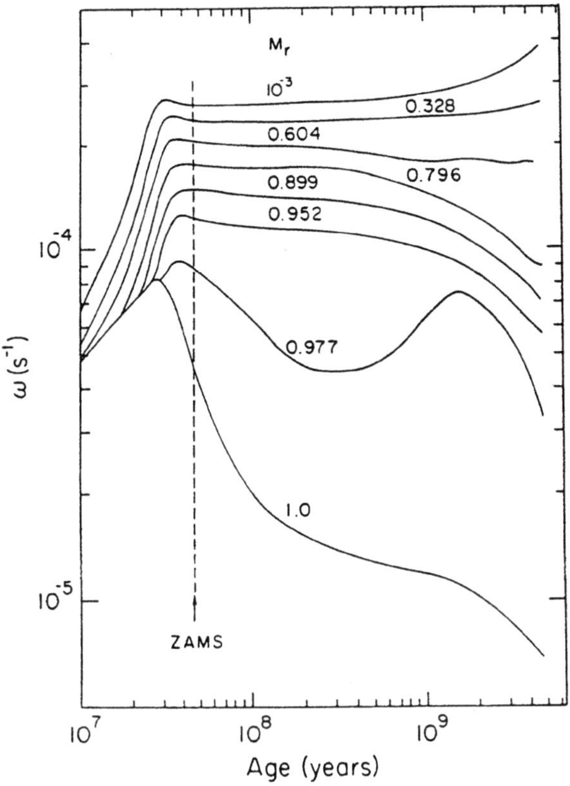

Figure 2. The angular velocities of selected mass layers (M_r) in the 1 $M_\odot$ model, as functions of time. From Endal and Sofia (1981).

3.2 Observed Rotation of the Main Sequence Stars

Recent observations of G dwarfs in open clusters (see Figure 4) confirm the Skumanich (1972) result and are in reasonable agreement with the solar models. Additional velocities are available for field stars and their ages can be estimated from their lithium abundances. However, there is a danger in this procedure in that atmospheric lithium depletion may be caused by rotationally driven circulation below the convective envelopes. In this case, the lithium abundance will depend on both age and initial rotation rate.

Some very interesting recent observations pertain to the K dwarfs in the Pleiades and α Per clusters. These stars have been found to rotate much more rapidly than the G dwarfs in the same clusters (see Figure 5). An interpretation of this result has been offered by Stauffer *et al.* (1985). They note that, in our solar models, the braking action due to the solar wind initially only affects the rotation of the outer (convective) layers, while the interior continues to rotate rapidly. This is clearly shown in Figure 2, above. Since the wind torque acts on a very small fraction of the total mass, the initial spin-down of the surface occurs very quickly. The K dwarfs, by contrast, have much deeper convective envelopes. Thus a given wind torque will require a much longer time to slow the rotation of these stars, in accord with the observations.

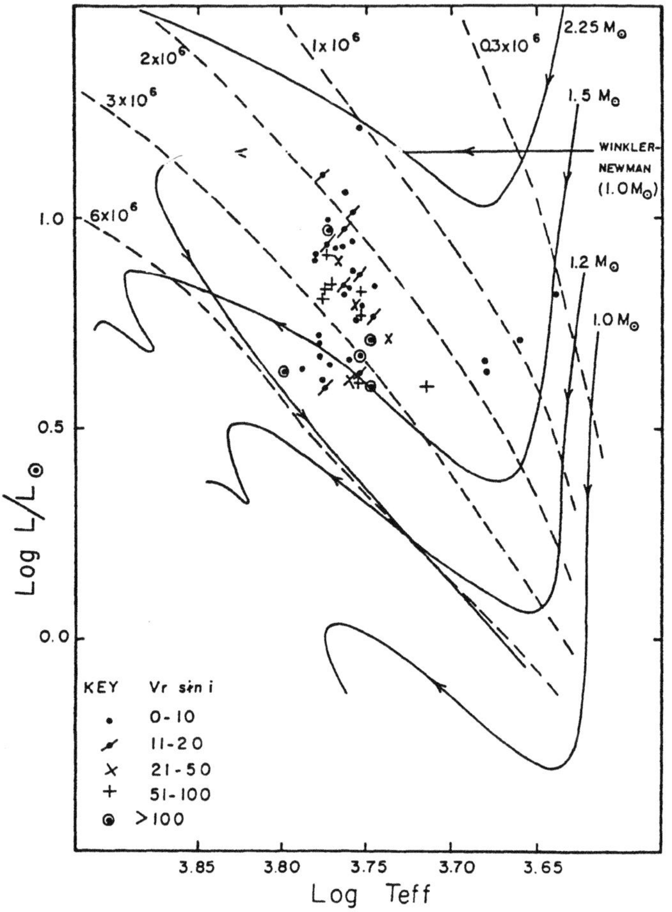

Figure 3. Vsini values measured for pre-main-sequence G stars in the Orion Ic cluster. From Smith, Beckers and Barden (1983).

If this explanation is correct, the clear implication is that a strong shear flow occurs near the bottom of the convective envelope. This provides a means to wind the poloidal magnetic field into a toroidal field in the solar dynamo cycle. Magnetic field amplification at the bottom of the convective envelope avoids the bouyancy problem encountered by dynamo models which have the field generated within the convective layers.

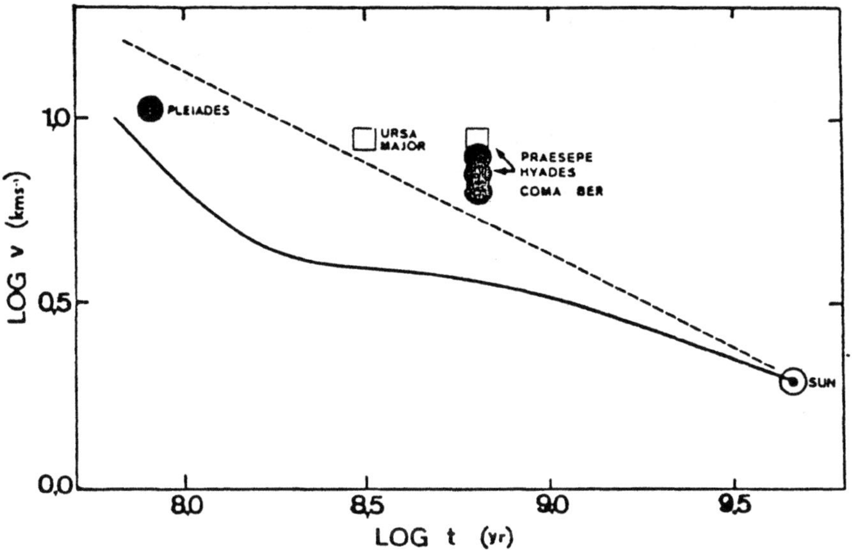

Figure 4. Rotation of G dwarfs as a function of age. Open squares show <v> from observations by Soderblom (1983); filled circles are from the observations of Benz *et al.* (1984). The dashed line is the Skumanich (1972) relationship. The solid line shows the model predictions of Endal and Sofia (1981). From Benz *et al.* (1984).

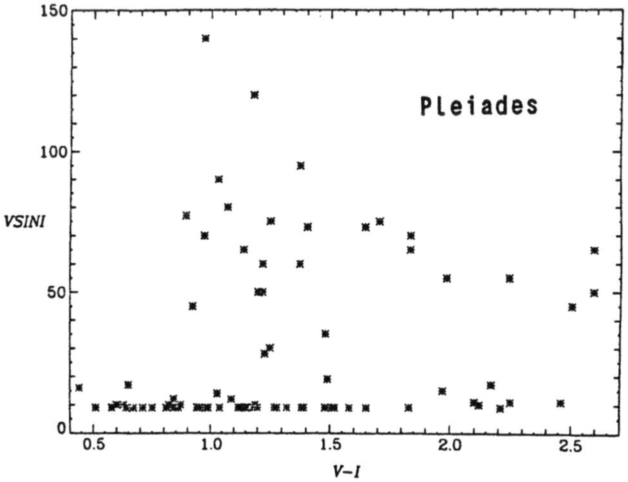

Figure 5. Observed rotational velocities (vsini) of G and K dwarfs in the Pleiades cluster. From Stauffer *et al.* (1985).

4. MASSIVE STARS

The stars with $M > 1.2\ M_{\odot}$ do not have radiative envelopes during the main sequence stage and, apparently, do not suffer any rotational braking until they move over toward the red giant branch and develop convective envelopes. Throughout the radiative-envelope stage, the predicted changes in the surface rotation rates from one time to another are in excellent agreement with the relative rates of rotation observed in the corresponding parts of the HR diagram (Endal and Sofia 1979; Endal 1982). These changes are caused by expansion/contraction of various parts of the star, with meridional circulation playing a significant role in redistributing the angular momentum within the star. The observations support the prediction that these stars develop significant differential rotation within the radiative envelopes during the rapid post-hydrogen-burning expansion/contraction. Winds do not have any observable effect on the surface rotation.

The situation changes dramatically when the stars evolve to the right side of the HR diagram and develop convective envelopes. This is illustrated in Figure 6. Our 3 $M_{\odot}$ models (without rotational braking) agree with the observed rotation rates of luminosity class III stars earlier than G5. At G5, there is a dramatic five-fold reduction in the observed rotation rates. This corresponds precisely to the stage where significant envelope convection first develops. If this initial spin-down is confined to the convective envelope, as seems likely considering how quickly the surface rotation is reduced, then a continued wind torque is required to prevent spin-up as the deepening convective envelope entrains excess angular momentum from the rapidly rotating interior. The required angular momentum loss rate is nearly constant during the evolution from G5 to K2 and exceeds the solar angular momentum loss rate by a factor of a million (Endal 1983).

Among the dwarfs (luminosity class V), stars of spectral type F2 and later show the effects of stellar wind braking, while the braking starts at G5 for the giants (luminosity class III). Is there evidence for a sharp onset of braking at the other luminosity classes? An affirmative answer for the subgiants (luminosity class IV) has been provided by the recent observations of Gray and Nagar (1985), displayed in Figure 7. As expected, the spectral type at which the effects of braking first appears (F8) is intermediate between F2 and G5. The location in the HR diagram of the boundary between fast and slow rotators coincides precisely with the boundary at which geometrically deep convective envelopes first appear. Finally, we do not expect that such a sharp boundary will be observed for luminosity classes I and II. By the time these massive stars reach the convective envelope stage, they have expanded sufficiently that, even without winds, the rotation rates will be at or below the threshold for current observational techniques.

5. SUMMARY

Most of the observations reviewed in the previous sections were obtained subsequent to publication of the main results from our calculations of the evolution of rotating stars. These observations often have not been in agreement with our predictions. This should have been expected since the calculations involve major assumptions regarding which processes are important and how these processes affect the observable surface rotation. In the above discussion, I have re-examined both the predictions and the observations to see what can be learned about the physical processes which control stellar rotation. Two themes which I would like to bring out in this summary are: evidence for stars rotating at different angular rates at different depths (differential rotation), and the one-to-one correspondence between convective envelopes and stellar wind braking.

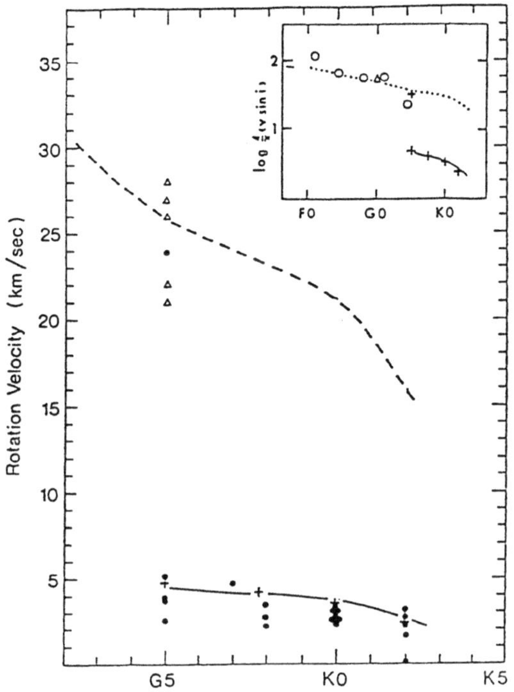

Figure 6. Comparison of predicted and observed rotation rates for giants crossing the Hertzsprung gap. The dashed and dotted lines are the predictions from a 3 $M_{\odot}$ model with no net angular momentum loss. In the main figure, filled circles and triangles refer to individual stars, while crosses are means for each spectral subclass. All symbols in the inset figure are means. From Gray (1982).

Magnetic-field pundits tend to dismiss the possibility of a long-lived radial shear by pointing out that any seed magnetic field will be quickly wound up and amplified by the differential rotation to a point where co-rotation of the various parts of the star will be enforced. While this seems a very compelling argument, the observational evidence seems to favor differential rotation. For instance, I have described observations of both main sequence and post-main-sequence stars which may most simply be interpreted in terms of differential rotation.

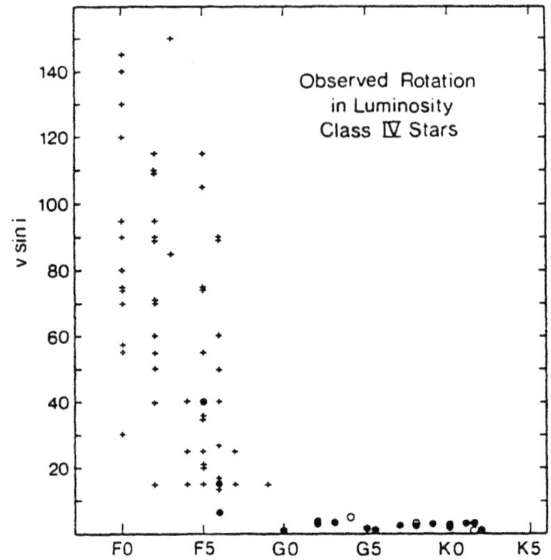

Figure 7. The observed rotation rates of luminosity class IV subgiants. From Gray and Nagar (1985).

While these particular observations may be open to alternate interpretations, there is one recent observational result which provides a most convincing case that a rapidly rotating core can survive for long periods within a slowly rotating envelope. I am referring to the observation by Peterson *et al.* (1983) and Peterson (1983) that a large fraction of the horizontal branch stars are rotating at surface velocities up to 30 km/s. These stars are the remnant cores of former red giants. During the ascent of the red giant branch, the rotation of the convective envelope slows dramatically due to the expansion and increased moment-of-inertia (and may be further slowed by the stellar wind). Co-rotation of the core with the envelope implies transfer of most of the core angular momentum to the envelope. In this case, the core would have insufficient remaining angular momentum to permit an observable surface rotation after the envelope is removed to form the horizontal branch star. In light of this observation, the relevant question seems to be what permits decoupling of the core from the envelope inspite of the magnetic field argument.

The second theme alluded to above concerns the close association between slow surface rotation and convective envelopes. This association has been known for some time for main sequence stars. The more recent work by D.F. Gray has extended this effect to the giants and subgiants. Many of these stars start their hydrostatic evolution in the blue part of the HR diagram, where rapid rotation is the rule, and evolve into the slow rotator regime. In this case, very strong braking is required by the rapid evolution across the Hertzsprung gap and the sharp demarcation between slow and rapid rotators. Such braking requires strong magnetic fields. These observations should permit us to examine the conditions for onset of the dynamo mechanism in stars which are very different from the Sun, in terms of mass, physical dimensions, and rotation rates, but which share the solar property of possessing geometrically deep convective envelopes.

The research described in this report was supported in part by NSF grant AST-8308035.

REFERENCES

Belcher, J.W., and MacGregor, K.B. 1976, *Ap.J.*, **210**, 498.
Benz, W., Mayor, M., and Mermilliod, J.C. 1984, *Astron. & Ap.*, **138**, 93.
Endal, A.S. 1982, in *Be Stars*, ed. M.Jaschek and H.-G.Groth (Dordrecht: Reidel), p. 299.
_____. 1983, in *Solar and Stellar Magnetic Fields: Origins and Effects*, ed. J.O. Stenflo (Dordrecht: Reidel), p. 493.
Endal, A.S., and Sofia, S. 1976, *Ap.J.*, **210**, 184.
_____. 1978, *Ap.J.*, **220**, 279.
_____. 1979, *Ap.J.*, **232**, 531.
_____. 1981, *Ap.J.*, **243**, 625.
Gray, D.F. 1982,*Ap.J.*, **262**, 682.
Gray, D.F., and Nagar, P. 1985, *Ap.J.*, **298**, 756.
Larson, R.B. 1969, *M.N.R.A.S.*, **145**, 271.
Peterson, R.C. 1983, *Ap.J.*, **275**, 737.
Peterson, R.C., Tarbell, T.D., and Carney, B.W. 1983, *Ap.J.*, **265**, 972.
Smith, M.A., Beckers, J.M., and Barden, S.C. 1983, *Ap.J.*, **271**, 237.
Skumanich, A. 1972,*Ap.J.*, **171**, 565.
Soderblom, D.R. 1982, *Ap.J.Suppl.*, **53**, 1.
Stauffer, J.R., Hartmann, L.W., Burnham, J.N., and Jones, B.F. 1985, *Ap.J.*, **289**, 247.
Winkler, K.-H.A., and Newman, M.J. 1980, *Ap.J.*, **236**, 201.

MAGNETIC FIELDS AND ROTATION IN LATE-TYPE STARS

L. Mestel, R. J. Tayler
Astronomy Centre
University of Sussex

D. L. Moss
Department of Mathematics
University of Manchester

1. INTRODUCTION (R. J. Tayler)

I wish to discuss three separate but related problems concerning the properties of magnetic fields in late type stars. It is generally accepted that the surface magnetic fields of late type main sequence stars are being maintained by a dynamo process in the outer convection zone which probably has its seat near the bottom of the zone. It is, however, uncertain whether or not there is a significant field in the radiative interior and, if so, whether the field is a relic of a dynamo-produced field from a pre-main-sequence phase or a genuine "fossil" field which was present when the star was formed. Two of the topics which I discuss are concerned with the properties of "fossil" fields, while the third is principally concerned with the influence of dynamo fields on convection and hence on the observed properties of stars. The second and third topics concern work which has already been published, while the first concerns work which is at present in progress, where I shall therefore stress the idea and not the details.

2. "FOSSIL" FIELDS AND PRE-MAIN-SEQUENCE MAGNETIC ACTIVITY

It is observed that the surface activity of late type main sequence stars is well correlated with the dynamo number which is believed to be related to the strength of the dynamo-generated magnetic field. The correlation is not so good for pre-main-sequence stars. Their surface activity is in some cases what might be expected from an extrapolation of the main sequence value but in other cases it is considerably higher. This has been discussed in a recent paper by Gilliland (1986). I suggest that the magnetic field near the surface of a pre-main-sequence star may not be purely a dynamo field but that it may be partly derived from a "fossil" field. If that is the case the variety in strength of surface activity may be related to the importance of the "fossil" field.

B. R. Durney and S. Sofia (eds.), The Internal Solar Angular Velocity, 143–157.

The argument goes as follows. Suppose that a star contains a large scale field at the time that it joins the fully convective Hayashi track; we will regard the strength of this field as arbitrary and we will not attempt to discuss what fraction of the original interstellar flux is likely to be retained at this stage. If the field has an energy density which is lower than that which would put it in approximate equipartition with the convective motions, it will be wound up and amplified by these motions until it is in such equipartition. If the field is stronger the motions will not be able to affect its basic structure. Instead the field will cause a substantial increase in the temperature gradient required for convection to carry most of the energy, but only for fields with an energy density comparable with the thermal energy density is convection likely to be replaced by radiation as the principal mechanism of energy transport.

If the "fossil" field is extremely weak ($\lesssim$ 1 gauss at a typical point in a solar mass star), its length scale will be reduced to such an extent that its magnitude may be limited by ohmic decay to a value less than the equipartition value while the star remains fully convective. For stronger initial fields, there will be an ultimate field strength which is of equipartition value or higher. This gives a field of order a few thousand gauss at a typical point in the interior of a solar type star approaching the end of its fully convective state. It is to be expected that such relatively strong or strong fields will be carried to the surface by convective motions and that as a result they will affect the magnetic activity of the star. The magnitude of the field will be rather insensitive to the initial field for a wide range of values of the latter, although the length scale of the field will be shorter for lower initial field strength.

As the star approaches the main sequence, it will develop a radiative interior. As the convection retreats outwards, a significant field may be left behind in the radiative interior. Although the existence and strength of such a field is not relevant to out present discussion of surface activity, we turn to its properties in the next section. Any field which remains in the outer convection zone will be confined to a region of lower temperature and it will have a length scale which is restricted to the thickness of the convection zone. Because the electrical resistivity, η, is proportional to $T^{-3/2}$, where T is the temperature, and the ohmic decay time is d^2/η, where d is the length scale of the field, it may be possible for the field to be influenced by ohmic decay close to the main sequence. If this happens the field in the outer layers will adjust to the standard dynamo value and the surface activity of the star will decline. It is therefore suggested that a spread in surface activity in the pre-main-sequence phase may be related to a spread in internal magnetic field strength and structure when stars joined the pre-main-sequence Hayashi track.

It would obviously be useful to have some idea of the magnetic field strengths in pre-main-sequence stars. Recently Johnstone and

Penston (1986) reported attempts to measure magnetic fields in three T Tauri stars. Two of these, R U Lup and G W Ori, are included in the list whose surface activity has been considered by Gilliland. Johnstone and Penston found a 2.3 sigma field of 550G in R U Lup and a 3 sigma upper limit of 1100G in G W Ori. If magnetic field strengths of this order could be confirmed in the stars, which appear to have surface activity in excess of that which would be generated by dynamo-maintained fields, this would be interesting evidence. It must of course be recognized that the observer measures an effective average field which will be less than the typical field strength, if the field has a length scale which is small compared with the radius of the star. An effective field of 550G could be accompanied by some very much larger positive and negative fields on different parts of the star's surface.

The detailed discussion which is contained in a paper in preparation is very approximate and in particular no attempt is made to discuss the influence of stellar rotation on both the magnetic field and the properties of convection. It is, however, believed that the basic idea does not really depend on the fine details of the discussion. The fine details will determine precisely how strong an initial field produces a significant surface field in the pre-main-sequence phase and precisely when the field will adjust to the dynamo value.

3. STABILITY OF MAGNETIC FIELDS IN RADIATIVE CORES

It has been suggested from time to time that the Sun might contain a very strong field in its radiative core; see e.g. Dicke (1979) for the suggestion of an extremely strong toroidal field. Strong internal solar fields have been suggested in connection with such matters as the resolution of the solar neutrino problem or with measurements of the solar oblateness. We have seen above that a relic of the initial solar field might have been left behind in the radiative core when the convection retreated. Such a field might have has a strength comparable with what we have called the equipartition field or have been even stronger if the primeval field was very strong. Even if the Sun does not possess a field of the strength suggested by Dicke, it could have an interior field much stronger than the average surface field. There are, however, arguments that the Sun cannot contain an interior field other than a very weak one, which are discussed elsewhere in this paper by Mestel. These are concerned with observations of solar oscillations which suggest that the radiative core is differentially rotating. Unless the field is extremely weak, it would be expected to enforce something very close to rigid body rotation. There is therefore a strong argument that the Sun does not at present possess a significant internal field.

It is therefore of interest to ask whether a strong internal field would be expected to survive for a main sequence lifetime. In a series of papers, I and various colleagues have discussed the stability of magnetic fields in non-rotating stars. In Tayler (1973) it was shown that a wide class of toroidal magnetic fields would be unstable to adiabatic perturbations and in Markey and Tayler (1973) a similar result was obtained for poloidal fields; in the latter case comparable results were obtained by Wright (1973). In the simplest cases the growth times for linear perturbations are comparable with the time taken by an Alfvèn wave to cross the region of interest. This linear growth time is very much shorter than the evolution time of the star unless the field is extremely weak; in the case of the Sun the relevant value of the field is about 10^{-5}G. For an extremely strong field of the type postulated by Dicke the instability growth time is not much greater than the dynamical timescale of the Sun.

This result, while of interest, does not immediately apply to fields with mixed poloidal/toroidal topology, which would a priori be more likely to be present. Nor does it take account of the rotation of the star. This is superficially the more important factor because, unless the field is very strong, the rotational energy is much greater than the magnetic energy or, equivalently, the rotation speed exceeds the Alfven speed. It is known from laboratory studies of hydromagnetic stability that helical magnetic fields are likely to be more stable than purely poloidal and toroidal fields and inside a star some remaining instabilities are suppressed by the strong gravitational field. Tayler (1980) showed that, although some mixed poloidal/toroidal fields would be stable, a wide class would be unstable even if with a reduced growth rate.

In a more recent paper Pitts and Tayler (1985) have discussed the influence of rotation on the instabilities. Although their results are suggestive rather than conclusive, it appears that rotation does not generally stabilize instability but that it leads to lower growth rates. These were found to be of order $\Omega_B^{\,2}/\Omega_s$, where Ω_B is a hydromagnetic frequency and Ω_s a sound frequency; $\Omega_B^{\,-1}$ and $\Omega_s^{\,-1}$ are the times taken by Alfvèn waves and sound waves to cross the relevant region. Although this result follows consideration of the influence of rotation, the rotation frequency does not appear directly in the growth rate. In fact, the growth rate is comparable to that obtained for mixed poloidal/toroidal fields without rotation. The linear growth time is of order $(4\pi P/B^2)$ times the dynamical timescale for the region of the star being considered. For a field stronger than about 10G the linear growth time is still less than the evolution timescale of the Sun.

None of this actually proves that magnetic fields significantly stronger than this would have decayed to a much lower strength by now. For one reason the non-linear behaviour of the instabilities has not yet been established. In addition, for the weaker fields which we have just been discussing, non-adiabatic effects might exert some damping influence but it is also known that there are some non-adiabatic instabilities of magnetic fields in stars which would also have to be taken into account.

As a result of these discussions, it is suggested that, even if a significant internal field survives the pre-main-sequence phase of stellar evolution, it may decline in strength thought instability during the main sequence phase. In particular any magnetic field in the centre of the Sun today may be significantly weaker than the field that the Sun possessed when it joined the main sequence. It seems very unlikely that the Sun could contain a field which is strong enough to affect the solar neutrino problem or to produce a significant oblateness or quadrupole moment. It is not however clear that the instability could reduce an initial field to such a low value that it would allow the solar core to be differentially rotating as seems to be required by the oscillation measures. If these results are confirmed and if there is no more efficient method of destroying a solar field, it may be necessary to deduce that no significant field was present when the Sun reached the main sequence.

4. INFLUENCE OF MAGNETIC FIELD AND ROTATION ON SURFACE CONVECTION ZONES

We now return to the question of the dynamo-maintained fields in late-type main sequence stars. In Section 2 we did not attempt to discuss how the convection might be affected by the magnetic field because this would not make any qualitative change in the properties we were discussing. Here we discuss one way in which convection is affected by both magnetic fields and rotation which might affect the comparison between theoretical and observed stellar properties.

It is known that a magnetic field changes the critical temperature gradient for the onset of convection from ∇_{ad}, the adiabatic value of PdT/TdP, to

$$\nabla'_{ad} = \nabla_{ad} + \lambda_1(B^2/4\pi P) \qquad (1)$$

where λ_1 is O(1) unless the field is primarily horizontal (Gough and Tayler, 1966). If the field is being maintained by a dynamo it is appropriate to write

$$B^2 = 4\pi\lambda_2\rho v^2_{conv}, \qquad (2)$$

where ρ is the density and v_{conv} is a typical value of the convective velocity. λ_2 may also be O(1) if the dynamo is very vigorous but it is much smaller than this in the present Sun. Observations and theoretical ideas suggest that λ_2 may have been much larger in the Sun when it first joined the main sequence. If we suppose that we can use the mixing length theory of convection and that the magnetic field does not change the actual value of the mixing length but simply replaces $\nabla-\nabla_{ad}$ by $\nabla-\nabla'_{ad}$, the convective flux takes the form

$$F_{conv} = c_p \rho T g^{1/2} \ell^2 (\nabla - \nabla'_{ad})^{3/2} / 4H_p^{3/2}$$

$$= c_p \rho T g^{1/2} \ell^2 (\nabla - \nabla_{ad})^{3/2} / 4H_p^{3/2} \left(1 + \frac{\eta_1 \rho^2}{4H_p^2}\right)^{3/2} \quad (3)$$

where $\eta_1 = \lambda_1 \lambda_2$ and c_p, g, ℓ and H_p are the specific heat at constant pressure, gravitational acceleration, mixing length and pressure scale height.

This expression for the convective flux can be interpreted in two different ways. It can be said that the magnetic field reduces the convective flux for a given temperature gradient, which is large enough for convection to occur, by the factor $(1+\eta_1 \ell^2/4H_p^2)^{3/2}$. Alternatively it can be said that the effective mixing length is not ℓ but $1/(1+\eta_1 \ell^2/4Hp^2)^{3/4}$. This effective mixing length is what would have to be used in calculations of stellar structure and evolution. If such calculations determined an appropriate value of the mixing length which would provide agreement between theory and observations, the true mixing length would be larger by the factor $(1+\eta_1 \ell^2/4H_p^2)^{3/4}$. This is discussed in slightly more detail in Tayler (1986).

Before we discuss the consequences of this result, it should be noted that a very similar effect can be produced by rotation. As in the case of a magnetic field, the influence of rotation is anisotropic but on average we can again write

$$\nabla''_{ad} = \nabla_{ad} + \lambda_3 (\rho \Omega^2 \omega^2 / P) \quad (4)$$

as the critical gradient for the onset of convection, where Ω is the angular velocity, ω the distance from the axis of rotation and λ_3 is probably O(1). It is again possible to write

$$\Omega^2 \omega^2 = \lambda_4 V^2_{conv} \quad (5)$$

but there is no reason why the rotation velocity and the convective velocity should be directly related which means that the magnitude of λ_4 is quite uncertain. For the minority of late type stars which rotate very rapidly, λ_4 would be significantly greater than unity.

Rotation then produces an effective mixing length $(1+\eta_1 \ell^2/4H_p^2)^{3/4}$, where $\eta_2 = \lambda_3 \lambda_4$. Of course in reality the simultaneous effects of magnetic fields and rotation would have to be considered in a single star.

Suppose we now consider the effect of a change of effective mixing length on the properties of late type stars. VandenBerg (1983) has produced a large series of models of such stars using different values of the mixing length. He decided that $\alpha \equiv \ell/H_p = 1.6$ gave the best agreement between theory and observations both on the main sequence and on the giant branch and he concluded that α should lie in the range 1.5 to 1.7. Most of the stars whose properties were fitted by VandenBerg probably do not have very active dynamo-generated magnetic fields and we will therefore tentatively assume that $\alpha = 1.6$ is the true value of the mixing length. Now consider a star with such an active dynamo so that $\eta_1 \gtrsim 1$. In this case the effective value of α corresponding to a true value of 1.6 is 1.02. VandenBerg has calculated models with $\alpha = 1$ as well as $\alpha = 1.6$. His results, for stellar effective temperatures for a star of $1.1M_0$, on the zero age main sequence and on the giant branch, are:

$\alpha = 1.0$	ZAMS	5300K	GB	4200K
$\alpha = 1.6$	ZAMS	5600K	GB	4700K.

These results suggest that a change in the mixing length by the amount predicted for $\eta_1 \gtrsim 1$ would lead to a change of a few hundred degrees in the stellar effective temperature. This would be significant on the pre-main-sequence track and when the star first reached the main sequence. It would clearly affect the theoretical interpretation of some of the observed properties of the stars discussed in section 2. As the surface activity of the star declined during its main sequence life, the effective mixing length would adjust to the true mixing length and the star would therefore move up the main sequence.

It should be stressed that both in this section and in section 2 we have ignored any influence of magnetic fields on the true value of the mixing length. It could be that magnetic fields will interfere with convective motions in such a way as to reduce the true value. If that is the case it might serve to reinforce the effect that we have just been discussing.

5. DIFFERENTIAL ROTATION AND MAGNETIC FIELDS IN THE SOLAR CORE (L. Mestel)

This section is mainly a summary of work done by L. M. with N. O. Weiss (submitted to Monthly Notices). The most recent observations of G stars in clusters of different ages (Van Leeuwen & Alphenaar 1982;Stauffer et al. 1984, 1985; Gray 1982; Lockwood et al. 1984) imply that a star like the Sun arrives on the main sequence with a surface angular velocity $\Omega \cong 60$ times the present value Ω_0. During the next 3×10^7 years Ω decreases quite rapidly; subsequently Ω decays more gradually, roughly like $t^{-1/2}$. Inferences from helioseismological data (Claverie et al. 1981; Duvall & Harvey 1984; Duvall et al. 1984, 1986; Brown 1985; Libbrecht 1986) imply a sharp inward rise in the internal solar rotation at a radius $r \cong .2R_0$ to a value of at least $2\Omega_0$, equal to the rotation rate of the Sun about 3×10^9 years ago. This imposes a constraint on any magnetic field within the inner radiative core. One of the earliest kinematic results in magnetohydrodynamics is Ferraro's (1937) law of isorotation: if motions in meridian planes are sufficiently slow (see below), a necessary condition for the existence of a steady magnetic field (symmetric about the rotation axis) is that the angular velocity should be constant along a line of force. Any rotational shear acting on a poloidal component $\underset{\sim}{B}_p$ generates a toroidal component $\underset{\sim}{B}_t$ which grows until the Lorentz force reacts back on the shear, generating torsional waves which travel along the field with the Alfvén speed $v_A = B_p/(4\pi\rho)^{1/2}$. This implies that a shear of scale D will be reversed in a time of the order of the Alfvén time $\tau_b = D/v_A$. It is well known that this time is very short compared with either a stellar evolution time or the inferred braking time through a magnetic wind. For example, with $B_p \sim 1$ gauss, $D \sim 0.2R_0$, $\rho \sim 30$ g/cm^3, $\tau_b \sim 10^4$ years: to increase τ_b to 3×10^9 years, B_p would need to be 3×10^{-6} gauss - no greater than the strength of the galactic background field.

For a system strictly symmetric about the rotation axis, one could imagine a quasi-steady state in which a rapidly rotating inner part of the Sun is magnetically uncoupled from the outer parts of the radiative core and the convective envelope, so that an initial rotation higher than that at the base of the convective envelope could persist. The point is that shearing of one axisymmetric field line relative to another makes no difference either to the density or to the magnetic field: the only consequence of isorotation rather than uniform rotation is a change in the (curl-free) electric field $\underset{\sim}{E}$ and hence of the polarization charge density (which is in any case a very small quantity). Departures from <u>isorotation</u> generate the magnetic tensions with moments about the rotation axis which drive the torsional Alfvén waves that redistribute angular momentum along a given flux tube. However, it is doubtful whether the limited freedom allowed by isorotation will survive when the field is even slightly non-axisymmetric, for significant departures from <u>uniform</u> rotation will then generate also toroidal magnetic <u>pressure</u> gradients which will interchange angular momentum between different flux tubes. Order of magnitude arguments yield again the estimate D/v_A for the characteristic time.

The Ω-$\underset{\sim}{B}_t$ coupling process just described is non-dissipative: τ_b is the time in which the shear is reversed rather than destroyed, and if no dissipation occurs the Ω-$\underset{\sim}{B}_t$ fields will oscillate indefinitely with typical periods of order τ_b. However, the outer part of the radiative core is probably coupled hydrodynamically to the highly dissipative convective envelope (Rosner & Weiss 1957); classical mhd then predicts that a poloidal field linking the inner and outer parts of the radiative core will ensure that any permanent non-uniform rotation in the core is at most of the order of the variations with latitude and longitude over the base of the convective zone, and that even a very weak field is sufficient for this quasi-steady state to be reached well within a stellar evolution time. The helioseismic data suggest, however, that near the solar center there are radial gradients in Ω that are two orders of magnitude greater than those observed at the solar surface. We are therefore driven to look for physical processes which might increase the characteristic time-scale for shear reversal over the estimate D/v_A. At time t, a shear of scale D will have generated a toroidal field

$$|B_\phi| \cong \tilde{\omega} B_p (\Omega/D)\tau \tag{6}$$

where $\tilde{\omega}$ is the axial distance. After the Alfvénic time $\tau_b = D/v_A$ (based on the poloidal field B_p),

$$\frac{|B_\phi|}{B_p} \cong \frac{\tilde{\omega}\Omega(4\pi\rho)^{1/2}}{B_p} = \left[\frac{(1/2)\rho\Omega^2\tilde{\omega}^2}{B_p^2/8\pi}\right]^{1/2} >> 1, \tag{7}$$

since the rotational kinetic energy is almost certainly much greater than the poloidal magnetic energy density. However it is known (Tayler 1973) that purely toroidal magnetic fields tend to be unstable to adiabatic perturbations, with growth times in the linear domain of the order of a local Alfvén time. A similar result holds for purely poloidal fields (Markey & Tayler 1973; Wright 1973). (Note that these instabilities involve the spontaneous conversion of magnetic energy into kinetic energy, and can occur in a stably stratified medium such as a sub-adiabatic radiative zone; the adiabatic stabilization is nullified by the displacements being perpendicular to gravity). It appears that a necessary condition for stability is that the fields must have poloidal and toroidal components of the same order of strength. This suggests that hydromagnetic instabilities will halt the linear growth (6) of $|B_\phi|$ at values much less than (7). In a purely toroidal field instabilities develop near the magnetic axis, where they are locally similar to the kink and sausage instabilities in magnetic pinches (Tayler 1973), with typical growth rates in the linear domain

$$\tau_\phi \cong \tilde{\omega}/w_A \equiv \tilde{\omega}(4\pi\rho)^{1/2}/|B_\phi| \tag{8}$$

i.e. determined by the local Alfvén speed based on the toroidal field. Thus long before $|B_\phi|$ reaches the value (7), $w_A >> v_A$ and $\tau_\phi << \tau_b$. There is no theory extant of the non-linear development of these instabilities, but it can be plausibly argued that the reduction of length-scales by a dynamically-driven instability leads to accelerated Ohmic destruction of toroidal magnetic flux. The dynamical time-scale τ_ϕ yields an approximate upper bound for the rate of toroidal flux destruction. We are interested in estimating an upper limit to the time for which a given shear can persist in spite of the action of the magnetic torques: we are therefore plausibly minimizing the reaction of the field if we assume that the increase in $|B_\phi|$ according to (6) is halted when $t = \tau_\phi$, implying that the time-scale for flux destruction has become equal to the time in which the shear would double $|B_\phi|$. This is equivalent to postulating a macroresistivity such that the effective magnetic Reynolds number is of order unity. From (6) and (8), $|B_\phi|$ is limited to the value $\cong B_p \{(\tilde{\omega}/D)(\tilde{\omega}\Omega/v_A)\}^{1/2}$, so that the associated torque density $\tilde{\omega}|B_\phi|B_p/4\pi D$ will reverse the shear in a new time

$$\tau_B \cong \left(\frac{D}{v_A}\right)\left(\frac{\Omega D}{v_A}\right)^{1/2} \cong 10^7\left(\frac{\Omega}{\Omega_o}\right)^{1/2}\left(\frac{B_p}{B_o}\right)^{-3/2} \text{ years,} \quad (9)$$

where typical values $D \cong 1.4\text{x}10^{10}$cm, $\rho = 30\text{g/cm}^3$, $\Omega_o = 3\text{x}10^{-6}$, $B_o = 1$ gauss have been adopted. From (8), the corresponding value of $\tau_\phi \cong 10\{(B_p/B_o)(\Omega/\Omega_o)\}^{-1/2}$ years, with $|B_\phi|/B_p \cong 1900$ $(\Omega/\Omega_o)^{1/2}/(B_p/B_o)^{1/2}$. Thus if we again require that the braking time exceeds $3\text{x}10^9$ years, (9) yields $B_p < 2\text{x}10^{-2}$ gauss, with $|B_\phi|/B_p > 1.4\text{x}10^4$ and $\tau_\phi \geq 70$ years.

This increase by a factor $\cong 10^4$ of the allowed upper bound may however be over-optimistic. Pitts & Tayler (1985) have shown that in a rotating star with $\Omega\tau_\phi >> 1$ (satisfied by $|B_\phi| << 10^6$ gauss), the growth time of the instability of a toroidal field is increased, typically by the large factor v_s/w_A where v_s is the sound speed. Repeating the previous calculation then yields

$$\tau_B \cong (D/v_A)\left[(v_A/v_s)\,(\tilde{w}/D)\,(\Omega D/v_A)^2\right]^{1/3} \quad (10)$$

$$\cong 10^5\,(\Omega/\Omega_o)^{2/3}\,(B_p/B_o)^{-4/3} \text{ years,} \quad (11)$$

with similar typical solar numbers inserted. A braking time $\tau_B > 3\text{x}10^9$ now requires $B_p < 4\text{x}10^{-4}$ gauss. This value is probably a more reliable upper bound to the poloidal field.

An equatorially symmetric shear acting on a dipolar poloidal field generates an equatorially anti-symmetric $\underset{\sim}{B}_t$, and the associated magnetic pressure gradient provides a pinching force towards the equator. Reconnection across the equator would likewise limit the growth of $\underset{\sim}{B}_t$ and the associated torque. However, even if there were a dynamical instability driving toroidal flux towards the equator, the rate of reconnection could not exceed the Petschek limit with its corresponding time-scale τ_ϕ (Priest 1985; Priest & Forbes 1986). More realistically, the reconnection is more likely to be caused by resistive instabilities of which the more important is the tearing mode (Furth et al 1963; Priest 1985). This occurs in a boundary layer of thickness εL, where L is the scale of variation of the field; the fastest growing modes have $\varepsilon \cong (\tau_A/\tau_d)^{1/4}$, a wavelength parallel to the equator $\cong 2\pi L(\tau_d/\tau_A)^{1/4}$, and a growth time $\tau_r = (\tau_d \tau_A)^{1/2}$, where the Alfvénic time $\tau_A = L/w_A$ and the diffusion time $\tau_d \equiv D^2/\eta$. The instability may very well develop with the formation of magnetic islands in the boundary layer and reconnection at the neutral points, but the associated rate of flux destruction (per unit length normal to B_ϕ) as estimated from the linear theory is $|B_\phi| L\varepsilon^2/4\tau_r = |B_d| L/4\tau_d$ - no greater than the very slow classical Ohmic diffusion rate. Probably the as yet undeveloped non-linear resistive instability theory will yield a higher rate. Other instabilities that rely on double-diffusive effects or on magnetic buoyancy (cf. Hughes 1985 a,b) might also help to limit $|B_\phi|$, despite the stabilizing effect of rotation. Nevertheless, we think it very unlikely that the time scale for $\underset{\sim}{B}_t$ reconnection at the equator could fall below the Petschek limit, so that again (9) sets an upper limit on the braking time, with a corresponding upper bound of 10^{-2} gauss for B_p.

To estimate a lower bound on the field in the radiative core we assume that any relic of the galactic field has been destroyed during the Hayashi phase, and that only an alternating dynamo (of period τ) has operated in the gradually retreating convective envelope. During each magnetic cycle, flux generated at the base of the convective zone diffuses inwards only through a skin depth, given by equating the diffusion speed of retreat of the convective zone boundary. Since there is imperfect symmetry in successive half-cycles -- measured by a difference δB- there will be a net remnant flux left after each cycle. A random walk argument then yields a net field on the main sequence $\bar{B} \cong (\tau/T)^{1/2} \delta B \cong 10^{-3} \delta B$. Because of the more rapid rotation, dynamo action in a T Tauri star should yield strong alternating fields: $B \cong 500$ gauss has been observed in one case (Johnstone & Penston 1986). If we take $\delta B/B < 1/10$, then $\bar{B} < .05$ gauss. Thus by combining the most pessimistic assumptions about the residual field in the radiative core with the most optimistic assumptions regarding dynamical instabilities of the toroidal field, one can suppose that magnetic torques are unable to enforce uniform rotation on the core. If (as we think more likely) the field in the Sun's interior is stronger, then we must conclude that either (1) the Ω-field in the core is time-dependent (Gough 1986); (11) the inner part of the core is magnetically decoupled from the outer part (Rosner & Weiss 1985); or (111) the inferred rapid rotation of the centre will not survive future observations.

6. A NOTE ON THE EFFECT OF MERIDIAN CIRCULATION (L. Mestel and D. L. Moss)

The discussion has ignored the thermally-driven Eddington-Sweet circulation in the solar core. This will be very slow even with an inner zone with $\Omega = 2\Omega_o$, though it could be larger in a region where the Ω-gradient is locally large. (The gradients in mean molecular weight arising during normal stellar evolution easily stifle the circulation, so that the flow must be limited to regions where virtually no hydrogen burning has occurred (Mestel 1953; Mestel & Moss 1986)). Intuitively one expects that in a steady axisymmetric state the diviation from isorotation will be small provided the circulation speed $v_p << v_A$ -- i.e. if the speed with which the gas advects angular momentum is much less than that with which torsional Alfvén waves act to restore isorotation. This is confirmed from the well-known axisymmetric integrals for a perfectly conducting, inviscid gas (Chandrasekhar 1956; Mestel 1961), which combine into

$$\underset{\sim}{v}_p = \kappa \underset{\sim}{B}_p, \qquad \rho\kappa = \eta, \tag{12}$$

$$\Omega = \frac{(\alpha + \eta\beta/\rho\tilde{\omega}^2)}{(1 - 4\pi\eta^2/\rho)}, \qquad \tilde{\omega}B_\phi = \frac{(\beta + 4\pi\eta\alpha\tilde{\omega}^2)}{(1 - 4\pi\eta^2/\rho)}, \tag{13}$$

where η, α and β are constants along poloidal field-streamlines. If $4\pi\eta^2/\rho \equiv v_p^2/v_A^2 << 1$, it follows that for Ω/α to differ much from unity, $|B_\phi/B_p|$ would need to be $\cong \alpha\tilde{\omega}/v_p$ which is at least 10^9. (As already noted, such a predominantly toroidal field is almost certainly ruled out on stability grounds). To get significant deviation from isorotation one needs v_p^2/v_A^2 to be at least comparable with unity.

The quantities α and β can be expressed in terms of the toroidal flux dF_t threading an infinitesimal poloidal flux tube, and the angular momentum dJ within the same tube. It is of some interest to study the evolution of the Ω, B_ϕ fields from an initial state with $\Omega = \alpha_o$ (isorotation) and $\tilde{\omega}B_\phi = \beta_o$ (zero torque - Lüst & Schlüter 1954), under a prescribed circulation field, taken to be of E-S quadrupolar form. (A fully self-consistent treatment would have to take account of the effect of variations in Ω and B_ϕ on the circulation field). To reach a steady state of the form (13), dissipative processes are required to damp out the torsional oscillations that will be set up. If one arbitrarily postulates that dF_t and dJ are conserved during the approach to a steady state, this will fix α and β in terms of α_o and β_o. More realistically, one can solve numerically the time-dependent equations for Ω and B_ϕ with viscous and resistive terms included; these terms will both damp out the oscillations and also yield some diffusion of dF_t and dJ between different poloidal flux tubes. Moss (1986) has calculated some illustrative examples for a medium taken for simplicity as incompressible, with the initial state uniformly rotating and with $\beta_o = 0$. The circulation speed v_p is taken typically as $10^{-2} v_A$ and $10^{-3}(\Omega\tilde{\omega})$; in most cases the poloidal magnetic field is assumed quadrupolar with field lines coinciding with the streamlines. When the magnetic (R_m) and hydrodynamic (R) Reynolds numbers are about unity, an approximately steady state is reached in about 12 Alfvén wave travel times τ_A, but the surfaces of constant α and β (defined by equations (13)) are not particularly close to the field-streamlines - not surprisingly, as $R_m \cong 1$ implies that field freezing (assumed in deriving (12, 13) is a poor approximation. When R_m and R are taken to be 10, the damping time is longer than $40\tau_A$, but the contours of constant α and β are not much closer to the field-streamlines. Similar results are found when $\underset{\sim}{B}_p$ is dipolar, so that finite resistivity is needed for the quadrupolar circulation to cross the $\underset{\sim}{B}_p$-lines near the equator.

In a realistic astrophysical problem, if the circulation is to be relevant over a stellar lifetime, then the Reynolds numbers based on microscopic dissipative processes will be still larger, with the asymptotic states correspondingly closer to the forms (13), but the time for the computations to reach a steady state becomes prohibitively long. The calculations confirm that when the circulation is strongly sub-Alfvénic the allowed departures from isorotation are small, but enough to generate from $\underset{\sim}{B}_p$ the small toroidal component $\underset{\sim}{B}_t$ needed for the magnetic torque to be of the order of the angular momentum advection term. (A dynamically stable magnetic field must have toroidal flux linking the $\underset{\sim}{B}_p$-lines, of the same order as the poloidal flux. In the absence of circulation this field will be torque-free, with $\underset{\sim}{B}_t$ maintained by currents parallel to $\underset{\sim}{B}_p$; the modification to $\underset{\sim}{B}_t$ generated by the circulation will then be a small perturbation).

These computations are counter-examples to the statement by Tassoul & Tassoul in a preprint (1985) that "no large-scale magnetic field can off-set the advection of angular momentum by the meridional flow and so enforce almost uniform rotation". This claim depends on an expansion procedure which manifestly breaks down after a time ≅ the Alfvén travel time. Their method leads to a toroidal field which is a solution of the Cowling decay equation; this is decoupled from the basic poloidal field, and so clearly is irrelevant to the problem.

REFERENCES

Brown, T. M., 1985. Nature, 317, 591.
Chandrasekhar, S., 1956. Astrophys. J., 148, 905.
Claverie, A., Isaak, G. R., McLeod, C. P., van der Raay, H. G. & Roca Cortes, T., 1981.
Dicke, R. H., 1979. Astrophys. J., 228, 898.
Duvall, T. L., & Harvey, J. W. 1984. Nature, 310, 19.
Duvall, T. L., Dziembowski, W. A., Goode, P. R., Gough, D. O., Harvey, J. W. & Leibacher, J. W. 1984. Nature, 310, 22.
Duvall, T. L., Harvey, J. W. & Pomerantz, M. A., 1986. Nature, 321, 500.
Ferraro, V. C. A., 1937. Mon. Not. R. Astr. Soc., 97, 458.
Furth, H. P., Killeen, J. & Rosenbluth, M. N., 1963. Phys. Fluids, 6, 459.
Gilliland, R. L., 1986. Astrophys. J., 300, 339.
Gough, D. O., 1986. Highlights of Astronomy, ed. J.-P. Swings, 7, 283.
Gough, D. O. & Tayler, R. J., 1966, Mon. Not. R. Astr. Soc., 133, 85.
Gray, D. F., 1982. Astrophys. J., 261, 259.
Hughes, D. W., 1985a. Geophys. Astrophys. Fluid Dyn., 32, 273.
Hughes, D. W., 1985b. Geophys. Astrophys. Fluid Dyn., 34, 99.
Johnstone, R. M. & Penston, M. V., 1986. Mon. Not. R. Astr. Soc., 219, 927.
Libbrecht, K. G., 1986. Nature, 319, 753.
Lockwood, G. W. Thompson, D. T., Radick, R. R., Osborn, W. H., Baggett, W. E., Duncan, D. K. & Hartmann, L. W., 1984. Publ. Astron. Soc. Pac., 96, 714.
Lüst, R. & Schlüter, A., 1954. Z. Astrophys., 34, 263.
Markey, P. M. & Tayler, R. J., 1973. Mon. Not. R. Astr. Soc., 163, 77.
Mestel, L., 1953. Mon. Not. R. Astr. Soc., 113, 716.
Mestel, L., 1961. Mon. Not. R. Astr. Soc., 122, 473.
Mestel, L., & Moss, D. L., 1986. Mon. Not. R. Astr. Soc., 221, 25.
Moss, D. L., 1986 (unpublished).
Pitts, E. & Tayler, R. J., 1985. Mon. Not. R. Astr. Soc., 216, 139.
Priest, E. R., 1985. Rep. Prog. Phys., 48, 955.
Priest, E. R. & Forbes, T. G., 1986. J. Geophys. Res., 91, 5579.
Rosner, R. & Weiss, N. O., 1985. Nature, 317, 790.
Stauffer, J. R., Hartmann, L. W., Burnham, J. N. & Hones, B. F., 1985. Astrophys. J., 289, 247.

Stauffer, J. R., Hartmann, L. W., Soderblom, D. R. & Burnham, J. B., 1984. Astrophys. J., 280, 202.

Tassoul, J.-L. & Tassoul, M., 1985 (preprint).

Tayler, R. J., 1973. Mon. Not. R. Astr. Soc., 161, 365.

Tayler, R. J., 1980. Mon. Not. R. Astr. Soc., 191, 151.

Tayler, R. J., 1986. Mon. Not. R. Astr. Soc., 220, 793.

VandenBerg, D. A., 1983. Astrophys. J. Suppl., 51, 29.

Van Leeuwen, F. & Alphenaar, P., 1982, ESO Messenger, 28, 15.

Wright, G. A. E., 1973. Mon. Not. R. Astr. Soc., 162, 339.

SOLAR ROTATION AND AGE

E.L. SCHATZMAN
Observatoire de Nice
B.P. 139
06003 NICE Cedex
France

ABSTRACT. The constraint on transport processes derived from chemical abundances lead to low values of the diffusion coefficient ($D \simeq 300$) whereas the constraint derived from the transport of vorticity lead to high values ($D \simeq 10^5$). A brief discussion of the meaning of this discrepancy is given.

1. INTRODUCTION.

It is necessary, in the present situation of the problem of solar rotation, to use our knowledge in the best possible way. Especially, due to the failure of the solar models with turbulent diffucion mixing (Schatzman, Maeder, 1981a, b; Berthomieu, Schatzman, 1985), it seems possible to inverse the problem and to derive from the data on transported quantities (chemical elements, vorticity) which constraints have to be applied to the transport processes.

Reviewing the problem, it seems necessary to start with the first question concerning rotation. Assuming that molecular clouds have the angular velocity produced by the local differential rotation, we can calculate the first constraint on rotation. With the local value of the Oort's constant,

$$\omega_{GAL} \simeq 2 \cdot 10^{-16}\ s^{-1}$$

we find that an interstellar cloud is stable if its density is large enough,

$$\rho_{CLOUD} \gtrsim 0.1 \text{ baryon cm}^{-3}$$

During contraction, if there is conservation of angular momentum, the critical density increases like $1/R$, R being the radius of a spherical cloud, whereas the actual density increases like $1/R^3$: the cloud density always overtakes the critical density and the cloud becomes rotationally unstable. This is the classical problem of the spin down of

B. R. Durney and S. Sofia (eds.), The Internal Solar Angular Velocity, 159–171.

interstellar cloud and, as everyone knows, it is not really solved.

In fact, loss of angular momentum takes place all the way, down to the main sequence, as has been shown recently for example by Bouvier et al. (1986). (Fig. 1). In this respect, there is a number of classical

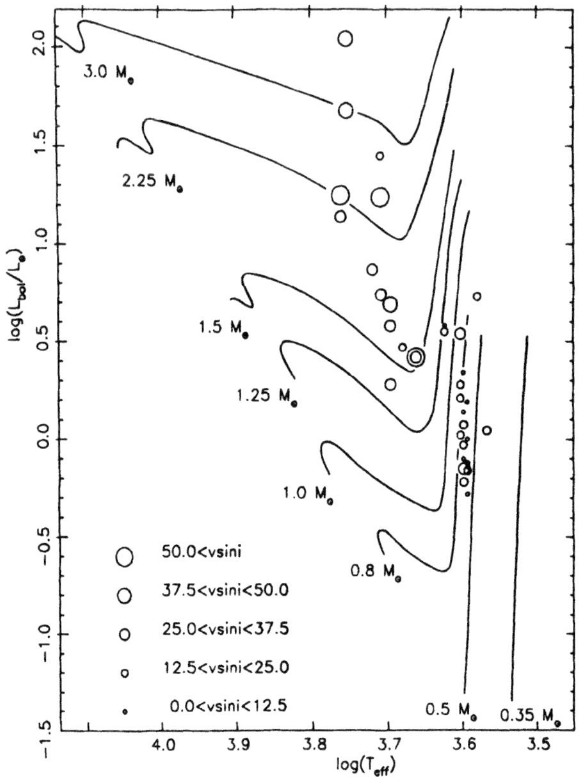

Figure 1. v sin i distribution of T Tauri stars in the HR diagram. The ZAMS is located at the end of the convection radiative tracks (Bouvier et al., 1986).

problems, especially (a) the distribution of v sin i on the main sequence which appeals to a strong relation between the dynamo and the loss of angular momentum (Schatzman, 1959, 1962), and (b) the evolution of v sin i with time. In connection with this last question, it is important to notice the work of Simon et al. (1985), where the evolution of the period of rotation P_{ROT} as a function of age has been derived from chromosphere activity as a function of $(P_{ROT}/\tau_c) = R_0$ (Rossby number) where τ_c is the characteristic time of overturn in the convective zone.

The result of Simon et al. (1985) (Fig. 2) gives the measure of the braking torque as a function of time. This torque increases with time, and the dependance on t can be written $(a+bt)^{-\frac{1}{2}}$, more or less in agreement with the relation found by Skumanich (1972). The point representing the present Sun falls on the extrapolated curve of Simon et al. (Fig. 2), if one assumes that the Sun rotates like a solid body, the core perhaps excepted (Duvall and Harvey, 1984; Brown, 1985; Libbrecht, 1986; Duvall et al., 1986).

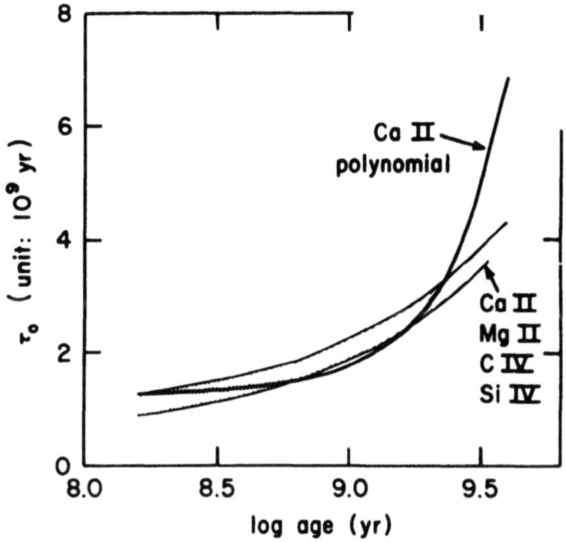

Figure 2. The braking parameter $\tau_o = -(\Omega/\dot{\Omega})$ as a function of age, as derived from individual ions and from Noyes et al. (1984) polynomial. It is measure of the torque which acts to decelerate the surface layers of a rotating star. For a law $\Omega = \Omega_o|1 + (2/\tau)t|^{-\frac{1}{2}}$, this would give $-(\Omega/\dot{\Omega}) = \tau_o = \tau + 2t$, which fits more or less the data (Simon et al., 1985).

In the following, we shall consider the constraints resulting from the abundance of light elements (Lithium, ^{3}He), from the solar neutrino problem, and from the transport of vorticity. As a conclusion, we shall discuss the problems concerning the solar core, and be inevitably brought back to the neutrino problem.

2. THE LITHIUM CONSTRAINT.

It is absolutely necessary when considering the Lithium abundance problem to avoid any ad hoc assumption and to try to rest on the best possible physics.

I would like first to stress that it does not seem possible to me to accept the idea of Lithium burning at the bottom of the convection zone. This bas been considered for the first time by Bodenheimer (1965), but if looking at an homogeneous sequence like the Hyades

(Cayrel et al., 1984), it is quite clear that this would mean a mixing length mass dependant or with $\alpha = (\ell/H_p)$, a function $\alpha(M)$. The same difficulty appears if one assumes that mixing is due to overshooting. I would then be necessary to assume a depth of overshooting which should be mass dependant. This is again an assumption which is not physically acceptable.

Slow turbulent mixing, as suggested by Schatzman (1969, 1977) seems to have the general properties which are necessary to explain the behaviour of Lithium abundance along the amin sequence, essentially for F - G type stars. But it is necessary to reach a complete consistency of the model. To that effect, let us consider the physical model proposed by Zahn (1983). The main idea of Zahn is to start from a 2-D turbulence generated on equipotentials by differential rotation. At a lower scale, 2-D turbulence upgrades into a 3-D turbulence, as soon as the inertia overtakes the Coriolis force, or as soon as the Rossby number is larger than one. The experiments with fluids on rotating tables (Hopfinger et al., 1982) show actually this transition from 2-D to 3-D turbulence, a fact from which stems the assumption of Zahn. Zahn then defines an efficiency factor

$$E = \frac{\Omega^2 r}{g} (\nabla_{ad} - \nabla_{rad})^{-1}$$

which is usually so small (in the Sun, away from the convection zone, $E \simeq 2 \cdot 10^{-6}$) so that the region where $\Delta\nabla$ goes to zero and E becomes larger than one, which is physically meaningless, is very small, $\delta r \simeq 0{,}2$ kilometers, and can be ignored.

The corresponding value of the turbulent diffusion coefficient is then

$$D_{turb} = K \frac{\Omega^2 r}{g} (\nabla_{ad} - \nabla_{rad})^{-1}$$

where K is the thermal diffusivity. According to Baglin et al. (1985), a reasonable fit to the N(Li) - M relation in the Hyades can be obtained with

$$D_{turb} = \alpha(\nabla_{ad} - \nabla_{rad})^{-1}$$

and

$$\alpha \simeq 700$$

which corresponds, for a solar model, in the region of Lithium burning ($T = 3 \cdot 10^6$ °K) to a value $D_{turb} \simeq 3500$. This corresponds to a Reynolds number

$$R_e^* = (D_{TDM}/\nu) \simeq 273$$

This has to be compared now to the value of α which fits with the solar abundance of Lithium.

The average value of α in a time-independant model of turbulent diffusion mixing is $\alpha \simeq 350$. Due to the Ω^2 dependance of α , it is necessary to compare average values. With a law

$$\Omega = \Omega_o \left| 1 + \frac{2t}{\tau_o} \right|^{-\frac{1}{2}}$$

we obtain for $\tau_o = 4 \cdot 10^9$y,

$$\frac{\langle D_{TDM} \rangle_\odot}{\langle D_{TDM} \rangle_{Hyades}} = 0.501$$

which is in full agreement with the average solar value. This would have not been true if comparing the present periods of rotation. This would have given, with $(\Omega_\odot/\Omega_{Hyades})^2 \approx 1/50$, $(\alpha_\odot/\alpha_{Hyades}) \approx 1/50$ instead of 1/2.

For earlier spectral types, the dip in the abundance of ^{7}Li found by Boesgaard and Tripico (1986) can be explained by gravitational settling and radiation pressure drag, as shown by Michaud (1986). The results of Michaud confirm that for the lower masses, depletion has to be taken over by another diffusive process, actually the turbulent diffusion mixing.

The ^{3}He constraint.

The abundance of ^{3}He at the surface of the Sun is given by

$$(X_3/X_4) = 4 \cdot 10^{-4}$$

Is this the combination of primordial ^{3}He abundance and Deutrium burning? In fact, Geiss (1971,1973) has shown that the abundance ratio (X_3/X_4) has increased by about 20% during the last 10^9 years. This can be explained by turbulent diffusion mixing, ^{3}He being carried by diffusion along the gradient of ^{3}He in the outer half of the Sun. From the results of Schatzman and Maeder (1981a, b) (Figure 3), it is possible to derive an estimate of the average turbulent diffusion coefficient in the outer half of the Sun (Table 1).

TABLE 1.-

R_e^*	$(^3He/^4He)_{today}$
30	$2 \cdot 10^{-4}$
100	10^{-3}
200	10^{-2}

The surface abundance of ^{3}He as a function of the pseudo Reynolds number R_e^*.

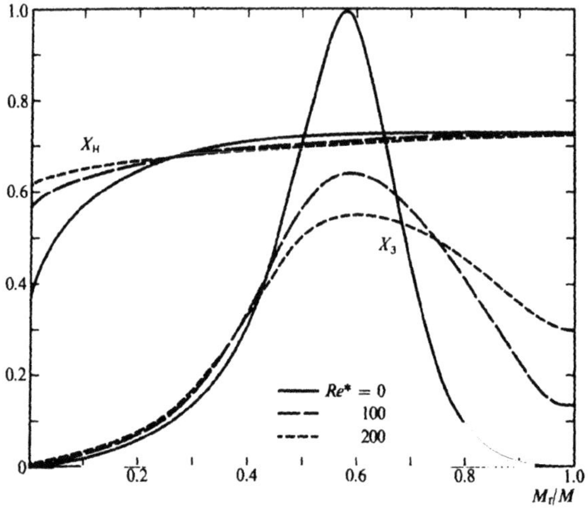

Figure 3. The effect of turbulent diffusion on the hydrogen concentration X_H, and on the ^3He concentration X_3. The curves are given for $R_e^* = 0$ (continuous line), $R_e^* = 100$ (dashed line) and $R_e^* = 200$ (points dashed line). (Schatzman and Maeder, 1981a).

It is quite clear that the present abundance of ^3He at the surface of the Sun does not allow a large turbulent diffusion coefficient, $D_{TDM} = R_e^* \nu$, with $R_e^* \lesssim 35$. The corresponding diffusion coefficient is $D_{TDM} \lesssim 300$.

It should be mentioned that this corresponds to a decrease of D_{TDM} at the level of Lithium burning, $D_{TDM} \simeq 750$ to a lower value, $D_{TDM} \simeq 300$ at the level of ^3He diffusion. This is in agreement with the picture given by Zahn (1983); if the driving force of the turbulence is finally the meridional circulation, the turbulent diffusion coefficient should roughly vary like ($r\ v_{mer}$), where v_{mer} is the velcoity of the meridional circulation (the Eddington Sweet currents). The product ($r\ v_{mer}$) decreases sharply when going inside the Sun, clearly in agreement with the values of D_{TDM} derived from the surface abundance of Li and ^3He.

The neutrino constraints.

Table 2 summarizes the values of the observed solar neutrino flux and in various standard and non-standard models.

Table 2.-

Source	ϕ_ν (SNU's)
Davis experiment	2
Standard model (a)	13
Standard model (b)	7.4
Mixed model (c)	3.5
Mixed model (d)	2

The solar neutrino flux for different models.

(a) Lebreton, Y. and Maeder, A.: 1986, Astron. Astrophys. (to be published); (b) Cahen, S. et al.: 1986b, Moriond, Massive Neutrinos, Tran Than Van ed., Editions Frontières, p. 83 ; (c) Lebreton, Y. and Maeder, A.: 1986b, Mixed model with mixing down to the center with $R_e^* = 100$; (d) Extrapolation from Lebreton and Maeder (1986c) to a fit of both the solar neutrino flux and present solar luminosity (see text).

The standard model of Lebreton and Maeder (1986a) assumes no mixing and is based on the last opacity tables of Los Alamos and takes into account the low temperature opacities (kT < 1 ev.) in computing the fitting of the atmosphere to the convective zone. The standard model of Cahen et al. (1986) differs of the preceding model only in the use of a different cross section for the $^7Be\ (p, \gamma)\ ^8B$ reaction.

The mixed model of Lebreton and Maeder (1986b) assumes mixing with $R_e^* = 100$. Mixing lowers the central temperature by bringing some extra-fuel, but this has to be compensated by an increase of temperature, related to an increase in Y (^4He concentration) in order to keep the present luminosity. The result is a smaller neutrino flux, but still too large: $\phi_\nu = 3.5$ SNU's.

From the models of Lebreton and Maeder (1986c) it is possible to extrapolate the values of Y and M_{conv} in order to fit both the present day luminosity of the Sun and the observed solar neutrino flux. This gives Y = 0.4 and $M_{conv} = 0.3$, and both values are astrophysically and physically unacceptable! Y = 0.30 seems to be an upper limit to the primordial Helium concentration, well in agreement with primordial nucleosynthesis and ^4He production in the Galaxy before the formation of the Sun. $M_{conv} = 0.3$ implies an overshooting which violates the most elementary dynamical law of conservation of momentum.

The conclusion is that it seems extremely difficult with non-standard models to fit all the solar constraints (other aspects of the problem will be given in the conclusion).

Further difficulties come from solar sismology. The difference $\delta\nu_{n\ell} = \nu_{n,\ell} - \nu_{n-1,\ \ell+2}$ implies the integral (Tassoul, 1980)

$$\int \frac{1}{r} \frac{dc}{dr} \, dr$$

where c is the sound velocity. In the standard model, c increases outwards from the center of the Sun, due to the decrease in the mean molecular weight outwards. In the mixed model (diffusion or convection) c decreases outwards from the center, due to the decrease in temperature outwards. There is then a change of the sign of the contribution of the core to the integral, shifting the $\delta\nu_{n\ell}$ from about 10 μs to 15 μs. This seems to be a very strong constraint on the mixing process in the center of the Sun: mixing in the center of the Sun leads to a model which is not in agreement with the splitting of the pressure modes.

Transport of vorticity.

With the transport of vorticity, we are meeting another kind of constraint. For my purpose, I shall assume solid body rotation of the Sun, which is kind of simplification of the results of Duvall and Harvey (1984), Brown (1985), Libbrecht (1986), Duvall et al. (1986). The aim of this simplification is that the Sun has been losing angular momentum since its formation and only some kind of internal friction can carry the vorticity outwards, keeping the Sun rotating like a solid body.

I shall not consider, for the time being, the nature of the friction, but I shall show that the viscosity associated with chemical turbulent mixing is too small by several orders of magnitude for doing the work.

Let us first consider orders of magnitude. We first recall here that the quantity which is transported inside the Sun is the vorticity and not the angular momentum. This implies a dissipative process; experiments on turn tables show that in non dissipative process angular momentum redistribution take place, whereas dissipative process leads to a distribution of vorticity.

Solid body rotation will be achieved if

$$R^2 \langle D \rangle \simeq t_\odot$$

This gives

$$\langle D \rangle \simeq 35000$$

$$\simeq 100 \; D_{TDM} \text{ (derived from } {}^3\text{He abundances)}$$

More precisely, if we have $D = D_o \, (\Omega/\Omega_o)^2$ and assume a law

$$\frac{\Omega^2}{\Omega_o{}^2} = (1 + (t/t_o))^{-1}$$

we are led in the diffusion equation to a change of variables,

$$d\tau = (1 + (t/t_o))^{-1} \; dt$$

or

$$\tau = t_o \log(1 + (t/t_o))$$

which defines a diffusion time expressed in the new available ,

$$\tau_D = R^2/D_o$$

Coming back to the real time t, we have

$$t_{diff} = t_o|-1 + \exp\left(\frac{R^2}{D_o t_o}\right)|$$

from which we derive

$$D_o = (R^2/t_o)\,(\log(1 + (t_{diff}/t_o)))^{-1}$$

and

$$\langle D\rangle = (R^2/t_{\odot})(\log(1 + (t_{\odot}/t_o)))\,(\log(1 + (t_{diff}/t_o)))^{-1}$$

As an example, if we take $t_{diff} \simeq \frac{1}{2}\, t_{\odot}$, $\Omega_{today} = (1/6)\,\Omega_o$, which gives $t_o = (1/35)t_{\odot}$ we obtain

$$D_o = 414000$$

$$\langle D\rangle = 39000$$

We can obtain a similar estimate by considering the diffusion equation for the vorticity,

$$\frac{\partial\Omega}{\partial t} = \frac{1}{r^2}\frac{\partial}{\partial r}\, r^2\, D\, \frac{\partial\Omega}{\partial r}$$

With

$$D = D_o(\Omega/\Omega_o)^2$$

we can separate the variables by assuming

$$\Omega = \Omega_O(t/t_o)^{-\frac{1}{2}}$$

With the change of variables

$$R = \alpha\xi$$

$$\Omega = \Omega_O\left|\frac{3}{2}\left(\frac{\alpha^2}{D_o t}\right)^{\frac{1}{2}}\right|\left(\frac{G}{\xi}\right)^{1/3}$$

we have to solve the differential equation

$$G'' + \xi^{2/3} G^{1/3} = 0$$

For ξ small, we can write

$$G = \xi - \frac{\xi^3}{3} + \frac{\xi^4}{144} + \ldots$$

The boundary conditions express that the loss of angular momentum is proportional to the rate of mass loss,

$$\frac{d}{dt} I\xi = - k^2 R^2 \dot{M}$$

kR being the distance of the Alfvenic point (Cowling, 1965).
Te separation of variables obliges us to assume that k^2M varies like $t(1 + (t/t_o))^{-\frac{1}{2}}$. We can only fit the boundary conditions to the present day value of $k^2\dot{M}$. With $k^2\dot{M} = 10^{-10}$ yrs^{-1}, and $R = R_{HCZ} = 6 \cdot 10^{10}$ cm we obtain the value of at the boundary,

$$\xi = 0.9$$

and

$$D(\text{to day}) \simeq 43000$$

$$\Omega(\xi) = 0.953\ \Omega(\Theta)$$

(see Fig. 4), the angular velocity of the center being 5% larger than the surface angular velocity. The solution provides also

$$D_o = 720000$$

and

$$\langle D \rangle = 133000$$

We can conclude from this section that the average diffusion coefficeint which is needed to carry the vorticity, in order to explain the present quasi-solid rotation of the Sun is of the order of 100 to 400 times the diffusion coefficient which is acceptable, according to the surface abundance of ^3He and to the constraint coming from solar sismology.

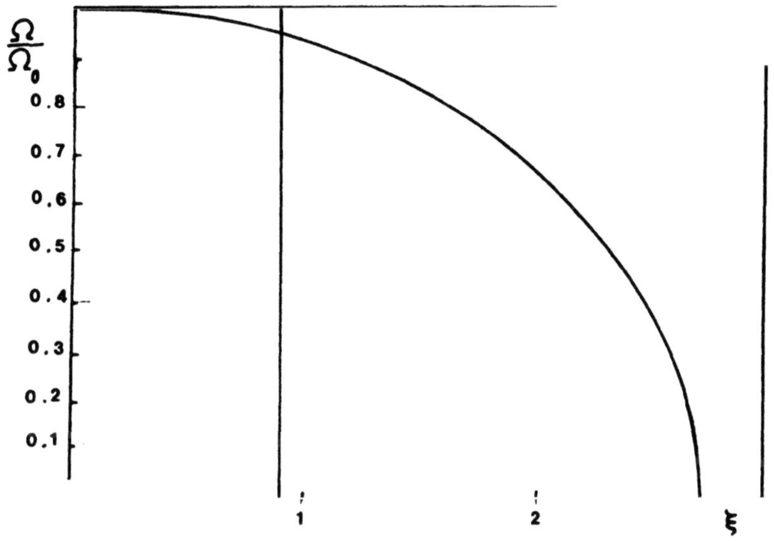

Figure 4. Solution of the diffusion equation with separation of the variables. The reduced variable ξ is defined by the boundary conditions. The vertical line corresponds to $\xi = 0.9$ and a present day torque $\tau_o = 10^{11}$ years (see text).

CONCLUSION.

Transport of vorticity implies a diffusion coefficient much larger than it is allowed by transport of chemical elements. The only other physical effect which we can consider is the friction due to the magnetic field.

We should notice first that the energy density which is present in the 2-D turbulence mentioned by Zahn is much larger than the energy density of the 3-D turbulence which carries the chemical elements. A dynamo mechanism related to this 2-D turbulence could easily provide the magnetic field which is necessary for the transport of vorticity. This implies a factor of the order of 100 and we are still much below the limit of the efficiency factor E = 1.

We can consider the model of Spruit (1986) as very attractive. The winding of the lines of force lead to an increase of the magnetic field and a magnetic buoyancy instability, which could provide the amount of chemical mixing which is needed by abundance data on ^{3}He. However, it would also be necessary to explain in the same way the surface abundance of ^{7}Li, that is to say to recover a law in $\Delta\nabla^{-1}$ for the diffusion coefficient.

The problem of the core raises a number of other questions. Without any magnetic field, mixing in the core has to be rejected, as a consequence of the interpretation of the pressure modes; this result is naturally in complete contradiction with the former interpretation of Schatzman and Maeder (1981a, b).

On the other hand, as shown by Ulrich and Rhodes (1983), the presence of a magnetic fiel in the core shifts the $\delta\nu_{n\ell}$ in the right direction. In principle, if the magnetic field is sufficiently large ($\sim 10^8$ gauss), mixing (and low neutrino flux) and pressure modes splitting are compatible. On the other hand, such a large magnetic field in the core does not seem to be compatible with any estimate derived from thermodynamical reasoning. Considering the energy production in the core, and a thermodynamic efficiency ($\Delta T/T$) where ΔT is the temperature difference between the center and the boundary of the core of radius r, I obtain an estimate

$$B^2 \simeq 8\Pi\rho\ (2.78\ \varepsilon\ T_\odot)^{2/3}\ (r/R_\odot)^2$$

For $\varepsilon = 20$ ergs g^{-1} sec^{-1}, $(r/R_\odot) \simeq 0.1$, this gives

$$B \simeq 10^5 \text{ gauss}$$

On the other hand, the mechanism considered by Spruit (1986) may mead to much larger fields, the maximum value of the field being given by the balance between the growth of the field and the magnetic buoyancy instability.

It is known now that in alternate to the mixed model is the WIMP model of Faulkner et al. (1986) which, by a supplementary heat transport by the weakly interacting passive particles, provides a lower central temperature to the Sun and agrees with the sismology data.

The answer to this question will be given probably by the new Gallex experiment (see Kirsten, 1986). The alternate to the guilt of the solar model being naturally the guilt of the neutrino. The consideration of the transport of massive neutrinos in the Sun (Mikheyev and Smirnov, 1986; Wolfstein, 1978) takes into account the different cross section

$$\sigma(\nu_e,\ e) \neq \sigma(\nu_\mu,\ e)$$

and provides a different interaction of the ν_e and ν_μ with the solar matter. With a small mass difference, $\Delta M^2 \simeq 10^{-6}$ $e.v^2$ and a proper mixing angle, it is easy to explain the low neutrino flux observed by Davis.

REFERENCES.

Baglin, A., Morel, P., Schatzman, E.: 1985, Astron. Astrophys. **149**, 309.
Berthomieu, G., Schatzman, E.: 1985, 2nd Gallex Meeting, Heidelberg.
Bodenheimer, P.: 1965, Astrophys. J., **142**, 451.
Boesgaard, A.M., Tripico, M.J.: 1986, Astrophys. J., **303**, 724.

Bouvier, J., Bertaut, C., Berez, W., Mayor, M.: 1986, Astron. Astrophys. **165**, 110.
Brown, F.M.: 1985, Nature **317**, 591.
Cahen, S., Doom, C., Cassé, M.: 1986, in 86 Massive Neutrinos, O. Fackler and J. Tran Thanh Van eds., Ed. Frontières, p. 83.
Cayrel, R., Cayrel de Strobel, G., Campbell, B., Däppen, W.: 1984, Astrophys. J. **283**, 205.
Cowling, T.G.: 1965, in Stellar and Solar Magnetic Fields, IAU Symposium n° 22, R. Lüst ed., p. 405.
Duvall Jr T.L., Harvey, J.W.: 1984, Nature **310**, 19.
Duvall Jr T.L., Harvey, J.W., Pomerantz, M.A.: 1986, Nature **321**, 500.
Faulkner, J., Gough, D.O., Vahia, M.N.: 1986, Nature **321**, 226.
Geiss, J.: 1971, in Solar Wind, NASA SP-308, p. 559.
Geiss, J.: 1973, in 13th International Cosmic Rays Conference, Denver, August 1975, **5**, 3375. Invited Paper: 'Solar Wind Composition and Implications about the Origin of the Solar System'.
Hopfinger, E.G., Broward, F.K., Gagne, Y.: 1982, J. Fluid Mech. **125**, 505.
Kirsten, T.: 1986, in 86 Massive Neutrinos, O. Fackler and J. Tran Thanh Van eds., Ed. Frontières, p. 119.
Lebreton, Y., Maeder, A.: 1986a, Astron. Astrophys. **161**, 119.
Lebreton, Y., Maeder, A.: 1986b, Astron. Astrophys. accepted.
Lebreton, Y., Maeder, A.: 1986c, private communication.
Libbrecht, K.G.: 1986, Nature **319**, 753.
Michaud, G.: 1986, Astrophys. J. **302**, 650.
Mikheyev, S.P., Mirnov, A.Yu.: 1986, Nuovo Cimento **9c**, 17.
Schatzman, E.: 1959, in The Hertzsprung Russell Diagram, J.L. Greenstein ed., IAU Symposium n° 10, Ann. Astrophys. Suppl. n° **8**, 129.
Schatzman, E.: 1962, Ann. Astrophys. **25**, 18.
Schatzman, E.: 1969, Astron. Astrophys. **3**, 331.
Schatzman, E.: 1977, Astron. Astrophys. **56**, 211.
Schatzman, E., Maeder, A., Angrand, F., Glowinski, R.: 1981a, Astron. Astrophys. **96**, 1.
Schatzman, E., Maeder, A.: 1981b, Nature **290**, 683.
Simon, Th., Herbig, G., Boesgaard, A.: 1985, Astrophys. J. **293**, 551.
Skumanich, A.: 1972, Astrophys. J. **171**, 565.
Spruit, H.: 1986, this meeting.
Tassoul, M.: 1980, Astrophys. J. **43**, 469.
Ulrich, R.K., Rhodes, E.J.: 1983, Astrophys. J. **265**, 551.
Wolfenstein, L.: 1978, Phys. Rev. D **17**, 2369.
Zahn, J.-P.: 1983, Astrophysical Process in Upper Main Sequence Stars, Saas Fe 1983, B. Hauck, A. Maeder, eds., Geneva Observatory, p. 253.

EVOLUTION AND INTERNAL ROTATION OF THE SUN

Sabatino Sofia
Yale University
P.O. Box 6666
New Haven, CT 06511

ABSTRACT: This paper describes how stellar evolution studies assist in determining the current internal rotation of the Sun, and understanding the physical processes that lead to it from a significantly different proto-Sun. It summarizes the formalism which converts solar rotation into an initial value problem in terms of parameters whose values can be inferred from observations of solar analogues. The main perceived source of uncertainty in the initial angular momentum distribution, and the angular momentum redistribution and loss processes are discussed. These uncertainties are then related to our inability to produce a model of the present Sun totally compatible with solar observations. The most promising next steps to be taken on this problem are then suggested.

1. INTRODUCTION

The internal rotation of the Sun can, in principle, be determined observationally from the solar oscillation spectrum, by means of the inversion technique (cf. Gough, in these Proceedings). In practice, because of observational and model limitations, the task has not been satisfactorily accomplished, since considerable disagreements exist on the results obtained by different groups. Moreover, to determine the rotation of the innermost solar interior in a robust way requires observation of g-mode oscillations, and these might not propagate through the solar convective envelope. Despite these difficulties, though, there is optimism that these obstacles will be ironed out, and that we will, in the near future, obtain $\omega(r)$, or even $\omega(r, \theta)$ for the current Sun.

In order to fully utilize the knowledge derived from helioseismology though, even $\omega(r, \theta)$ is not the end of the road. It alone will not tell us how such a rotation state was established in the Sun, and what it will be like in the future, and even more unfortunately, what is the rotation state and its effects on other stars.

B. R. Durney and S. Sofia (eds.), The Internal Solar Angular Velocity, 173–183.

Evolution affects rotation, and viceversa. During the evolution of a star, differential contraction/expansion takes place, and conservation of angular momentum leads to differential rotation with cylindrical symmetry. Often, the differential rotation leads to fluid-dynamical instabilities that re-distribute angular momentum and mix the stellar material. Of course, this mixing affects the distribution of the elements in the star, and consequently, evolution. The interaction between rotation and evolution is highly non-linear, and consequently, it has to be studies in a coupled way, i.e., we must be able to evolve rotating stars.

It is desirable to convert the problem of the evolution of the rotating Sun into an initial value problem, and then identify an early stage of solar evolution when the physical conditions are sufficiently simple so that all uncertainties can be reduced to one or a few parameters. As we shall see subsequently, such a stage exists. If evolution could be computed with no uncertainties, the present solar conditions could be used to determine the appropriate parameter values. For example, for non-rotating solar models, the appropriate mixing length and initial chemical composition are those required to produce a 1 M_0 model which, at an age of 4.7 billion years (the estimated age of the Sun) has a radius and luminosity equal to the current radius and luminosity of the Sun. As will be shown, the fluid-dynamical processes occurring in the solar interior are sufficiently uncertain that a variety of possible scenarios exist from the initial state to the current Sun. In order to be able to address the problem of initial value uncertainties in the presence of evolution uncertainties, we need to make use of redundant information. This redundancy is provided both by diverse observational information of the current Sun, and by comparison with solar analogues of different ages and/or chemical composition.

Obviously, tackling the problem of the evolution of the rotating Sun is a greatly more complex undertaking than its non-rotating counterpart. Many more physical uncertainties must be considered, and many more observations must be used in order to establish the solar model that more closely represents the Sun. On the other hand, as we know, there is a number of solar observations that are unreconcilable with the non-rotating "standard solar model". These difficulties are well known, and include among others the solar-neutrino problem, the recently detected intermediate timescale solar variability, as well as anomalies of the surface solar chemical composition. It is likely that some, if not all, of these problems will be improved, or even solved, by considering rotation, a feature that the real Sun undoubtedly possesses. In this paper I will describe the techniques we currently have for treating this problem, the uncertainties that we have identified, and the areas of current research that offer promise for significant advance.

2. EVOLUTIONARY CODES FOR ROTATING STARS

Rotating stars undergo deformations and generate circulation currents whose nature is clearly in the three-dimensional regime. In fact, if adequate computational capabilities existed, the "brute force" technique of solving the structure and fluid-dynamical equations in 3D is simple in principle, and more realistic than any other numerical alternative we may propose. However, such a solution is not realistic at the present time, and in fact, even a full 2D version of this problem exceeds current computational capabilities. Obviously, it is necessary to develop a compromise method which is able to display the significant features of the rotating stars while remaining in the numerically treatable (in terms of realistic computational resources) realm. The first such techniques was the perturbation method, used for example by Chandrasekhar (1933) and Sweet and Roy (1953). This technique has been applied only to very simple stellar models (eg. polytropes) and it is not easily applicable to realistic stellar models. Next, the double-approximation method was developed (Roxburgh et al., 1965; Faulkner et al, 1968; Strittmater et al., 1970; Sackmann and Anand, 1970, etc.) In this method, the star is divided in two parts, a slowly rotating core and a massless envelope. Only first-order expansion in some rotation parameter is used in the core, and the Laplace potential is used in the envelope. This method will not be very accurate when differential rotation becomes significant. Other techniques are the J^2 method (described in detail by Papaloizou and Whelan 1973, for example) in which the Roche approximation is used for the rotating stellar potentials. As described in Endal and Sofia, 1976 (Paper I) the regime of validity of this approximation is also much too restrictive. Finally, at the other extreme of the spectrum, there is the self consistent field (SCF) method (Ostriker and Mark, 1968). This method is a very accurate way of describing the stellar potential, but it is extremely time-consuming for implementing in evolutionary codes. In my estimation, the best compromise at the present stage is the technique proposed by Kippenhahn and Thomas (1970), as implemented in Paper I. We will describe hereafter the main points of this technique, which allows a realistic treatment of rotation even for strongly differentially rotating stars, while requiring tractable (if not negligible) numerical resources. In many of these methods, we replace the role played by spherical surfaces in the (spherically symmetric) non-rotating case by equipotential (in the sense of gravitational plus rotational) surfaces. The differences arise in the evaluation of ψ.

Endal and Sofia derived in Paper I a set of stellar structure equations equivalent to the spherically symmetric case, which are valid if the total potential is conservative (the procedure is similar that used by Kippenhahn and Thomas, 1970). These equations are:

Hydrostatic equilibrium

$$\frac{\partial P}{\partial M_\psi} = - \frac{G M_\psi}{4\pi r_\psi^4} f_p \tag{1}$$

Conservation of Mass

$$\frac{\partial r_\psi}{\partial M_\psi} = \frac{1}{4\pi r_\psi^2 \rho} \tag{2}$$

Conservation of Energy

$$\frac{\partial L_\psi}{\partial M_\psi} = \varepsilon - \frac{\partial E}{\partial t} - P \frac{\partial (1/\rho)}{\partial t} \tag{3}$$

and Energy Transport

$$\frac{\partial \ln T}{\partial \ln P} = \min \left[\nabla_{ad}, \nabla_{rad} \frac{f_T}{f_P} \right] \tag{4}$$

where ∇_{ad} ad ∇_{rad} are the usual (spherically symmetric) adiabatic and radiative gradients. The subscript ψ denotes quantities contained within the equipotential ψ, and r_ψ is defined as

$$r_\psi = \left(\frac{3 V_\psi}{4\pi} \right)^{1/3}$$

The distortion terms are f_p and f_T, which are defined in Paper I, and derived in the appendices thereof for an approximation that has been shown to be valid up to a large fraction of the critical rotational velocity.

The main point of this method is that the equations are sufficiently similar to their spherical counterpart that conventional stellar evolution codes can be used by only adding the computation of the distortion terms f_p and f_T.

Our technique of computation of f_p and f_T, which in turn requires computation of ψ, divides the total potential in three components, namely the gravitational component (spherically symmetric), the rotational component (cylindrically symmetric) and a cylindrically symmetric component due to the distortion of the star. The approximations made apply to the third component which, for a given point in the star, is computed as if the point were on a polytrope rotating with constant ω and having the same ratio of mean interior density and central density as the model. To compute this component, then, it is necessary to know $\omega(\psi)$.

As stated earlier, for the solar case it is at least in principle, possible to obtain the current values of $\omega(\psi)$, and thus the structure of the current Sun can be derived from the equations given above. This, however, will not solve all our problems, since the current rotational solar velocity is a consequence of both the initial rotation conditions (established in the pre-main sequence stage) and the mechanisms for angular momentum redistribution and loss, which are themselves functions of ω. To know how the Sun got to rotating the way it does, and how it will in the future, and what the structural consequences are, we must be able to formulate the problem of solar structure and evolution as an initial value problem. This requires identifying a solar evolutionary stage where the initial conditions are sufficiently simple that they can be characterized in terms of a few parameters, and to formulate the processes in such a way that uncertainties can be identified and parameterized. The hope is that there are enough redundant observations of the current Sun to be able to uniquely and separably narrow-down the range of uncertainty of all parameters to be able to reliably reproduce the past solar history, and confidently predict its future. In this sense, helioseismology is extremely valuable as a diagnostic of our model, in the sense that, once we obtain a solar model by using a given set of parameters, we perturb this model and derive the oscillation spectrum. We can now compare the agreement or disagreement of predicted and observed oscillations even for a mode sample that would be insufficient to allow solution of the inverse problem.

3. INITIAL CONDITIONS

As we know from many past studies, for non-rotating stars it is sufficient to begin evolutionary sequences at the so-called zero-age main sequence (ZAMS), since that is the latest stage in which the star is chemically homogeneous, and all other stellar parameters can be uniquely determined from the stellar mass and this composition. In the case of rotating stars, however, this stage is much too late to begin evolving, since by the ZAMS stage an unknown angular momentum distribution has already been established in the star. This point is discussed in detail in Endal and Sofia (1981), (Paper III) but it can be summarized as follows: if the Sun goes through a Hayashi track, as conventional non-rotating star formation predicts, then, during this stage, it is fully convective, and the detailed angular momentum distribution will quickly evolve into the equilibrium rotation law for a rotating fluid, whatever that law may be. Both current numerical simulations (Chan, reported elsewhere in these Proceedings) and current helioseismology results (Brown, these Proceedings) suggest that solid-body rotation may be a good approximation for this state. In this case, only the total angular momentum is unknown, and a good start can be made by assuming that the total angular momentum of the forming Sun can be obtained by extrapolating Kraft's (1970) rotation law for more massive stars. This curve suggests that the Sun was born with a total angular momentum $J_0 = 2 \times 10^{48}$ g cm^2 sec^{-1}, or 61.5 times the angular momentum of the current Sun if it were rotating as a solid body. Since the Sun did not necessarily begin with an angular momentum equal to the mean value of Kraft's relation, we must allow values of the initial solar angular momentum larger or smaller than this value. In any event in the Hayashi picture, all possible rotational scenarios can be expressed in terms of one parameter, J_0 initial.

The problem with the very simple picture presented above is that protostellar material necessarily has angular momentum, and that very likely star formation occurs by accretion from a disk. In this case, the Hayaski picture is not valid, and the proto-Sun is never fully convective. The initial rotation stage of the proto-Sun then must be derived from much more sophisticated physical considerations (Larson, personal communication).

4. ANGULAR MOMENTUM REDISTRIBUTION AND LOSS

The problem of redistribution of angular momentum is one of the most difficult parts of these studies, and fraught with most uncertainties, although it could, in principle, be the easiest part. It all depends on the role played by the ubiquitous magnetic fields. Let us first look at limiting cases. If the Sun were a perfect conductor, magnetic fields would be forever frozen in, and any amount of differential rotation would increase a seed field strength to such a degree as to kill the differential rotation even within a small fraction of the age of the Sun. This conclusion is not greatly modified if we include magnetic field diffusion by known mechanisms in a realistic solar model (Fox and Bernstein, these Proceedings). In these cases differential rotation would always be negligibly small, the structure of these stars could be computed accurately by any of the proposed techniques, and the deformation and circulation would play a negligible role in stellar evolution.

On the other hand, there is indirect evidence that rotational effects are not negligible, and that differential rotation does develop and remain for some time. These matters will be discussed in a subsequent paper, but can be summarized as follows. Since a white dwarf is but the core of a giant, if differential rotation were destroyed during the giant phase, all white dwarfs should have rotational periods in excess of hundreds of days contrary to observations. There are ways of spinning up white dwarfs in close binary systems, but not in isolated systems. It appears, then, that the frozen in condition does not apply to magnetic field in the stellar interiors, and that field magnification must be limited, either by more efficient diffusion than currently estimated or by magnetic field line breaking and reconnecting, as shown in the numerical experiments of Weiss (1966).

If the effects of magnetic fields are negligible, a large variety of circulations and instabilities come to operate when differential rotation is established. The most complete treatment of these instabilities is given in Endal and Sofia, 1978, (Paper II) and it will not be repeated here. I will only summarize the main ideas. If extreme differential rotation conditions operate, there will be instabilities to adiabatic perturbations with consequences that typically occur on free-fall timescales. These mechanisms are called dynamical instabilities, and they include convection, dynamical shear and the Solberg-Hoiland instability. Because of the fast timescales involved, the code assumes instantaneous redistribution whenever a dynamical instability operates. The magnitude of the redistribution assumed is the smallest amount required to remove the instability, and a corresponding chemical mixing is computed.

An angular momentum gradient that is stable to an adiabatic perturbation may be unstable when the stabilizing effects of entropy gradients are removed by virtue of radiative diffusion. In this case, secular instabilities set in which distribute angular momentum and mix stellar material. Since the timescales on which secular instabilities operate are comparable to evolutionary timescales, their effects must be computed in terms of coupled non-linear diffusion equations of the form

$$\frac{\partial \omega}{\partial t} = \frac{1}{\rho r^4} \frac{\partial}{\partial r}\left(\rho r^4 D \frac{\partial \omega}{\partial r}\right)$$

for angular velocity, and

$$\frac{\partial X_i}{\partial t} = \frac{1}{\rho r^2} \frac{\partial}{\partial r}\left(\rho r^2 D \frac{\partial X_i}{\partial r}\right)$$

for chemical composition. The mechanisms included in this category are secular shear instability, Goldreich-Shubert instability, and the special case of Eddington circulation. Again, the details for implementing these criteria, and for computing the diffusion coefficients D are given in Paper II.

The loss of angular momentum is assumed to be via the solar wind. This is another element of uncertainty, but at the moment the model of Belcher and MacGregor (1976) is included in our codes although, as recognized by these authors, significant limitations exist in this model.

5. A TEST CASE

I will describe here the results of calculations carried out by Endal and Sofia, in Paper III. In addition to summarizing the results of paper III, I will point out the shortcomings of that work both in the simplifications made and in the comparisons with observations. The elements currently under investigation will be pointed out.

As stated earlier, the starting solar model was a fully convective, chemically homogeneous proto-Sun in the Hayashi track. Magnetic fields were not included. The initial angular momentum was obtained from an extrapolation of the Kraft rotation curve for more massive stars. Future work on the initial phase includes following the details of star formation through accretion, and examining the role of magnetic fields. It is my current assessment that these new ideas will not produce modifications that cannot more simply be mimicked by varying the initial total angular momentum parameter, although this statement has not been documented.

As the proto-Sun evolves through the Hayashi track and afterwards, wind torquing according to the Belcher-MacGregor model removes angular momentum from the convective envelope which is assumed to rotate as a solid body. The Belcher - Mac Gregor model assumes a one-component, steady state, axisymmetric flow, a magnetic field proportional to the rotation rate, and a total energy for the wind equal to the thermal energy input of the current Sun, plus a time-dependent rotational energy input. Of course one may have quarrels with any of these assumptions, but it is my feeling that this is not the critical element of current solar research.

As the Sun evolves, non-homologous contraction/expansion of different radiative layers produces differential rotation. The current practice is to ignore magnetic fields, and to compute the redistribution according to the mechanisms described in Section 4.. The redistribution aspect of the problem is the most questionable not only because of the absence of magnetic fields, but also because: (a) the completeness of the catalog of instabilities is not known; (b) the efficiency of the mechanisms is only known approximately; (c) only local effects are included, and (d) we do not know, in some cases (e.g. meridional circulation and convection) what rotation law the mechanism tries to enforce. I believe this is the aspect of the problem in most urgent need of improvement. Obviously, we are not going to be able to solve all our problems on first principles and with complete certainty. We must make extensive use of observations to test a variety of plausible scenarios.

6. OBSERVATIONAL TESTS

There are two types of observations against which we can test our assumptions. On the one hand, we have observations of the Sun itself. The negative aspects of solar observations are that they relate to only one object, and that its physical conditions (e.g. possible degree of differential rotation) are very moderate, when compared to stars in later stages of evolution for example. On the positive side is that its nearness to us allows us to observe the Sun with a degree of detail not possible for any other star. On the other hand, stellar observations, while not as detailed as the solar ones, allow us to determine global effects of varying one of the stellar parameters at a time, such as age.

Our experience has shown that at the present time the solar observations are more difficult to satisfy, although new developments on the Fourier technique for measuring stellar rotation to a very high accuracy has made this stellar parameter also a sensitive test for our ideas.

The solar observations of relevance to this work are: the oblateness, oscillations, surface chemical composition of trace-elements (such as Li, Be, He^3), and the neutrino rate, which is also affected by rotation although not sufficiently to solve the solar neutrino problem. The results of paper III are a solar model whose rotational state is strongly incompatible with observations both of the solar oblateness and of the rotation rates of the outer solar regions (thus presumably more definitive) obtained by inverting p-mode oscillations (Gough, these Proceedings). Specifically, the model oblateness ranges from 6.4×10^{-5} to 9.5×10^{-4} depending on the total angular momentum assumed for the model, whereas the observational values are $(5.0 \pm 0.7) \times 10^{-5}$, as obtained by Dicke and Goldenberg (1967) or $(9.6 \pm 6.5) \times 10^{-6}$, as obtained by Hill and Stebbins, 1975. Clearly a disagreement of this size indicates that the rotational calculations are still far from realistic, and this same conclusion is obtained when one considers the oscillations data. Although we may eventually find a number of things that are wrong in the scheme of angular momentum redistribution and loss currently used, there is a particular feature, common to most of the redistribution mechanisms that most obviously produce trouble, and this is the local nature of the diffusion coefficient D. In most mechanisms $D \propto \omega^2$. Since the radiative region just below the convection zone is forced to co-rotate with the latter (as a consequence of convective overshoot), the diffusion coefficient there is very small. This provides a very efficient barrier to prevent the flow of angular momentum from the solar interior to the surface. However, this barrier is an artifact of our approximation, and probably has no physical basis. Before any significant advance can be made in this area, the effect of the local approximation in the diffusion processes must be removed.

7. SUMMARY AND CONCLUSIONS

To study the evolution of the rotating Sun is a very important and necessary undertaking of current research. It is necessary, since there are many solar observations which either disagree or cannot be addressed by the non-rotating "standard solar model". It is important because it bears on exciting fluid-dynamical problems and it has significant consequences on the evolution of all stars, especially at the late stages. It is timely to do this now because of the increased availability of powerful computers which are essential to tackling this problem, the development of techniques, and the development of new powerful observational tools such as helioseismology.

In a way, we cannot expect to take the next step in solar modelling and ignore rotation. The most pressing elements of solar physics is solar variability, and this is powered by a dynamo which arises in a rotating turbulently convective fluid. Even though the problem is difficult, it is now tractable, and the next decade should answer many of the questions that we have in this meeting. A judicious effort combining the direct technique of evolution of the rotating Sun, using the steady advances in our understanding of solar internal fluid dynamics, and the diagnostic tools of helioseismology will not only have a profound effect on solar physics, but it will also expand the field of stellar evolution by another dimension where the payoff is expected to be exciting.

REFERENCES

Belcher, J. W., and MacGregor, K. B. 1976, Ap.J., 210, 498.
Chandrasekhar, S. 1933, M.N.R.A.S., 93, 390.
Dicke, R. H., and Goldenberg, H. M., 1967, Phys. Rev. Letters, 18, 313.
Endal, A. S., and Sofia, S. 1976, Ap.J., 210, 184.
__________. 1978, Ap.J., 220, 279.
__________. 1981, Ap.J., 243, 625.
Faulkner, J., Roxburgh, I. W., and Strittmatter, P. A. 1968, Ap.J., 152, 203.
Hill, H. A., and Stebbins, R. T. 1975, Ap.J., 200, 471.
Kippenhahn, R., and Thomas, H.-C. 1970, in Stellar Rotation, ed. A. Slettebak (Dordrecht: Reidel), p. 20.
Kraft, R. P. 1970, in Spectroscopic Astrophysics, ed. G. H. Herbig (Berkeley: University of California Press), p. 385.
Ostriker, J. P., and Mark, J. 1968, Ap.J., 151, 1075.
Papaloizou, J. C. B., and Whelan, J. A. J. 1973, M.N.R.A.S., 164, 1.
Roxburgh, I. W., Griffith, J. S., and Sweet, P. A. 1965, Zs. f. Ap., 61, 203.
Sackmann, I.-J, and Anand, S. P. S. 1970, Ap.J., 162, 105.
Strittmatter, P. A., Robertson, J. W., and Faulkner, D. J. 1970, Astr. and Ap., 5, 426.
Sweet, P. A., and Roy, A. E. 1953, M.N.R.A.S., 113, 701.
Weiss, N. O. 1966, Proc. Roy. Soc. London, Ser. A 293, 310.

ANGULAR MOMENTUM TRANSPORT IN THE RADIATIVE INTERIOR OF THE SUN

H.C. Spruit
National Solar Observatory
Sunspot, New Mexico 88349, USA
and
Max Planck Institut für Physik und Astrophysik
Karl-Schwarzschildstr. 1
8046 Garching, West Germany

ABSTRACT. The known mechanisms for transporting angular momentum in the core of the Sun are reviewed. All hydrodynamic mechanisms are insufficient to account for the present low rotation rate with the possible exception of internal gravity waves. Several mechanisms are likely to be far too effective in depleting Lithium, however. It is argued that weak magnetic fields can easily solve both problems. Their stability properties are reviewed and a scenario is presented for the evolution of such fields. It predicts that the present field in the Sun has poloidal and toroidal components of similar strength, on the order of a few Gauss. The differential rotation in the core, if due to such a field alone, would be extremely small.

1. INTRODUCTION

The observations reported in these proceedings (see also Duvall and Harvey, 1984 and Brown, 1985) establish as one of the most remarkable properties of the solar rotation its *uniformity* with depth; $\Omega(r)$ being within 30% of its equatorial surface value down to $r/R_{sun} = 0.3$, perhaps even throughout the entire Sun. The continous braking of the Sun by the solar wind is compatible with such uniformity only if a very effective mechanism of angular momentum transport exists inside the Sun, keeping the interior rotating at nearly the surface rate. Such a mechanism is easily found in the convection zone, where turbulent flows produce such a high effective viscosity that virtually no differential motion is expected except that produced by the dynamics of the turbulent motions themselves (cf. Durney, this volume, Glatzmaier, this volume). Thus the more interesting task is finding the mechanism that keeps the radiative core rotating nearly uniformly. I will review briefly the mechanisms that have been proposed so far. This will lead us to one particular mechanism: torques exerted by (relatively weak) magnetic fields. As I shall argue, this is at present the most promising candidate because a) most of the other mechanisms have serious weaknesses, and b) magnetic fields with the required strength are easily accounted for and have just the right transport properties (section 3). There are some loose ends, though, so that the case for

B. R. Durney and S. Sofia (eds.), The Internal Solar Angular Velocity, 185–200.

magnetic fields can not be considered definitive. The kind of field configurations and strengths that can exist in the core are determined crucially by their stability properties. Section 4 reviews some important results on stability. On the basis of these a scenario is presented in section 5 of how the magnetic field and rotation of the interior could have evolved together under the influence of the solar wind torque. This scenario predicts that at present a nearly poloidal field exists in the interior with a strength of the order of a few Gauss. The level of differential rotation accompanying this field would be very low.

Some highly effective mechanism has to be present at *every* depth in the Sun, down at least to where present observations provide information ($r/R_{sun} \approx 0.2$). I will therefore skip over certain processes that work only in very restricted depth ranges (such as convective overshoot and baroclinic instability, which work only very close to the base of the convection zone) and concentrate mainly on the more general processes.

2. NONMAGNETIC MECHANISMS

I discuss here the three classes of nonmagnetic mechanisms proposed so far: 1) hydrodynamic turbulence generated by some form of shear instability, 2) meridional circulations and 3) waves. Since hydrodynamic turbulence ideas have been reviewed elsewhere in greater detail (Zahn, 1983, Spruit, Knobloch and Roxburgh, 1983 and Spruit, 1984) I limit myself here to some general remarks, and concentrate instead on the remaining two classes. Before doing so it is useful to discuss two constraints that appear to have the greatest effect in restricting the possibilities.

2.1. Mixing versus angular momentum transport

Most mechanisms do not just transport angular momentum (i.e. exert torques) but also have the effect of physically mixing material between different depths in the Sun. Both effects can be described by effective diffusivities (one for momentum and one for constituents of the gas) and the ratio of these diffusivities is characteristic for the mechanism. Observations of the Lithium depletion of the Sun and other stars (Soderblom, 1983, Cayrel et. al. 1984, Boesgaard and Tripicco, 1986) give a good estimate of the rate of mixing over the first scaleheight below the convection zone, averaged over the past of a star (Thévenin, Vauclair and Vauclair, 1986). If expressed in terms of a diffusivity which is constant in time and in depth in the Sun, it is of the order of 1000 $cm^2 s^{-1}$, or about 100 times the microscopic diffusion rate (Baglin, Morel and Schatzman, 1985). To account for the very small variation of rotation rate below the convection zone, the diffusivity for momentum (viscosity) has to be much higher, at least a factor of 20 (Law et. al. 1984; Schatzman 1984, private communication). Since the Sun is still being braked by the solar wind, this number is not just an average over the Sun's history but also applies to its recent (1 Gy, say) past. The mixing that burned the Lithium on the other hand may have taken place at any time in the past, and its is not neccessary to assume that it is still effective now. Thus we can be a bit more precise: the effective viscosity in the outer radiative core has most of the time including recently been at least 20 times larger than the average effective diffusivity of Lithium over the history of the Sun.

2.2. Energetics of mixing

Mechanisms in which angular momentum transport is associated with physical mixing have to account for the energy that is neccessary for mixing against the stable gradient in the radiative core. This energy depends on the scale on which the mixing takes place. On very small scales, radiative exchange effectively removes the stabilizing effect of the thermal buoyancy force. Since it is usually plausible that the mixing can indeed take place on such small scales (in the cases of hydrodynamic turbulence and wave mixing for example) the main concern is the stabilizing effect of a composition gradient. The energy required to mix the interior against the stable gradient of Helium that builds up during the evolution is substantial. To decrease the presently existing composition gradient by a factor of the order unity an amount of the order of a few percent of the total gravitational binding energy of the Sun is needed (the heavy Helium being a substantial part of the mass of the Sun, and most of it being located at the bottom of the gravitational potential well). This is more than the rotational energy the Sun has ever had (assuming that it never rotated faster than 150 $km\ s^{-1}$), and five orders of magnitude more than the present rotational energy. This shows that mixing mechanisms that ultimately draw upon rotation as an energy source can presently operate only in the outer part of the core, where the stabilizing Helium gradient is sufficiently weak. This makes them plausible candidates for the mixing implied by the Li-depletion, but not for lowering the neutrino flux or for explaining the low rotation in the inner half (by radius) of the Sun.

2.3. Shear instabilities

With the added complications of rotation, spherical geometry, thermal and compositional stratification, and thermal diffusion, the subject of shear instability cannot be considered fully understood at present. Various forms of instability have been studied in great detail (see the monographs by Monin and Yaglom, 1971; Joseph, 1976; Pedlosky, 1979; Drazin and Reid, 1981). The essentials of these instabilities from the astrophysical point of view have been summerized recently by Zahn(1983, see also this volume). Two instabilities thought to be of importance for astrophysics have been considered in some detail recently by Knobloch and Spruit(1983) and Spruit and Knobloch (1984).

A difficulty with shear flow stability is that no general stability conditions exist that would allow easy application to astrophysical problems. Changes in boundary conditions or the addition of ingredients like rotation or stratification all change the results in essential ways. Another difficulty is that in many cases a linear stability analysis is not enough to determine whether a flow is unstable or not. Nonlinear results are much harder to get, and can sometimes only be obtained numerically (e.g. Orszag and Kells 1980). The results for the astrophysically most relevant cases of *stratified* flows however all seem to have an intuitively obvious property: A shear flow can produce mixing in the direction of the stratification *only* if the envisaged mixing motions take more energy out of the shear than they spend against the buoyancy force due to the stratification. This handwaving argument is thus sufficient to exclude mixing by hydrodynamic instabilities in some cases. For example, the Richardson criterion for adiabatic stability in a stratified flow was originally derived this way. If the stratification is stronger than the shear flow in this sense, the surviving instabilities must all be such that they do not mix against the stratification, for example by having nearly horizontal motions (as in the ABCD instability, see Shibahashi, 1980; Knobloch and Spruit,

1983). Whether instability actually takes place in situations where it is allowed by this argument is far harder to establish. For the astrophysical case of a rotating stratified star however, it is generally felt that Zahn's criterion (Zahn, 1983) gives a reliable estimate for the occurrence of instability (see however Kippenhahn and Thomas, 1981). It includes the effects of composition and thermal stratifications, as well as viscosity and thermal diffusion. Its application to the Sun shows that instability driven by differential rotation is presently possible in a rather thin layer just below the convection zone but that in the past (at higher rotation rates) it was possible over a much deeper layer. Shear instability may therefore be involved in the Li-depletion process.

2.4. Circulations

2.4.1. Eddington-Sweet-Vogt and Ekman circulations

The most familiar of meridional circulations in a star is the Eddington-Sweet-Vogt (ESV) circulation. I refer to other texts (e.g. Zahn, 1983) for a description; its circulation (turnover) time scale is of the order

$$t_E = \frac{g}{\Omega^2 R}\frac{R^2}{\kappa} \tag{1}$$

where g is the surface gravity, Ω the rotation rate, R the stellar radius and κ the thermal diffusivity (in $cm^2 s^{-1}$). For the present Sun this time is about 10^{12} yr. ESV circulation is therefore presently not important and cannot explain the low internal rotation. It was much more vigorous in the past however and may therefore have done some Lithium depletion.

A second process is the Ekman circulation. In the stellar context the Ekman problem can be formulated as follows. The core is treated as a barotropic fluid (pressure a function of density only), rotating initially at a fixed rotation rate Ω. It is bounded by the base of the convection zone, which is treated as a solid boundary rotating slower by an amount $\Delta\Omega$. Near this boundary a small amount of core fluid rotates with it, due to viscosity. Since this fluid rotates at a different rate than the rest of the core, a Coriolis force acts on it and while the fluid is barotropic this force can not be taken up by internal forces in the core. Instead a circulation takes place. Along the boundary a layer of fluid of thickness $\delta_E = (\nu/\Omega)^{1/2}$ moves horizontally. This flow causes a circulation which penetrates through the entire core, eventually bringing it into corotation with the boundary (see Zahn, 1983 for a derivation).

Though there is no doubt that Ekman circulations can form in stars by this mechanism, I feel that the above picture is nevertheless very misleading, because the core of a star is not barotropic. As I will argue presently, this has the effect of confining the Ekman circulation to a very thin layer below the convection zone, so that its effect on the rotation of the core is negligible.

The reason why the Ekman flow can penetrate so freely through a barotropic fluid is that no buoyancy forces exist in such a fluid. In convectively stable layers of a star however buoyancy forces always dominate the rotation forces. The ratio of these is of the order N^2/Ω^2 where N is the buoyancy frequency. This ratio is 10^5 in the solar interior. Since the forces driving the Ekman flow are too weak to overturn the core, they produce instead a circulation that is deflected back into a horizontal return flow at the depth where the buoyancy acceleration is of the order $\Omega^2 r$. N^2 varies linearly over about 1 scaleheight below the base of the convection zone:

$$N^2 \approx \frac{x}{H}\frac{g}{H}(\nabla_a - \nabla) \tag{2}$$

where $x = r_b - r$ is the depth below the base of the convection zone r_b, and ∇ , ∇_a are typical values of the logarithmic temperature gradient and its adiabatic value in the core. The buoyancy acceleration on a fluid parcel starting at $x=0$ equals $\Omega^2 r$ when

$$\Omega^2 r = 1/2 \frac{x^2}{H^2} g\,(\nabla_a - \nabla). \tag{3}$$

Since $\Omega^2 r/g$ is always less than 1, it follows from (3) that the penetration depth x of the Ekman circulation is always less than a scaleheight. *Ekman layers play no role in the spindown of a star.* For the Sun, we get $x \approx 500km$, roughly equal to the depth of the overshoot layer. This is large compared with the Ekman layer thickness $\delta_E \approx 20m$, but negligible compared with the size of the core (for the applied mathematicians among the readers: the radius of the core is $5\,10^5$ km).

2.4.2. "Baroclinic" circulation

A more important kind of circulation associated with differential rotation is due to *thermal diffusion* and is related in character to the ESV circulation. Let us ignore viscosity (since it only produces the thin Ekman layer). To obtain order of magnitude estimates, I assume that Ω is a function of r only. Unless the rotation is uniform, there are Coriolis forces that cannot be balanced by the pressure gradient. They lead to displacements of the fluid which produce restoring buoyancy forces. When the system is in equilibrium the two balance, this state is called *baroclinic* and is governed by the relation

$$2r\,\sin\theta\cos\theta\,\Omega\Omega' = g\,\mathbf{n}.\nabla\,lnT \tag{4}$$

where $\mathbf{n}$ is a unit vector in the θ direction along a surface of constant pressure and $\Omega' = d\Omega/dr$ (this follows directly from the equations for equilibrium and our assumption $\partial_\theta\Omega = 0$). Integrating in θ direction Equation (4) yields the equator-pole temperature difference:

$$\Delta lnT \equiv \ln T_e - \ln T_p = r^2\Omega\Omega'/g. \tag{5}$$

In the absence of radiative diffusion, this would be a valid equilibrium (except for possible baroclinic instabilities, at high rotation rates, cf. Spruit and Knobloch, 1983). Since the quantities in the RHS of (4) vary with r, this temperature distribution is not in *thermal* equilibrium, however, and ΔlnT diminishes in time. The equilibrium equation (4) is then not satisfied any more, and fluid displacements take place to restore the value of ΔlnT. Thus, a circulation is set up similar to the ESV circulation. I call this a baroclinic circulation. The main difference is that its pattern is now determined by the details of $\Omega(r)$. Let us assume that Ω varies by a small amount:

$$\Omega' = \Delta\Omega\, f(r), \tag{5}$$

where $f(r)$ is of order unity. Assume further that f and its derivative vary on a radial scale L which is small compared with the scale on which the other variables change. The steady state energy equation is

$$\rho T\mathbf{v}.\nabla S = -div\mathbf{F}_{rad}.$$

With the assumed length scale for Ω, this yields the following relation for the radial velocity component v:

$$v \approx \frac{\kappa}{L} \frac{rH}{L^2} \frac{\Omega \Delta\Omega r}{g(\nabla_a - \nabla)} \tag{6}$$

and the circulation time scale $t_C = L/v$ is

$$t_C = \frac{L^4}{r^3 H} \frac{\Omega}{\Delta\Omega} t_E. \tag{7}$$

For Ω varying smoothly through the star, this time scale is of the same order as the the ESV time scale or longer. Baroclinic circulation can therefore not account for the low internal rotation of the Sun. But it may be important for depleting Lithium. For example a minor wiggle in Ω, $\Delta\Omega/\Omega = 0.1$ over one scaleheight ($L = 6\,10^4 km$) below the convection zone would at present ($\Omega = 3\,10^{-6}$) yield a circulation time scale of only $2\,10^8$ years, and much less in the past. The Lithium depletion due to such a circulation would be much larger than observed.

2.4.3. Zahn's mixing scheme

Zahn (1983, and these Proceedings) oberves that even without torques exerted on the star by a stellar wind, the ESV circulation will slowly produce, by advection of angular momentum, some variation of rotation rate on equipotential surfaces. These are likely to be unstable to shear instabilities, since there are no stabilizing forces to motions taking place entirely on such a surface. The turbulence produces a cascade to smaller scales, and on sufficiently small scales the motions become 3- dimensional, resulting in mixing in the radial direction. Zahn shows that on a scale of the star as a whole this mixing has a time scale comparable to the ESV time scale. On smaller scales, such as the gap between the Lithium burning layer and the base of the convection zone, the turbulence produces much faster mixing. Zahn finds that the mixing for the present Sun is large enough to deplete Lithium, which implies that in the past this process may well have been too efficient.

2.5. Wave induced mixing and momentum transport

Waves can transport angular momentum in two ways: a) by dissipation or reflection, and b) by gravitational interaction. The first of these takes place if at some point in the star waves are generated with a preferred direction with respect to the rotation. Such a preference may itself be brought about by rotation, through the Coriolis force acting on the wave's velocity field. Such waves may travel to another part of the star and dissipate there, depositing their momentum. An asymmetry with respect to the direction of rotation then implies that angular momentum is transferred as well. A likely source of waves is the convection zone, which generates p-modes as well as internal gravity waves. Since the p-modes have short periods, they are not strongly influenced by the Coriolis force. Also, their damping takes place mostly in the upper convection zone. Together, this suggests that they may not be important transporters of angular momentum, but no quantitative estimate seems to exist.

Internal gravity waves are excited at the base of the convection zone, by individual cells penetrating its lower boundary. This has been proposed by Press (1981) as a source of energy for mixing the core. He came up with a very optimistic estimate of the energy flux, much higher than we believe is realistic. An upper limit on the wave flux can be found by assuming that just below the convection zone the core moves at an amplitude equal to that of the boundary, and that all of this motion is in the form of propagating (rather than standing) gravity waves.

The flux F then follows from the typical vertical displacement x of the boundary, its horizontal wavenumber k_h, its frequency ω, and a typical value for the buoyancy frequency N in the core (cf. Press, 1981):

$$F = \rho(N^2-\omega^2)^{1/2}(\omega x)^2/k_h. \tag{8}$$

Inserting $N\approx10^{-3}$, $x\approx200km$, $\rho\approx0.2$, and $\omega\approx2\,10^{-6}$, $k_h\approx3\,10^{-10}$ (valid for the energy carrying convective scale near the base of the convection zone), we have $F\approx3\,10^7$ $erg\ cm^{-2}s^{-1}$. Integrated over 1 Gyr say, this corresponds to an amount of energy much larger than the rotational energy of the Sun. The question is however how much of it actually reaches places where mixing and momentum transfer are needed. Because of their low frequency the waves excited directly by the convection zone have a very small radial wavelength, and propagate almost horizontally. These waves would be damped by radiative exchange long before they reach substantially into the core. Press argues however that by their strong tendency for nonlinear interactions, the gravity waves will convert into higher frequency waves with longer vertical scales which are less damped and have a higher radial propagation speed, so that a substantial fraction of the initial energy can spread through the core. If these waves also have a substantial east-west asymmetry, their disspation by radiative damping could produce substantial internal torques in the core. Whether they would also mix (for which they were invoked originally by Press) is less clear because this would have to take place on such small scales that the buoyancy force is strongly reduced by radiative exchange. These scales would have to be excited from the larger scales in the wave field (presumably by some form of shear instability). This implies that intermediate scales, where radiative damping is maximally effective, would also be strongly present, and it is not clear that any of the wave energy would survive to sufficiently small scales to produce mixing.

The second way in which momentum can be transferred is by the gravitational perturbation associated with the wave. Since this perturbation extends through the entire star, momentum can be transferred to distances far from the place where the waves are excited or dissipated. The energy carried by the gravitational perturbation however is likely to be small in the case of waves generated by the convection zone.

2.6. Conclusions on nonmagnetic mechanisms

A number of candidates has been found for the mixing implied by the observed Lithium depletion, but none that can explain the low internal rotation (with the possible exception of internal gravity waves generated by the convection zone, which need further study). These mechanisms share the property that their efficiency increases rapidly with rotation rate (as Ω^2). This limits their usefulness somewhat because most of the Li-depletion would then have taken place early in the life of a star, contrary to the rather smooth variation of Li-depletion with age (Duncan, 1981). An exception may be again internal gravity waves generated by the convection zone, which do not depend on rotation, and therefore would produce a rather steady mixing. In addition some mechanisms (Zahn's scheme, baroclinic ciculation) may be far too efficient early in the life of the star.

3. MAGNETIC FIELDS

From the above it appears that we need a mechanism that is much more effective at transporting angular momentum. Also, we need something to prevent the mechanisms discussed in section 2 from becoming *too effective* at depleting Lithium. Magnetic fields can do both.

3.1. Effect of magnetic fields on mixing

A moderate magnetic field may be enough to interfere with the turbulence on which some of the mixing mechanisms of section 2 are based. Consider for example shear turbulence driven by differential rotation. In accordance with Zahn's estimate (Zahn, 1974,1983) such turbulence will occur on length scales not larger than a certain maximum L_{max}, such that the shear rate equals the radiative diffusion rate (at this length the buoyancy force is reduced enough by radiation to make instability energetically possible). For a rotation rate varying like

$$\Omega \sim r^{\alpha} \tag{9}$$

this yields

$$L_{max}^{2} \approx \kappa/(\alpha\Omega). \tag{10}$$

The magnetic forces start becoming important when the Alfven speed equals the turbulent velocity at this scale:

$$\frac{B}{(4\pi\rho)^{1/2}} \approx L_{max}\alpha\Omega = (\kappa\alpha\Omega)^{1/2}. \tag{11}$$

For α of order unity this yields, just below the convection zone ($\rho = 0.2$) a critical field strength of 7 Gauss for the present Sun, and 200 Gauss for a star rotating 30 times as fast (as the early Sun may have done). If magnetic fields large compared with this value are present (in the next section I will argue that this may well be the case), shear turbulence may be strongly suppressed. This would affect mostly the mixing by shear instabilities driven by differential rotation, and Zahn's mixing scheme.

In addition to their stabilizing effect on turbulence, magnetic fields would also indirectly reduce the mixing by turbulence and circulations, by reducing the differential rotation that drives the processes.

3.2. Torques exerted by a weak field

Assume that the star initially has a weak magnetic field, in the sense that the Alfven speed in the core is much less than the rotation speed. Consider small amplitude perturbations on the equilibrium in the form of displacements in the azimuthal (rotation) direction. These propagate through the star in the form of torsional Alfven waves, so that a typical oscillation time is r/V_A. This oscillation period also holds when the azimuthal displacements are *not small*. So it is also the time scale on which the differential rotation is braked by the torques that develop as the weak field is wrapped around into a strong field by differential rotation. To see this quickly (see also Mestel, Moss and Tayler, these Proceedings), assume that azimuthal flow is the only motion taking place (thus ignoring nonazimuthal motions due to magnetic instabilities for example). The induction equation $\partial_t \mathbf{B} = \nabla\times(\mathbf{u}\times\mathbf{B})$ then yields after integration in time:

$$B_\phi = \frac{1}{r}[\partial_r(rB_r\xi_\phi)+\partial_\theta(B_\theta\xi_\phi)], \tag{12a}$$

$$B_r = const., \tag{12b}$$

where ξ_ϕ is the fluid diplacement in the ϕ direction $(\partial_t \xi_\phi = \Omega r)$. Equations (12a,b) show that B_ϕ and the stress component $B_r B_\phi / 4\pi$ are linear in the displacement ξ_ϕ and so is the corresponding torque, even for azimuthal displacements corresponding to many rotations around the Sun. This yields the following well known estimate for the initial field needed to brake the internal differential rotation within the age of the Sun:

$$r/V_A < 4 10^9 \, yr \rightarrow B > 10^{-6} \, G, \tag{13}$$

where I have taken $\rho \approx 3$ as a typical density in the core. If the field is strong compared to (13), many oscillations will take place within the life of the Sun.

3.3. Evolution of the field (ignoring instabilities)

Though magnetic instabilities definitely cannot be ignored during the winding-up by differential rotation (see Section 4), it is important to understand first how the field and the differential rotation would evolve without such instabilities.

Each individual magnetic surface (a surface outlined by adjacent field lines) can support its own set of oscillation modes, independent of all other magnetic surfaces (in general these oscillations are not pure Alfven waves, but are waves from the coupled Alfven and slow continua, cf. Goossens, 1986, but this does not reduce the validity of what follows). Due to the inhomogeneity of the Alfven speed the oscillation frequencies vary continuously across the magnetic surfaces. The initial differential rotation therefore sets up a *continuum* of oscillation frequencies, each corresponding to oscillation of a different surface. Oscillating over many periods, these get out of phase, and arbitrarily high gradients develop in the oscillating field. This is known as phase mixing and it leads to an extremely rapid disspation of the oscillation by magnetic diffusion (see, e.g. Heyvaerts and Priest, 1983). The time scale for the differential rotation to be damped by this process can be estimated as follows. Assume that the Alfven speed based on the B_r component varies through the star on a length scale r, the radius of the core. Then oscillations on surfaces separated by a distance L will be out of phase after $n = r/L$ oscillation periods. Damping of the oscillation becomes effective when the oscillation period r/V_A equals the magnetic diffusion time scale L^2/η, where η is the magnetic diffusivity $(cm^2 s^{-1})$. Eliminating L from these relations yields

$$n = (r V_A / \eta)^{1/2} \tag{14}$$

as the number of oscillations needed for damping the oscillation by phase mixing. With the 1 μG field (13), this is $n \approx 1.6$ with parameter values as before, and $\eta \approx 10^4$. An initial field strength of only 3 μG would be enough to bring the damping time within the age of the Sun. The end result of this procedure would be a nearly uniform rotation with a field strength similar to the initial value.

4. MAGNETIC INSTABILITIES

It would be nice if we could make a more realistic estimate of how a magnetic field would evolve than I just did in section 3. The real difficulty here is that any proposed scenario must include a demonstration that the field as it evolves is also *stable*, on time scales at least as long as the evolution time scale, or else must include the effect of instabilities. Since the time scales involved are of the order of the age of the star, this is a very strong constraint.

Consider first dynamical stability, that is the stability of the field ignoring all disspative effects (radiative exchange, viscosity and magnetic diffusion). General methods exist for determining the stability of a given field configuration (energy method, normal mode analysis, see e.g. Roberts, 1967) but in spite of this it is at present unknown whether *any* stable field configuration can exist in a star (that is, no stable ones have ever been found, see discussions in Pitts and Tayler, 1985 and Mestel, 1984 and references therein). For special classes of fields, namely the purely poloidal and the purely toroidal fields fairly conclusive results exist. For fields with both poloidal and toroidal components some limited results were derived by Markey and Tayler (1973,1974), and Tayler (1980).

4.1. Poloidal fields

The easiest is the case of a purely poloidal field, that is a field in which all field lines lie entirely within meridional planes (with respect to some axis). There are two possibilities: the field outside the surface of the star is zero or it is nonzero. In the first case, all field lines must be closed loops inside the star. Since field lines do not cross, these loops are organized as concentric nests within the meridional planes. It follows that there is (at least one) line inside the star on which the field strength vanishes (namely the line connecting the centers of the nests). Near this line the field looks like a stack of circular rings (the field of a linear pinch). This field is always unstable (Markey and Tayler, 1973, 1974 ; van Assche, Tayler and Goossens, 1982). The unstable motions move the rings in adjacent planes sideways with respect to each other. Work done against the buoyancy force vanishes because the motions can be entirely on equipotential surfaces. If the star also rotates, the situation is more complicated but since Coriolis forces do not do any work, they probably can change only the growth rate of adiabatic instabilities, not the stability conditions. This was also made plausible by explicit calculations by Pitts and Tayler (1985). In the second case (external field nonzero) note that the vacuum outside the star supports only potential fields. The external field therefore is like that of ball shaped laboratory magnet. If we cut such a magnet in half along a meridional plane, we know that the two halves will not stay in their original positions with respect to each other because the external field exerts stresses on the surface of the magnet which are not taken up by the solid state forces of the material any more. In the case of a star, a lower energy state can be reached by rotating the two halves with respect to each other. Such a displacement does no work against the buoyancy force. It gains energy from the external field because on opposite sides of the dividing line we have (almost always) equal polarities of magnetic field which repell each other (this is visualized most easily for a pure dipole field, the rotation is then 180°). This argument has also been applied in other areas (e.g. Flowers and Ruderman, in the case of neutron stars). It works only for purely poloidal fields, because the star has to be "divided" along a magnetic surface. Such a surface is warped for non-poloidal fields so that the two parts then can not be rotated without doing work against magnetic or buoyancy forces.

Concluding, poloidal fields are always dynamically unstable, with the minor caveat that the effects of rotation have not been fully investigated. It is interesting to note that the destruction process is likely to involve again a kind of "phase mixing" process. Starting with a dipole configuration a rotation by 180° transforms the external field into a quadrupole. This can again release energy by transforming into an octupole (by rotating along a different plane) and so on. At each step however the energy released is less than in the previous step, the total energy that can be extracted being limited to the total magnetic energy of the external field. This implies that the growth rate of the instability goes to zero as the instability proceeds, and that some quasi-stable field configuration might be reached. This configuration would have exceedingly small length scales in it because the complication of the field increases exponentially with the number of steps. Thus in practice such a configuration would rapidly destroy itself by even a small amount of magnetic diffusion. For the case where the external field is zero, another such length scale decreasing mechanism works near the neutral line. Thus for practical purposes it is adequate to say that a poloidal field will decay in a few times the growth time of the instability.

At high rotation speeds the Coriolis force reduces the growth rates of dynamical instabilities typically by a factor σ/Ω where σ is the growth rate without rotation (for examples see Pitts and Tayler, 1985). Hence a good estimate of the decay time of poloidal fields by dynamical instability is

$$t_{decay} = r^2\Omega/V_A^2 \qquad (\Omega \gg V_A/r). \tag{15}$$

4.2. Toroidal fields

These are fields in which the only nonzero component is in the azimuthal direction (with respect to some axis). They are of special interest because they are the fields that result from differential rotation acting on a weak initial field. Their stability has been studied by Tayler and coworkers (see Tayler 1980; Goossens and Tayler 1980; Goossens, Biront and Tayler, 1981; Pitts and Tayler 1985 and references therein). In the absence of rotation, all toroidal fields are dynamically unstable (e.g. Goossens and Tayler, 1980). The key to the instability is the behavior near the pole. Here the field has the form of a stack of circles on top of each other. This configuration is unstable to sideways displacements of the rings with respect to each other (the same configuration as in the poloidal case with field lines closed inside the star). The buoyancy force is stabilizing for most unstable disturbances, but its effect can be made arbitrarily small by choosing the wavelength in the vertical direction small enough (in the poloidal case the unstable disturbances are different). The effect of rotation (Pitts and Tayler, 1985) seems to be a reduction of the growth rates, but no change in the instability condition. That is, all toroidal fields are dynamically unstable (with again the caveat that the effect of rotation has not been fully explored).

4.3. Stability of toroidal fields with diffusion

On time scales like the age of the Sun the diffusive effects, viscosity, thermal diffusion and magnetic diffusion are likely to become important. I discuss here briefly a new result on the stability of toroidal fields including such effects (a full account will be given in Spruit and Tayler, 1987).

The introduction of dissipative effects in general introduces *additional* instabilities, which often set in before the adiabatic ones (see Acheson, 1978). These are the double diffusive instabilities. Since we know already that toroidal fields are dynamically unstable, it would seem superfluous to add the complications of diffusion. Diffusive effects however can also be stabilizing, and it turns out that this does indeed happen in the present case at low field strength. (At higher field strengths dynamical instability and a number of double diffusive instabilities occur). There is therefore a *minimum field strength* for instability of toroidal fields.

An approximate condition for stabilization by dissipative effects can be obtained by equating the growth rate without diffusion to the decay rate due to the most effective diffusion mechanism. I ignore viscosity, which is so small that it has no practical effect. Let η be the magnetic diffusivity and κ the thermal diffusivity, as before (both in $cm^2 s^{-1}$). Let

$$\Omega_{B\phi} = V_A/r\,, \tag{16}$$

the growth rate without rotation and diffusion. For large Ω the growth rate with rotation but without diffusion is of the order

$$\sigma = \Omega_{B\phi}{}^2/\Omega \qquad (\Omega \gg \Omega_{B\phi}). \tag{17}$$

A corollary of the work of Goossens (1980) is that the adiabatic instability is a local one. That is, there is a local stability condition such that if violated, there exist unstable disturbances confined to a small region near the point of violation. Under such circumstances, the correct stability condition can be obtained in a much quicker way by a local analysis, in which the medium is reduced to its minimum necessary degree of inhomogeneity, and the disturbances are plane waves. This approach can also be used for the diffusive case. It allows the inclusion of several diffusive mechanisms without severely complicating the analysis. Of course we may overlook possible instabilities that cannot be captured this way, and Fearn (1983) has in fact found such an instability. It is however related to the presence of a solid boundary, which is absent in the case of a star. In any case I think that the nature of the argument below is sufficiently general that it is not limited to local instabilities.

Consider the instability near the pole (as opposed to Acheson's 1978, 1979 analysis near the equator). Ignore first the thermal diffusion (this omission can be repaired easily at the end of the analysis). Let k_z , k_h be the wavenumbers in the vertical and horizontal directions. The fluid velocities then satisfy $v_z/v_h = k_h/k_z$. The condition that the work done against the buoyancy force be smaller than that gained from the instability is then

$$\frac{k_h}{k_z} N^2 < \sigma^2 \;\rightarrow\; k_z > k_h \frac{N^2 \Omega^2}{\Omega_{B\phi}{}^4}. \tag{18}$$

For a given horizontal length scale, the vertical length scale of the instability therefore has to be sufficiently small. If the vertical scale becomes too small however, magnetic diffusion becomes effective, destroying some of the nagnetic energy gained in the process. The magnetic diffusion rate must therefore be smaller than the adiabatic growth rate:

$$\eta k_z{}^2 < \sigma \;\rightarrow\; k_z{}^2 < \frac{\Omega_{B\phi}^2}{\eta\Omega}. \tag{19}$$

For a given k_h there exist values for k_z that satisfy both (18) and (19) only if

$$\frac{\Omega_{B\phi}^2}{\eta\Omega} > k_h^2 \frac{N^4\Omega^4}{\Omega_{B\phi}^8}. \tag{20}$$

The smallest value k_h can have is limited by the size of the star to a value of the order $1/r$. Thus for instability one must have

$$\Omega_{B\phi}^{10} > \Omega^5 N^4 \frac{\eta}{R^2}. \tag{21}$$

Due to their small vertical scale, the buoyancy force acting on the perturbations is reduced by thermal diffusion. This is an important effect because the thermal diffusivity is larger than the magnetic diffusivity. The efect can be taken into account by reducing the buoyancy frequency in (18) by the ratio of the adiabatic growth rate to the thermal diffusion rate $\sigma/(k_z^2\kappa)$. The same calculation then yields as our final estimate

$$\Omega_{B\phi}^{10} > \Omega^5 N^4 \frac{\eta}{R^2}\frac{\eta^2}{\kappa^2} \tag{22}$$

for instability, or

$$B_\phi^{10} > r^8(4\pi\rho)^5\Omega^5 N^4 \frac{\eta^3}{\kappa^2} \tag{23}$$

(Note that this holds only as long as Ω is much larger than $\Omega_{B\phi}$). For the present Sun we find

$$B > 10^4 \;\; G$$

for instability. Because of the various order of magnitude estimates made, this number should not be taken too exactly. It does however show that a surprisingly large toroidal field strength can still be stable. This is due mostly to the strongly stabilizing effect of the thermal stratification, which forces the instability to adopt such small vertical length scales that magnetic diffusion becomes important.

The importance of this result is that it predicts that kilogauss fields fields generated in a winding up process due to differential rotation may in principle be present in stars.

5. EVOLUTION OF THE FIELD (WITH INSTABILITIES)

Suppose the Sun started its life with a field in which the radial and azimuthal components were of the same order of magnitude. Depending on its initial strength the subsequent evolution of the field will follow different paths. If the field stays nearly poloidal, it will decay by instability on the timescale given by (15). The solar wind torque induces a toroidal field component. If this becomes strong, the stability of the field is governed by (23). In the following a scenario for the evolution is given which takes these factors into account.

The condition that the magnetic torque exerted on the surface of the core equals the rate of change of its angular momentum yields

$$<sin^2\theta B_\phi B_r> = -\frac{4\pi}{3}k^2\rho\dot{\Omega}r^2, \tag{24}$$

where k^2 is the gyration constant $(I = k^2MR^2)$, ρ the mean denstity of the core and the average is over the surface of the core. Suppose that the torques exerted by the solar wind depend on Ω only, such that

$$\dot{\Omega} = -\beta\Omega^n. \tag{25}$$

For $n=3$, $\beta = 5 10^{-7}$ this yields Skumanich's (1972) $t^{-1/2}$ law for $\Omega(t)$ with $\Omega(0) \equiv \Omega_0 = 30\Omega_{sun}$. Inserting into (24) this yields, neglecting factors of order unity:

$$B_\phi B_r \approx \beta\rho\Omega^{-n}r^2. \tag{26}$$

This determines the strength of the field that is needed to transmit the solar wind torque. If the initial field strength is less than this, it will be wrapped up by differential rotation until B_ϕ is large enough to satisfy (26). Since the rotation is always fast compared with the spindowm time scale, the wrapping up is for all but the weakest initial fields practically instantaneous, and (26) holds to a good approximation at all times. At t=0 ($\Omega = 10^{-4}$, $\rho = 3$) we have

$$B_\phi B_r \approx 4\ 10^3\ G^2. \tag{27}$$

At t=0 we get $B_\phi < B_r$ if

$$B_r(0) > B_1 \approx 60\ G. \tag{28}$$

No significant wrapping up takes place in this case. B_r decreases however because the field, being mainly poloidal, is unstable. Its decay rate is (cf. section 4)

$$\frac{\dot{B}_r}{B_r} \approx \frac{B_r^2}{4\pi\rho r^2\Omega}. \tag{29}$$

This can be integrated with the known $\Omega(t)$ to get $B_r(t)$:

$$B_r^{-2} = B_r^{-2}(0) + (2\pi\rho r^2\Omega_0)^{-1} t_0 \frac{n}{n-1}[(1+t/t_0)^{n/(n-1)}-1]. \tag{30}$$

One can then verify that (26) yields a B_ϕ that initially is less than B_r, and asymptotically reaches a value similar to B_r during the spindown. Thus for an initial field above about 100 G the field is never wound up significantly.

From the form of Equation (30) one sees that the present value of the field is *independent* of $B_r(0)$ provided that

$$B_r(0) \gg B_2 \equiv (8\pi\rho r^2\Omega_0)^{1/2}(t_{sun}\frac{n}{n+1})^{-1/2}(t_{sun}/t_0)^{-1/2\ n/(n-1)} \approx 3\ G\ , \tag{31}$$

where $t_0 = (2\Omega_0\beta)^{-1}$ and t_{sun} is the age of the sun. The present value of the field strength in this case would also be of the order B_2 i.e. a few Gauss. For lower initial field strengths the present value would be lower. For initial fields less than $B_1 \approx 60 G$ however, we found that a toroidal field will develop first which is larger. It is not clear that our estimate of the decay time of B_r by instabilities is valid for such a field, so the only thing we can say on the basis of the above is that *for all initial field strengths above about 100G the field decays to a configuration with roughly equal toroidal and poloidal components, with a present strength of a few Gauss.* Note that this is about the value where the field starts interfering with the hydrodynamic processes (cf. section 3.1).

It takes a finite amount of time for the torque exerted at the surface to travel through the core (with the Alfven speed). Thus there is a time delay between the surface rotation rate and that in the center. From this we can easily predict the difference in rotation rate:

$$\Delta\Omega \approx \frac{d\Omega}{dt}\frac{r}{V_{Ar}}. \tag{32}$$

For $B_r = 3\ G$ and the $t^{-1/2}$ law this yields for the present Sun:

$$\Delta\Omega/\Omega \approx 10^{-6}.$$

This illustrates how effective even a field of a few Gauss is in transporting angular momentum.

The scenario becomes more complicated if the initial field strength is less than 100G. If the initial field is sufficiently weak such a strong toroidal field may develop that at some point in time the instability condition (23) is satisfied. During the subsequent spindown this instability may persist (because torques amplifying the toroidal field are still present). The present Sun could then have a mainly toroidal field in its core which would be in a state of marginal stability, of the order of $10^4\ G$. The initial field needed to create this situation is very low however (on the order of a few μG), and this scenario is perhaps not so realistic.

References

Acheson, D.J.: 1978, *Phil. Trans. R. Soc. London* **289A**, 459

Acheson, D.J.: 1979, *Solar Phys.* **62**, 23

Baglin, A., Morel, P.J. and Schatzman, E.: 1985 *Astron. Astrophys.* **149**, 309

Boesgaard, A.M., and Tripicco, M.J.: 1986, *Astrophys. J.* **302**, L49

Brown, T.M.: 1985, *Nature* **317**, 591

Cayrel, R., Cayrel de Strobel, G., Campbell, B., Däppen, W.: 1984, *Astrophys. J.* **283**, 205

Drazin, P., and Reid, W.H.: 1981, *Hydrodynamic Stability*, Cambridge University Press, Cambridge, U.K.

Duncan, D.K.: 1981 *Astrophys. J.* **248**, 651

Duvall, T.L., and Harvey, J.W.: 1984, *Nature* **310**, 19

Fearn, D.R.: 1983, *Geophys. Astrophys. Fluid Dyn.* **27**, 137

Flowers, E., and Ruderman, M.A.: 1977, *Astrophys. J.* **215**, 302

Goossens, M.: 1980, *Geophys. Astrophys. Fluid Dyn.* **15**, 123

Goossens, M. Biront, D. and Tayler, R.J.: 1980, *Astrophys. Sp. Sci.* **75**, 521

Goossens, M., Poedts, S. and Hermans, D.: 1986, *Solar Phys.* **102**, 51

Goossens, M. and Tayler, R.J.: 1980, *Mon. Not. R. Astron. Soc.* **193**, 833

Heyvaerts, J. and Priest, E.R.: 1983, *Astron. Astrophys.* **117**, 220

Joseph, D.D.: 1976, *Stability of Fluid Motions*, Vols. I and II, Springer, New York

Kippenhahn, R., and Thomas, H.-C.: 1981, in *Fundamental Problems in the Theory of stellar Rotation*, eds. D. Sugimoto, D.Q.Lamb and D.N. Schramm (IAU Symp.**93**), Reidel, Dordrecht, Netherlands, p.237

Knobloch, E., and Spruit, H.C.: 1983, *Astron. Astrophys.* **125**, 59

Law, W.Y., Knobloch, E., Spruit, H.C.: 1984, in A. Maeder and A. Renzini (eds.), *Observational Tests of stellar Evolution Theory*, D. Reidel, Dordrecht, p.523

Markey, P. and Tayler, R.J.: 1973, *Mon. Not. R. Astron. Soc.* **163**, 77

Markey, P. and Tayler, R.J.: 1974, *Mon. Not. R. Astron. Soc.* **168**, 505

Mestel, L.: 1984, *Astron. Nachr.* **305**, 301, Erratum **306**, 260 (1985)

Monin, A.S. and Yaglom, A.M.: 1971, *Statistical Fluid Mechanics*, MIT Press, Cambridge, Mass. Chapter 1

Orszag, and Kells, 1980, *Phys. Fluids* **96**, 159

Pedlosky, J.: 1979, *Geophysical Fluid Dynamics*, Springer Verlag, New York

Pitts, E. and Tayler, R.J.: 1985, *Mon. Not. R. Astron. Soc.* **216**, 139

Press, W.H.: 1981, *Astropys. J.* **245**, 286

Roberts, P.H.: 1967, *An Introduction to Magnetohydrodynamics,* American Elsevier Publishing Comp., New York

Shibahashi, H.: 1980, *Publ. Astron. Soc. Japan* **32**, 341

Skumanich, A.: 1972, *Astrophys. J.* **171**, 565

Soderblom, D.R.: 1983, *Astrophys. J. Supp.* **53**, 1

Spruit, H.C.: 1984, in T.D. Guyenne (ed.) *The Hydromagnetics of the Sun* (ESA SP-220), European Space Agency, Noordwijk, Netherlands, p.21

Spruit, H.C. and Knobloch, E.: 1984, *Astron. Astrophys.* **132**, 89

Spruit, H.C., Knobloch, E., and Roxburgh, I.W.: 1983, *Nature,* **304**, 320

Spruit, H.C., and Tayler, R.J.: 1987, in preparation

Tayler, R.J.: 1980, *Mon. Not. R. Astron. Soc.* **191**, 151

Thévenin, F., Vauclair, S. and Vauclair, G.: 1986, *Astron. Astrophys.* **166**, 216
van Assche, W., Tayler, R.J. and Goossens, M.: 1982, *Astron. Astrophys.* **109**, 166
Zahn, J.-P.: 1974, in P. Ledoux, A. Noels and R.W. Rodgers (eds.) *Stellar Instability and Evolution*, (IAU Symp. 59), Reidel, Dordrecht, p.185
Zahn, J.-P.: 1983, in A.N. Cox, S. Vauclair and J.-P. Zahn (eds.), *Astrophysical Processes in upper Main Sequence Stars*, Geneva Observatory, Switzerland, p.255

TURBULENT TRANSPORT IN THE RADIATIVE ZONE OF A ROTATING STAR

J.-P. Zahn
Observatoires du Pic-du-Midi et de Toulouse
14, avenue E. Belin
31400 Toulouse
France

ABSTRACT. Hydrodynamical instabilities that occur in the radiation zone of a differentially rotating star may produce the mild turbulent mixing which is required to explain various observations. We discuss here a consistent scheme in which the energy which sustains the turbulent motions originates in the thermal imbalance caused by the rotation. An approximate expression is derived for the vertical diffusion coefficient and the consequences of this process are compared with those of the classical meridional circulation.

1. INTRODUCTION

Many stars display composition anomalies which seem to indicate that their radiative interior is the seat of some mild mixing process. Such is the underabundance of ^{7}Li in solar-type stars, whose smooth variation with spectral type cannot be explained only by convective undershoot from the outer convection zone (Cayrel *et al.* 1984, Baglin *et al.* 1985).

The secular increase of the ^{3}He/^{4}He ratio in the solar wind, as demonstrated by the comparison of old (aubritic) meteorites with both recent lunar samples and the present solar wind (Geiss 1971, Geiss *et al.* 1972), is also hinting that ^{3}He is transported towards the surface from the deep interior of the Sun.

Massive stars too show an overabundance of ^{13}C and N as they evolve off the main sequence (Lambert *et al.* 1980), and those elements must have been brought to the surface from the stellar core.

Schatzman (1969, 1970) was the first to interpret such observations by invoking a mild mixing of the stable radiation zone, which he ascribed to small amplitude turbulent motions. He later estimated that a turbulent diffusivity about 100 times larger than the molecular diffusivity would be adequate to explain the solar abundance anomalies (Schatzman 1977). This implies that the Reynolds number characterizing those turbulent motions is of order 100, and in Schatzman's original treatment that number was assumed constant throughout the star, an unknown parameter to be adjusted by comparison with the observations. Further work by

B. R. Durney and S. Sofia (eds.), The Internal Solar Angular Velocity, 201–212.

Genova and Schatzman (1979), by Schatzman and Maeder (1981a), and by Bienaimé *et al.* (1983) proceeded along the same line.

If one seeks a physical and quantitative description of this turbulent mixing, one is led immediately to ask the following questions:

- 1. which are the hydrodynamical instabilities responsible for this turbulence?
- 2. which is the energy source that maintains it?

The observations cannot answer directly those questions, and therefore any theoretical explanation will be necessarily based on assumptions, whose plausibility may only be established *a posteriori* - a familiar procedure in astrophysics. To be considered as such, a theory must consistently derive the properties of the turbulence from the assumptions made, and in particular its vertical diffusivity. It is the effects of that diffusivity, on the structure and on the evolution of the star, which will ultimately be compared with the observations, thus permitting the validation (or the invalidation!) of the theory.

In the scheme that we have proposed (Zahn 1983), the energy source is the thermal imbalance due to the rotation of the star, and the instabilities are those which are caused by the differential rotation. Let us examine those processes, in that order.

2. THERMAL IMBALANCE IN A ROTATING STAR

As it is well known since Von Zeipel (1924), there is a conflict, in the radiative zone of a rotating star, between the isotropy of the transfer of energy and the oblateness imposed by the centrifugal force. The result is that the energy flux is no longer conservative, as it is in a non rotating, spherical star. (Exceptions exist for some specific distributions of the angular velocity, but we shall not consider them here.) Heat sources and heat sinks appear, in which the energy is exchanged on a rate defined by the divergence of the energy flux; per unit volume it is thus $\nabla \cdot (\chi \nabla T)$, where χ is the radiative conductivity, and ∇T the local temperature gradient. Given the rotation law $\Omega(r, \theta)$ (r is the radial coordinate and θ the colatitude), and the internal structure of the star, one knows well how to calculate this rate at every point (Sweet 1950, McDonald 1972). For the present purpose, it will suffice to give a rough estimate

$$\epsilon = \left| \frac{1}{\rho} \nabla \cdot (\chi \nabla T) \right| \sim K \Omega^2, \tag{1}$$

where the energy rate is now expressed per unit mass, ρ being the density, and $K = \chi / \rho C_P$ the radiative diffusivity; Ω is a typical value for the angular velocity. In the classical theory, which we owe to Eddington (1925) and Sweet (1950), these energy sources and sinks give rise to a meridional circulation, whose (very slow) velocity $\mathbf{U}$ satisfies the following equation, expressing the conservation of heat in a stationary regime:

$$\rho T \, \mathbf{U} \cdot \nabla S = \nabla \cdot (\chi \nabla T) \, . \tag{2}$$

Here ∇S is the gradient of specific entropy, whose direction is very close to the vertical, and whose magnitude may be written $g(\nabla_{ad} - \nabla_{rad})/T$, in terms of the logarithmic gradients which are commonly used in stellar structure theory

($\nabla = \partial \ln T / \partial \ln P$). An approximate expression for the vertical component of **U** is therefore

$$U_r \sim \left(\frac{K\Omega^2}{g}\right)(\nabla_{ad} - \nabla_{rad})^{-1}, \tag{3}$$

which will serve later on (g is the local gravity). The time scale of this circulation is the Eddington-Sweet time $t_{ES} = R/U_r = (R^2/K)(\Omega^2 R/g)^{-1}$, R being the radius of the star.

Such a meridional circulation will advect angular momentum, and it will thus create differential rotation in a star which rotates initially as a solid body. It is this differential rotation that we invoke as the cause of the instabilities mentioned above.

3. INSTABILITIES DUE TO DIFFERENTIAL ROTATION

Differential rotation gives rise to various instabilities, and a complete description of these is out of the cope of this presentation. We refer the reader instead to some recent reviews on the subject: Tritton and Davies (1981), Knobloch and Spruit (1982), Zahn (1983, 1984). Here we shall just recall some salient properties of these instabilities.

They can be classified into two broad families, according to the type of energy that feeds them. The first is that of the Rayleigh-Taylor and related instabilities; they take their energy from the potential energy of the medium, which is stratified under the combined action of the gravity and and of the centrifugal force. The second family is that of the shear instabilities, which draw their energy from the kinetic energy of the flow (here the non-uniform rotation).

3.1. Baroclinic instabilities

The baroclinic instabilities are, in the first family mentioned above, the most likely to occur in a differentially rotating star, provided that the rotation rate is not constant on cylinders. In such a star, the surfaces of constant pressure (i.e. the level surfaces) do not coincide with the surfaces of constant density, a property which is called baroclinicity. Consequently, there is also an angle between the level surfaces and the surfaces of constant specific entropy, which in a region of uniform chemical composition is given by

$$\sin\beta = N^2 \frac{\partial \varpi \Omega^2}{\partial z}, \tag{4}$$

ϖ and z being the cylindrical coordinates in the radial and axial directions. N is the buoyancy frequency, which measures the strength of the entropy stratification (and hence its stability):

$$N^2 = \frac{g}{H_P}(\nabla_{ad} - \nabla_{rad}), \tag{5}$$

H_P being the pressure scale height.

At first view, it seems that when a blob of stellar matter is displaced within the small wedge between two such surfaces, it will acquire kinetic energy: for example, when it rises sidewards, it will be hotter than its surroundings, and hence lighter, and will thus keep accelerating. One is thus tempted to conclude that a baroclinic star is always unstable.

However, in such a gedankenexperiment one cannot choose an arbitrary perturbation: it must necessarily ensure the conservation of angular momentum, and furthermore it must satisfy the boundary conditions. These constraints are the most severe for axisymmetric perturbations, which in a realistic star are found to be unstable only if one includes the dissipative effects. Since stellar matter diffuses heat on a much faster rate than momentum, double diffusive instabilities arise, which in the astrophysical context have been described by Goldreich and Schubert (1967), and Fricke (1968). More recently, Knobloch and Spruit (1983) have included the diffusion of chemical species; their generalization leads to the conclusion that a baroclinic star of moderate differential rotation is indeed liable to linear, axisymmetric instabilities, whose growth rate is related to the local thermal time scale. But the nonlinear fate of all these multidiffusive instabilities is still a matter of debate, and their efficiency is questionable; it is likely that in most cases they will be obliterated by other instabilities, proceeding on a much faster time scale.

For non-axisymmetric perturbations, the constraint imposed by the conservation of angular momentum is much milder. The theory has been worked out in detail for an ideal fluid (no thermal diffusion, no viscosity). In such a fluid, the linear instability criterion involves both the strength of the differential rotation in the vertical direction, and its profile, much as in the case of inviscid shear instabilities; for a complete account on this problem, see Pedlosky (1979). Some specific profiles have been tested by Spruit *et al.* (1983), and by Zahn (1983). They conclude that in a stellar radiation zone, linear baroclinic instabilities occur only in the vicinity of its border with the adjacent convection zone, because the stabilizing effect of the entropy stratification is there sufficiently weakened.

Let us stress that the theory of the baroclinic instabilities has barely gone beyond its linear phase, in spite of the efforts made by many geophysicists, who were the first to explore the subject. It may well be that such instabilities, but of finite amplitude, are as likely to occur as the nonlinear shear instabilities which will be considered next.

3.2. Shear instabilities

Those instabilities may be characterized by their ability of extracting vorticity from a mean laminar flow, to build turbulent vortices. As stated above, they draw their energy from the kinetic energy of the mean flow. We shall recall here only their main properties (for references on this vast subject, see for instance Zahn 1983).

Whereas the linear instability requires some specific flow profile, for instance one presenting an inflexion point (in the inviscid limit), one observes that all shearing flows become unstable as soon as the Reynolds number exceeds some critical value Re_c. This number compares the rate at which a perturbation draws energy from the mean flow with the rate at which it is dissipated by viscosity; it is evaluated in terms of the typical length L and velocity U characterizing the mean flow: $Re = LU/\nu$, ν being the viscous diffusivity. The critical Reynolds number at which the instability appears depends somewhat on the profile of the flow and on the boundary conditions; typically, it is of order 1000.

In a rotating medium, the Coriolis force, although it acts as a restoring force, is unable to inhibit the shear instability. This is well established both by theoretical work (Johnson 1963) and by laboratory experiments (Rabaud and Couder 1983).

Consequently, a differentially rotating flow will become unstable, at least for finite amplitude perturbations, under conditions which are similar to those applying on a plane-parallel shear flow.

In the absence of magnetic field, the only force which can stabilize a shear flow is the buoyancy, acting in a stratified medium (a density stratification in an incompressible fluid, or an entropy gradient in a compressible one). But this stabilizing effect can only operate on a vertical shear; furthermore, it is seriously weakened by thermal dissipation, since in stars heat diffuses much faster than momentum . One finds that the instability sets in when the vertical gradient of the angular velocity exceeds a critical value given by

$$\cos\theta \frac{\partial \ln \Omega}{\partial \ln r} \geq \left(\frac{\nu}{K} Re_c\right)^{\frac{1}{2}} \left(\frac{N}{\Omega}\right), \tag{6}$$

which can be much lower than that given by the classical Richardson criterion, namely $2(N/\Omega)$ (Zahn 1975).

The most important conclusion we shall draw at this stage is that a flow which is sheared in the horizontal direction is unstable whenever its Reynolds number is supercritical. This condition is easily achieved in a star, due to its large size. Thus even a slight departure from uniform rotation, on horizontal surfaces, will be liable to shear instabilities.

Those instabilities generate turbulent motions whose properties we shall now examine.

4. PROPERTIES OF THE TURBULENCE

4.1. Anisotropy

The turbulent motions produced by the instabilities arising in a horizontal shear are predominantly horizontal, due to their origin, and to the fact that such motions are not hindered by the stratification. (The same would occur if the most efficient instabilities were of the baroclinic type considered above.) But this two-dimensional character will presumably hold only for the large scales at which the energy is injected from the mean flow. Due to nonlinear interaction, a whole spectrum of scales will unfold in the stationary state, from the largest allowed by the size of the star, to the smallest at the viscous cut-off.

The largest scales will result from the pairing of vortices, as is observed in two-dimensional turbulence; they will play a major role in the transport of vorticity over level surfaces, and they will keep those in a state of moderate differential rotation.

We shall assume that the smallest scales are three-dimensional, having lost the memory of their origin, and that all the energy drawn from the mean flow will ultimately be channelled into a classical inertial cascade, to be dissipated by viscous friction at the Kolmogorov scale. Assuming this, we neglect the energy which is possibly emitted and transported elsewhere in the form of waves (of the Rossby type, for instance).

The flow diagram of the energy is sketched in Fig. 1. A fraction of the thermal energy appearing in the heat sources and sinks, due to the thermal imbalance of the rotating star, serves to drive the meridional circulation and hence to create some

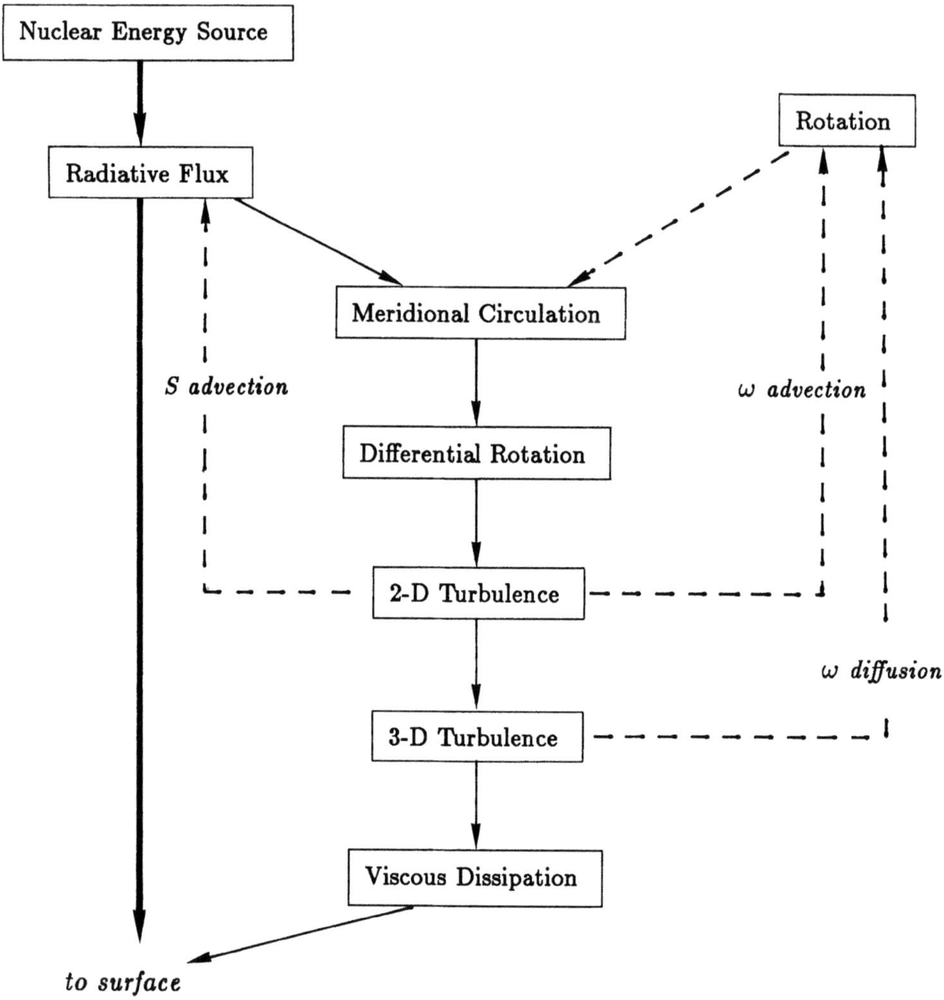

Figure 1. Flow diagram of the energy, also showing the main feedback mechanisms discussed in section 6 (transport of entropy S and of vorticity ω).

amount of differential rotation; it is thus converted into kinetic energy. This energy is injected into the turbulent motions, and it is dissipated at the Kolmogorov scale.

4.2. Strength

The strength of the turbulent motions is determined by the energy injection rate ϵ_t, which may be evaluated as follows. Its upper limit is obviously the thermal energy available in the heat sources/sinks, for which we gave earlier a rough estimate (Eq. 1). Therefore we shall write it in the form

$$\epsilon_t = \eta K \Omega^2, \tag{7}$$

where the coefficient η measures the efficiency of the conversion of heat into kinetic energy, and is thus by definition smaller than unity. Let us estimate this key factor. In principle, it should be possible to derive it from the second law of thermodynamics, but it is not too obvious here to define the closed system required for its application. We prefer to use an explicit scenario: from the conservation of angular momentum

$$\frac{\partial}{\partial t}\varpi^2\Omega + \mathbf{U}\cdot\nabla\left(\varpi^2\Omega\right) = \Gamma, \tag{8}$$

where Γ is the torque exerted per unit mass by the turbulent motions, we can calculate the rate of variation of the rotational energy:

$$\frac{\partial}{\partial t}\frac{1}{2}\left(\varpi\Omega\right)^2 + \Omega\,\mathbf{U}\cdot\nabla\left(\varpi^2\Omega\right) = \Omega\,\Gamma. \tag{9}$$

In a stationary state, the advection term balances the right hand side, which is precisely the energy injection rate into the turbulence; approximating ∇ by $1/r$ and ϖ by r, using the crude estimate (Eq. 3) for the meridional velocity, and averaging over level surfaces, we get the following expression for the efficiency factor η:

$$\eta = \min\left[1, \frac{\Omega^2 r}{g}\left(\nabla_{ad} - \nabla_{rad}\right)^{-1}\right]. \tag{10}$$

where we have imposed in a rather artificial way that η not exceed unity.

Therefore the velocities v and the scales l of the three-dimensional tail of the turbulence follow the similarity law

$$v^3/l = \eta K \Omega^2, \tag{11}$$

which is the transposition in the physical space of the Kolmogorov power law, usually written in Fourier space (see for instance Landau and Lifschitz 1953).

4.3. Vertical diffusivity

It remains to evaluate the vertical diffusivity of the turbulence, which according to this law is due mainly to the largest of the three-dimensional scales. We need therefore to identify the scale below which the motions are predominantly three-dimensional, and will therefore have a vertical component. Having achieved this, we may evaluate the turbulent diffusivity as $D_v = \frac{1}{3}vl$.

When the cause of anistropy is the stratification, the scale at which the transition occurs, known as the Ozmidov scale, is such that the turn-over rate of the eddies, v/l, is of the order of the buoyancy frequency N (Eq. 5). In a homogeneous star, however, the restoring effect is suppressed through the very effective thermal diffusion; the Péclet number vl/K of the relevant scales is indeed much smaller than unity.

Therefore we shall invoke the other restoring force which is present in rotating stars, namely the Coriolis force, to play this crucial role of inhibiting the three-dimensional motions. The existence of such a transition from three to two-dimensional turbulence, when v/l is of the order of Ω, has been demonstrated in a very nice laboratory experiment by Hopfinger *et al.* (1982). Numerical simulations in progress seem to confirm that the same transition occurs in the opposite sense, from two to three dimensions, when the energy is injected in large two-dimensional eddies (Herring 1986).

Combining $v/l = \Omega$ and the power law (Eq. 11), we get the following value for the vertical turbulent diffusivity

$$D_v = \eta K = K \min \left[1, \frac{\Omega^2 r}{g} \left(\nabla_{ad} - \nabla_{rad} \right)^{-1} \right]. \tag{12}$$

Let us stress that this expression is a very crude estimate, since in addition to the approximations made, we have neglected several factors of order unity (such as the usual coefficient $\frac{1}{3}$ in the formula above).

4.4. Suppression by a chemical composition gradient

In the presence of a molecular weight gradient $\nabla_\mu = \partial \ln \mu / \partial \ln P$, the buoyancy force partly remains, and its strength is measured by a modified buoyancy frequency N_μ, where $N_\mu^2 = g\ \nabla_\mu / H_P$. The μ-gradient is assumed to point downwards (i.e. $\nabla_\mu > 0$); otherwise it would be liable to a double diffusive instability related to the Rayleigh-Taylor instability (see Turner 1974).

Such a composition gradient will hinder the vertical motions if $N_\mu > v/l$, and the vertical turbulent diffusion will therefore be completely suppressed if this condition holds even for the Kolmogorov scale, where $vl = \nu$. This is the case when $N_\mu^2 > \epsilon_t / \nu$, or equivalently

$$\frac{\partial \ln \mu}{\partial \ln r} > \frac{K}{\nu} \frac{\Omega^2 r}{g} \min \left[1, \frac{\Omega^2 r}{g} \left(\nabla_{ad} - \nabla_{rad} + \nabla_\mu \right)^{-1} \right]. \tag{13}$$

Due to the production of helium in its core, in an evolving star this condition is easily achieved, if its rotation rate is not too large; such a star will therefore develop an inhomogeneous structure, and it will follow the classical evolution track to the red giant branch. However, fast rotators may well be mixed by the process considered here; they may thus evolve homogeneously, and be observed as so-called "blue stragglers."

4.5. Possible improvements

Refinements of this scheme are quite possible, and they are certainly desirable, but we shall see that one cannot completely avoid yet some sort of parametrisation.

For instance we can use the correct value of the meridional velocity $\mathbf{U}$, instead of the rough approximation of Eq. 3, and calculate the local input of kinetic energy into the differential rotation, and hence into the turbulence. For this purpose, we need the local value of the angular velocity, which itself evolves, due both to vertical diffusion and to the advection of angular momentum by the meridional circulation.

In a stationary regime, the angular momentum advected into a horizontal layer serves only to modify its mean angular velocity Ω_0, since the horizontal turbulent motions will keep the rotation close to a spherical one. Neglecting the oblateness of the star, and splitting the angular velocity into its spherical and non spherical parts, $\Omega(r,\theta) = \Omega_0(r) + \delta\Omega(r,\theta)$, we may write the conservation of angular momentum of a spherical layer as

$$\frac{\partial}{\partial t}\Omega_0(r) \iint \varpi^2 dS = -\iint \mathbf{U} \cdot \nabla \left(\Omega_0(r,\theta)\varpi^2\right) dS. \tag{14}$$

This expression permits to calculate the secular variation of Ω_0 due to the meridional circulation. For completeness, another term should be added to take into account the vertical diffusion of momentum. One has also to check if the vertical shear remains stable, according to the criterion given in Eq. 6; otherwise, the gradient of Ω_0 must be replaced by the critical one.

Subtracting the kinetic energy of the mean flow from that which is advected locally into the layer, we get an improved expression for the turbulent energy input, which to first order is

$$\epsilon_t(r) = -\int_0^1 \delta\Omega(r,\theta)\, \mathbf{U} \cdot \nabla \left(\Omega_0(r)\sin^2\theta\right) d(\cos\theta). \tag{15}$$

The meridional velocity U is easily derived from the mean rotation rate $\Omega_0(r)$ through the classical theory; for a spherical rotation law its vertical component takes the simple form $U_r(r,\theta) = -U_2(r)P_2(\cos\theta)$, $U_2(r)$ being a positive function of r when the star is homogeneous and the departures from solid rotation not too large (McDonald 1972).

To proceed further, we need a better knowledge of the transport of angular momentum through the horizontal turbulence, which determines the differential rotation $\delta\Omega/\Omega$. For want of anything better, we may assert $\delta\Omega = -C\,\Omega_0 P_2(\cos\theta)$, the coefficient C being left as an adjustable parameter, with plausible values not much greater than 1/10. The sign is chosen such that C is positive when the equator rotates faster. One sees that we have made no attempt to describe the latitudinal variation of ϵ_t; it will be considered here as a function of depth only.

A further improvement is to take into account the heat flux carried by the large scale horizontal motions, which does not vanish here because of the baroclinicity. Although only the largest scales, for which the thermal diffusion is negligible, will contribute to this transport, it may provide the feedback which prevents the efficiency factor η from exceeding unity.

Using the value of the baroclinic angle β (Eq. 4), one can estimate the horizontal heat flux in the diffusion approximation:

$$F_h \sim vl\, H_P \frac{\partial\Omega^2}{\partial z}. \tag{16}$$

This flux is directed from pole to equator in an almost spherically rotating star, and it tends to restore the thermal equilibrium. It will fully succeed if it is able to balance the divergence of the radiation flux, hence if $F_h \sim \varpi\rho K\Omega^2$, or

$$\frac{vl}{K} H_P \frac{\partial \ln\Omega^2}{\partial z} \sim 1. \tag{17}$$

Since the differential rotation is of order C (the coefficient introduced above), we see that only the largest scales, for which the Péclet number $Pe = vl/K$ is significantly larger than one, will contribute to this thermal feedback.

5. COMPARISON WITH THE OBSERVATIONS

Following Schatzman (1977), one can define a Reynolds number associated with the vertical diffusivity estimated above : $Re^* = D_v/\nu$. Applying to the Sun the expression (12) for D_v, and taking for the angular velocity Ω the average of that which is observed on the surface, one gets $Re^* = 15$ and 52 at the depths where respectively the beryllium and the lithium are destroyed. These values are of the order of those required to explain the observed abundance anomalies, according to Schatzman and Maeder (1981a). However, due to the crudeness of the derivation of the turbulent diffusivity, this agrement was taken merely as an encouragement to further proceed along these lines (Zahn 1983).

Since then, Baglin *et al.* (1985) have performed several evolution calculations of lower main sequence stars, in order to interpret the observed depletion of lithium in the Hyades and in the Sun. They test both Schatzman's prescription of a constant Re^* and the present proposal, and conclude that the strong increase of the vertical diffusivity predicted by the latter near the convection zone is needed to explain the observations. On the other hand, the comparison between the Sun and the Hyades stars indicates that there is a milder dependence of the turbulent viscosity on the rotation than that given by a Ω^2 law (Eq. 12, for $\eta < 1$). The fact that they did not cut the efficiency factor η at 1 near the convection zone may well account for this discrepancy.

Another series of calculations by Lebreton and Maeder (1986) dealt mainly with the helium content of the Sun. They make the assumption that the profile of the spherical rotation $\Omega_0(r)$ is everywhere just at the threshold of the multidiffusive baroclinic instabilities described by Knobloch and Spruit (1983); in that case, and according to Eq. 10, Re^* follows the simple law

$$Re^* \sim \alpha^{-2} \frac{8r}{H_P} \qquad \text{where} \qquad \alpha = -\frac{\partial \ln \Omega_0}{\partial \ln r}. \tag{18}$$

They adjust this parameter α to best represent the observations.

They conclude that the enrichment of ^{3}He suggested by Geiss (1971) is well accounted for by such a turbulent diffusion, with $\alpha \sim 1$. They verify also that the vertical diffusion is inhibited by the μ-gradient in the central region of the Sun; this agrees with the heliosismic observations, but cannot account for the low neutrino flux, as was suggested earlier (Schatzman and Maeder 1981b).

Further work with this same expression for Re^* concerned the evolutionary tracks of low mass stars (Lebreton 1986). Those tracks were shown to better fit the observed cluster isochrones when they were computed with a vertical diffusion obeying the law above (Eq. 18), rather than with a constant Re^* or no mixing at all.

Finally, another positive indication comes from the observations of the rotational splitting of the solar p-modes. Although the question is far some settled, it is now believed that the most plausible model is one in which the differential rotation of the Sun resembles that seen at the surface throughout the convection zone, but becomes constant on spherical surfaces within the radiative interior (Brown and Morrow 1986).

6. DISCUSSION

One may wonder how the turbulent diffusion considered here differs from the classical Eddington-Sweet circulation. Indeed, they have the same characteristic time scale $t_{ES} = (R^2/K)(\Omega^2 R/g)^{-1}$, which is prohibitively large in the case of the Sun, and can compete with the nuclear evolution time only in fast rotating stars (Sweet 1950).

However, diffusion operates in a quite different way than a circulation, and its effects may be sensed in a time which is much shorter than t_{ES}, since they strongly depend on the steepness of the relevant gradient (of chemical composition, or of angular velocity). The detailed calculations mentioned above are demonstrations of this.

No selfconsistent model of the Eddington-Sweet circulation has been worked out yet, since none, so far, has taken the secular advection of angular momentum in account. One exception is that of Tassoul and Tassoul (1978), which postulates a very strong turbulent diffusivity to keep the rotation almost constant throughout the star.

In the model proposed here, the differential rotation over horizontal surfaces is kept at a moderate level by the horizontal turbulence; whereas the depth-dependent mean angular velocity $\Omega_0(r)$ is modified both through turbulent diffusion, and through advection by the meridional circulation.

It is often argued that solid rotation is imposed by the presence of the slightest magnetic field, which will wind up until its azimuthal component becomes strong enough to react back on the differential rotation (Mestel 1953, Mestel and Moss 1986). This may indeed be the case, but to settle this important question, it needs to be checked whether the hydrodynamical instabilities which are triggered by the differential rotation would not destroy the magnetic field by ohmic dissipation before it reaches a sufficient strength. Further heliosismic observations may soon provide the answer.

In the classical theory, the major feedback is due to the μ-gradients built up by nuclear burning (Mestel 1953). In the present diffusion model, this feed-back is due primarily to the horizontal advection of heat, which restores the thermal equilibrium, even in the absence of inhomogeneities.

Whereas in the classical model the main effect of a vertical μ-gradient is to cancel the vertical component of the meridional velocity, and thus to split the circulation in superposed cells, it is to suppress the vertical diffusion in this new model.

We thus see that the differences between circulation and diffusion are significant enough to be used as observational checks, and the first of those which have been carried out look indeeed rather encouraging, as we have just seen. We are fully aware, however, that in its present state the scheme presented here is far from satisfactory, due mainly to the large uncertainties concerning the turbulent motions. But are not astrophysicists constantly obliged to use such phenomenological models, which like this which is presented here involve one or more adjustable parameters, whenever they dare to deal with fluid dynamical problems?

REFERENCES

Baglin, A., Morel, P., Schatzman, E.: 1985, *Astron. Astrophys.* **149**, 309
Bienaimé, O., Maeder, A., Schatzman, E.: 1983, *Astron. Astrophys.* **131**, 316
Brown, T.M., Morrow,C.A.: 1986, (preprint, and comm. at this coll..)
Cayrel, R., Cayrel de Strobel, G. Campbell, B., Däppen ,W.: 1984, *Astrophysical J.* **283**, 205
Eddington, A.S.: 1925, *Observatory* **48**, 78
Fricke, K.J.: 1968, *Z. Astrophys.* **68**, 317
Geiss, J.: 1971, In Solar Wind, NASA SP-308, 559
Geiss, J., Bühler, F., Cerutti, H., Eberhardt, P., Filleux, C.: 1972 , Apollo 16 Preliminary Science Report, NASA SP-315, 14-1
Genova, F., Schatzman, E.: 1979, *Astron. Astrophys.* **78**, 323
Goldreich, P.A., Schubert, C.: 1967, *Astrophys. J.* **150**, 571
Herring, J.: 1986, (private communication))
Hopfinger, E.J., Browand, F.K., Gagne, Y.: 1982, *J. Fluid Mech.* **125**, 505
Johnson, J.A.: 1963, *J. Fluid Mech.* **17**, 337
Knobloch, E., Spruit, H.C.: 1982, *Astron. Astrophys.* **113**, 261
Knobloch, E., Spruit, H.C.: 1983, *Astron. Astrophys.* **125**, 159
Lambert, D.L., Downing, J.F., Siverton, S.: 1980, *Astrophys.J.* **235**, 114
Landau, L., Lifschitz, E.: 1953, Hydrodynamika (English translation 1959; Pergamon edit., London)
Lebreton, Y.: 1986, Thèse de doctorat (Univ. Paris)
Lebreton, Y., Maeder, A.: 1986, *Astron. Astrophys.* **161**, 119
McDonald, B.E.: 1972, *Astrophys. Space Sci.* **19**, 309
Mestel, L.: 1953, *M.N.Roy. Astron. Soc.* **113** 716
Mestel, L., Moss, D.L.: 1986 *M.N.Roy. Astron. Soc.* **221**, 25
Pedlosky, J.: 1979, Geophysical Fluid Dynamics (Springer edit.; Berlin, Heidelberg, New York), p.482
Rabaud, M., Couder, Y.: 1983 , *J. Fluid Mech.* **136**, 291
Schatzman, E.: 1969, *Astrophys. Lett.* **3**, 139
Schatzman, E.: 1970, *CERN Lecture Notes, Physics and Astrophysics*
Schatzman, E., Maeder, A.: 1981a, *Astron. Astrophys.* **96**, 1
Schatzman, E., Maeder, A.: 1981b, *Nature* **290**, 683
Schatzman, E.: 1977, *Astron. Astrophys.* **56**, 211
Spruit, H.C., Knobloch, E., Roxburgh, I.W.: 1983 *Nature* **304**, 520
Sweet, P.A.: 1950, *M.N. Roy. Astron. Soc.* **110**, 548
Tritton, D.J, Davies, P.A.: 1981, Hydrodynamical Instabilities and the Transition to Turbulence, edit. Swinney, H.L. and Gollub, J.P. (Topics in Applied Physics; Springer; Berlin, Heidelberg, New York)
Turner, J.S.: 1974, *Ann. Rev. Fluid Mech.* **6**, 37
Von Zeipel, H.: 1924, *M.N. Roy. Astron. Soc.* **84**, 665
Zahn, J.-P.: 1975, *Mém. Soc. Roy. Sci. Liège, 6e série* **8**, 31
Zahn, J.-P.: 1983, Astrophys. Processes in Upper Main Sequence Stars (Publ. Observatoire Genève), p.253–329
Zahn, J.-P.: 1984, Proc. 25th Liège Intern. Astrophys. Coll., (Univ. Liège edit.), p. 407

THE GENERATION OF MAGNETIC FIELDS IN THE SUN

Peter A. Fox and Ira B. Bernstein †
Center for Solar and Space Research
Yale University
P.O. Box 6666
New Haven, CT 06511
USA

ABSTRACT. In this paper results of a preliminary investigation into the existence of large scale magnetic fields in the Sun are given. Using a kinematic model with prescribed internal rotation and a standard solar model, the poloidal and toroidal components of the magnetic field are calculated. The basic decay time is of the order of the age of the Sun. In addition the fields are quite sensitive to slight variations in the internal rotation. The boundary condition at the solar surface also does not seem to influence the inner regions where large scale fields seem possible.

1. INTRODUCTION

In this paper, the results of an initial investigation into the stength and configuration of large length scale magnetic fields in the Sun are given. Some of the more basic questions to be addressed are :

(1) What is the Sun's internal magnetic field?
(2) Is there any remnant of the primordial magnetic field?
(3) Where and how is magnetic field regeneration acting?
(4) How is the magnetic field affected by the internal rotation, the dynamics, and the evolution of the Sun and vice versa.
(5) Are there large fields confined to certain regions (like directly below the convection zone or in the core)?
(6) What field strengths are likely to be generated?

Some of these questions have been addressed by other authors but with varying degrees of success. Since the formation, evolution and dynamics of stars involves magnetic fields, answers to these questions may provide valuable information. In many cases the magnetic field is neglected, usually based on an order of magnitude estimate for the governing equations, or by consideration of the timescale over which the magnetic field may be acting.

Previous work on solar activity, apart from phenomological models, falls into two categories (i) kinematic models - which have had moderate success but nevertheless suffer many drawbacks and criticisms (see Krause and Radler 1980, Kuklin 1971, Ruzmaikin 1985, Stix 1981, Yoshimura 1983, Yoshimura et al 1984)

† also Section of Applied Physics

B. R. Durney and S. Sofia (eds.), The Internal Solar Angular Velocity, 213–220.

and (ii) dynamic models , which are limited by present computing resources and the understanding or parameterization of turbulent flow (see Gilman and Miller 1981, Gilman 1983, Glatzmaier 1984,1985a,b). There is thus still a long way to go.

Many phenomological models assume that large toroidal fields exist in certain regions, for example (i) just below the convection zone, or (ii) in the lower part of the convection zone. So far few attempts have been made to establish if such configurations satisfy the basic equations of magnetohydrodynamics or if they are stable. Dynamic models appear to be the proper way to proceed but since such a program is fraught with other difficulties the kinematic approach, keeping in mind its many limitations, may be used initially.

The aim here is to look at large length scale magnetic fields and their evolution. Thus a large scale velocity field is considered. There are many aspects which distinguish this from previous attempts. A key point is the use of a standard solar model including variations over the entire depth of the Sun, instead of replacing thermodynamic and transport quantities by averages. Some other studies have only treated surface fields (DeVore et al 1985, Kuklin 1971) or not integrated over the entire depth of the Sun (Nakagawa 1971).

2. EQUATIONS

The sun is modelled, in this work, as an axisymmetric plasma, using the Maxwell equations in cgs units, neglecting displacement current,

$$\nabla \cdot \mathbf{B} = 0 \tag{2.1}$$

$$c\nabla \times \mathbf{B} = 4\pi \mathbf{J} \tag{2.2}$$

$$c\nabla \times \mathbf{E} = -\frac{\partial \mathbf{B}}{\partial t}. \tag{2.3}$$

It is convenient to employ spherical polar coordinates r, θ, ϕ and introduce two scalars the poloidal stream function $\psi(r,\theta,t)$ and a toroidal part $F(r,\theta,t)$ such that

$$\mathbf{B} = (\nabla\psi) \times (\nabla\phi) + F\nabla\phi. \tag{2.4}$$

The scalar product of (2.2) with $\nabla\phi$, after some manipulation, yields

$$\nabla \cdot \left[\frac{(\nabla\psi)}{R^2}\right] = \frac{-4\pi J_\phi}{Rc}, \tag{2.5}$$

where $R = r\sin\theta$ is the cylindrical radius (note that $\nabla\phi = \mathbf{e}_\phi/R$). Let S be a circle of radius R in a plane z = constant, centered on the z axis which is the pole of the coordinate system. Then if we integrate the z component of (2.3) over S there results

$$cRE_\phi = -\frac{\partial\psi}{\partial t}. \tag{2.6}$$

Of particular interest is the Sun's internal rotation. If the large scale velocity is considered to be made up of differential rotation characterized by an angular speed $\omega(r,\theta,t)$ and a meridional circulation characterized by a function $m(r,\theta,t)$ then

$$\mathbf{v} = \omega\, R^2\nabla\phi + (\nabla m) \times (\nabla\phi)\,. \tag{2.7}$$

The particular form of ω and m would be self consistently computed in a dynamic model, but must be assumed here. In this way various hypothetical cases can be investigated to see what magnetic field configurations result. Here the meridional circulation will be neglected and the differential rotation functions are as given by Yoshimura et al (1984), appropriately scaled, will be used.

The Sun rotates comparitively slowly ($1.8\,km\,sec^{-1}$ at the equator c.f. $125\,km\,sec^{-1}$ for an A5 ($2M_\odot$) star) which makes it an interesting case, and it also has a deep and highly turbulent convection zone.

On adopting the Ohm's law

$$\mathbf{J} = \sigma\left(\mathbf{E} + \frac{\mathbf{v}\times\mathbf{B}}{c}\right) = \sigma\left(\mathbf{E} + \frac{\omega}{c}\times\nabla\psi\right). \tag{2.8}$$

Thus

$$J_\phi = \sigma E_\phi. \tag{2.9}$$

In this investigation an isotropic molecular electrical conductivity $\sigma = 9\times10^6\,T^{\frac{3}{2}}\,sec^{-1}$ is used. It is a function of radius only and varies from $10^{19}\,sec^{-1}$ to $5\times10^{12}\,sec^{-1}$ from center to surface. Because of the convective zone, magnetic diffusion is likely to be enhanced but this will be neglected at present. The validity of the use of a scalar σ can be monitored at each time instant in the calculation.

Thus, on introducing $\mu = cos\,\theta$ and $\eta = c^2/4\pi\sigma$ (the magnetic diffusivity), it follows on employing (2.5) and (2.9) that one can write (2.6) as

$$\frac{\partial\psi}{\partial t} = \eta\left[\frac{\partial^2\psi}{\partial r^2} + \frac{1}{r^2}(1-\mu^2)\frac{\partial^2\psi}{\partial\mu^2}\right]. \tag{2.10}$$

Similarly using (2.8) and taking the ϕ component of (2.3) the equation for F is

$$\frac{\partial F}{\partial t} = \frac{\partial}{\partial r}\left(\eta\frac{\partial F}{\partial r}\right) + \frac{\eta}{r^2}\left(1-\mu^2\right)\frac{\partial^2 F}{\partial\mu^2} + Q \tag{2.11}$$

where

$$Q = R(1-\mu^2)\left(\frac{\partial\omega}{\partial r}\frac{\partial\psi}{\partial\mu} - \frac{\partial\psi}{\partial r}\frac{\partial\omega}{\partial\mu}\right). \tag{2.12}$$

Since $\eta = \eta(r)$, equations (2.10) and (2.11) can be solved easily by separation of variables. Let us in (2.10) seek solutions of the form

$$\psi = f(r)M(\mu)exp^{-\nu t}. \tag{2.13}$$

If we introduce the integer index n equal to the number of nodes of the radial eigenfunction, and index ℓ to represent the latitudinal wave number, then the radial eigenfunctions are determined from the Sturm-Liouville equation,

$$\frac{d^2 f_{\ell n}}{dr^2} + \left[\frac{\nu_{\ell n}}{\eta} - \frac{\ell(\ell+1)}{r^2}\right] f_{\ell n} = 0, \tag{2.14}$$

and the latitudinal eigenfunctions satisfy a Legendre type equation.

If the solar surface is modelled as a perfect conductor one must require that ψ vanish there, in which case $f_{\ell n}$ must vanish there. If the boundary is viewed as matching on to an exterior vacuum region where the current vanishes, then it follows from (2.5) and (2.13) that in the vacuum outside the sun one has the separated solutions $\psi_{\ell n} = M_\ell(\mu) r^{-\ell}$. On the solar boundary the radial eigenfunction satisfies $(R_\odot / f_{\ell n})(df_{\ell n}/dr) = -\ell$, where $R_\odot$ denotes the solar radius. Since ψ and its radial derivative must be continuous, this fixes the boundary condition on the radial eigenfunction on the inside.

The general solution of (2.10) may be written as

$$\psi(r,\mu,\phi,t) = \sum_{\ell=1}^{\infty} \sum_{n=0}^{\infty} \alpha_{\ell n} M_\ell(\mu) f_{\ell n}(r)\, exp^{-\nu_{\ell n} t}, \tag{2.15}$$

the constants $\alpha_{\ell n}$ are determined from the initial conditions on using the orthogonality of the eigenfunctions via

$$\int_{-1}^{1} d\mu \int_{0}^{R_\odot} dr\, \psi(r,\mu,\phi,0) M_\ell(\mu) f_{\ell n}(r) = \alpha_{\ell n} \int_{-1}^{1} d\mu \int_{0}^{R_\odot} dr\, M_\ell^2(\mu) f_{\ell n}^2(r). \tag{2.16}$$

To fully constrain the solution $\psi(r,\mu,\phi,0)$ must be known, perhaps from models of star formation (Nakano and Umebayashi 1986, Phillips 1986). Equation (2.11) too is conveniently dealt with by separation of variables. Since (2.10) is independent of F it may be solved for ψ alone and used in (2.11) for F.

As the Sun evolves so does the way in which the magnetic field behaves. The central temperature increases, the radius changes, the convection zone depth decreases and the surface rotation rate decreases (solar wind etc.). The standard solar model will be taken from the Yale stellar structure code but in this investigation is frozen in time. In addition the boundary conditions at the solar surface may alter as well. In the early stages of formation a vacuum boundary may be the best choice, whereas currently a perfectly conducting boundary is plausible. These unfortunately represent two extreme cases with the true condition somewhere in between. Hopefully the choice of boundary condition may not influence the inner magnetic field, and this will be discussed later.

3. RESULTS

Solving (2.14) as an eigenvalue problem gives the decay rates for each modal component (ℓ, n) of the poloidal field. For the fundamental component $(\ell = 1, n = 0)$ the e-folding decay time, T_d, is $2\,T_{sun}$ for vacuum boundaries and $1.7\,T_{sun}$ for perfectly conducting boundaries, where T_{sun} is the solar age. For higher order modes like $\ell = 10, n = 0$; $T_d = 0.07\,T_{sun}$ and $\ell = 1, n = 9$; $T_d = 0.04\,T_{sun}$ for vacuum boundaries. Similar values hold for perfectly conducting boundaries. Keeping in mind the restriction of a static solar model, these decay times are still very long. Also it is worth keeping in mind that current models of star formation including magentic fields (Nakano and Umebayashi 1986, Phillips 1986) suggest dipolar type magentic fields with only small quadrupole components. This suggests that the magnetic field evolution, below the turbulent convection zone, is dominated by the lower order components of ℓ and n.

In this preliminary investigation only the stationary state configuration will be considered. The aim is to get an idea of field strength enhancements, given the obvious restriction of a fixed solar model.

Enhanced (toroidal) field strengths depend on the differential rotation and its gradient, the strength and configuration of the poloidal field and the length scale of the field.

Figures 1 a) and b) represent the fundamental ($\ell = 1, n = 0$) stationary poloidal field configurations for vacuum and perfectly conducting boundaries respectively. In each case the influence of the boundaries is restricted to the outer regions which has implications for surface phenomena but not so much for the study of large fields in the inner regions.

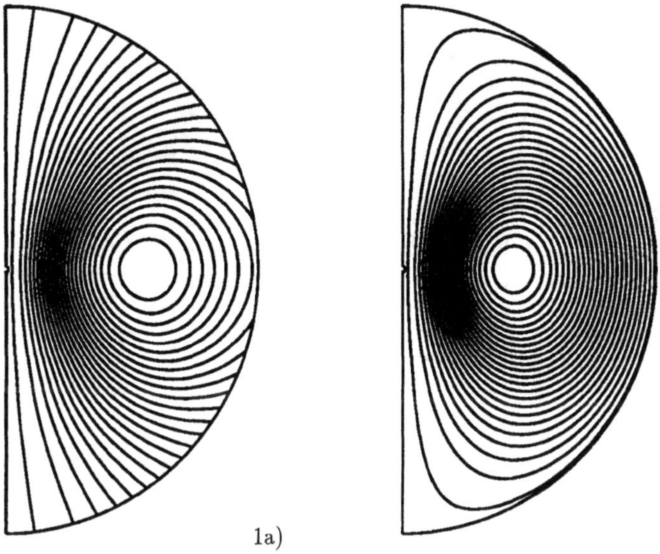

1a) 1b)

Figure 1. Streamlines of poloidal magnetic field ($\ell = 1, n = 0$) for a) vacuum and b) perfectly conducting boundaries.

The variation of ℓ and n provide no surprises except for the location of nodal points in radius, which tend to be more central as n increases. Because of limited space only a small sample of the low order components are presented. Instead, many configurations due to $\omega(r, \theta)$ are examined and the sign of $\nabla\omega$ can be tested.

Figures 2 a) - f) give the toroidal field configurations corresponding to the poloidal field in figure 1 a) for different differential rotation profiles. These figures emphasize the sensitivity of the field configuration to ω. In particular many of the fields are restricted to the inner regions (certainly below the convection zone). This is especially true for perfectly conducting boundaries, but is not shown here. Since the equations are linear and not constrained by an initial magnitude at present, only the ratio of toroidal to poloidal field strength is significant. For the cases in figure 2, the average ratio is 10^{10} i.e. a $1\mu G$ poloidal field would produce a $10\,kG$ toroidal field, in the stationary state. At this particular level of intensity the magnetic field would have an effect on the dynamics and on the differential rotation. In all cases the field strength increases with depth but for cases a) and f) the field strength is at least a factor of 10 lower than for other cases.

Intense core fields are suggested by some cases, however, the true form of ω in the inner regions of the Sun may remain an unknown for a long time. The assumption of isotropic diffusion seems acceptable, however, if the turbulent contribution in the upper layers is considerable this situation may change.

It is possible that some remnant of the primordial field may exist in the lower order modes. Such studies require a time dependent calculation coupled with the evolving Sun and this work will be reported elsewhere.

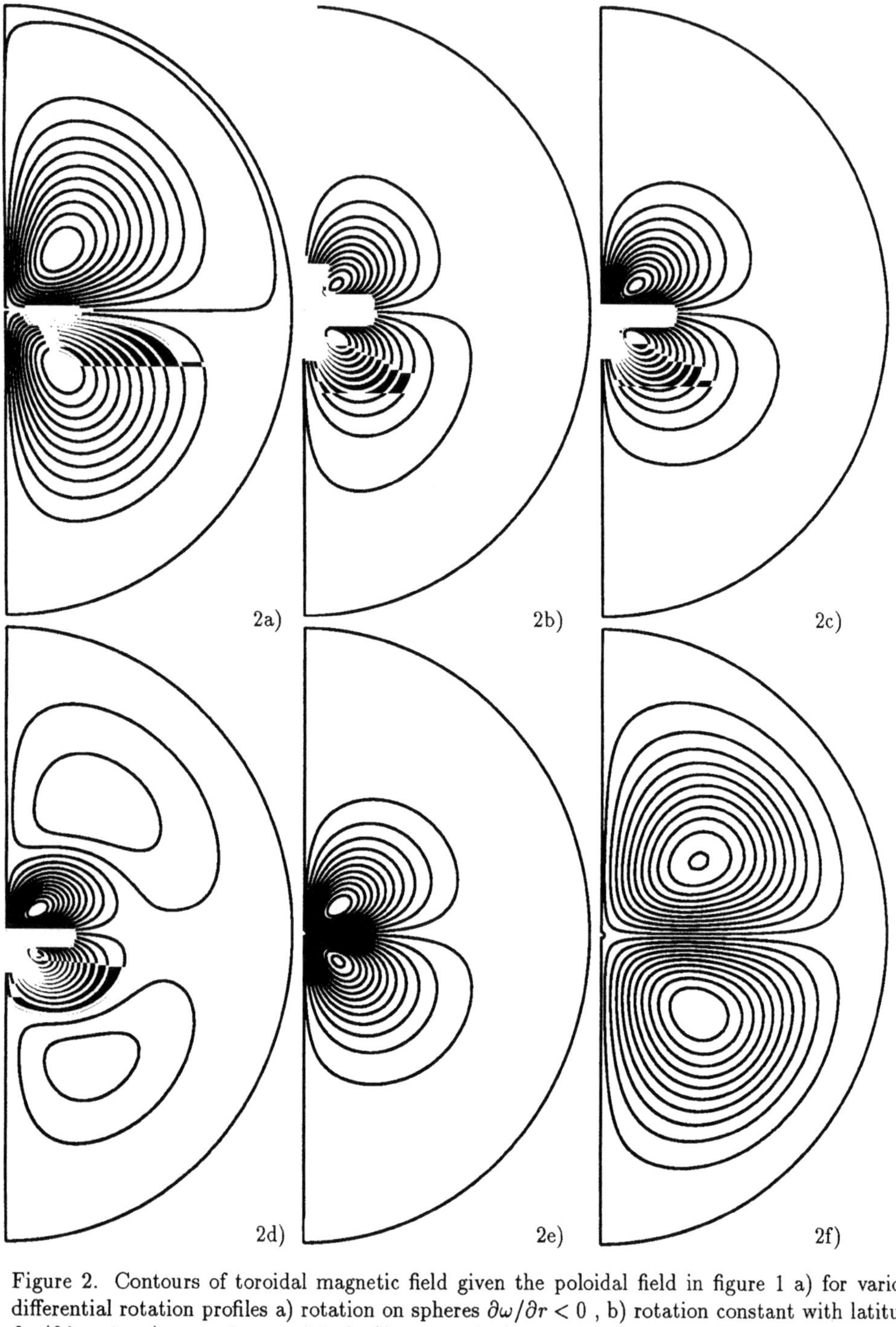

Figure 2. Contours of toroidal magnetic field given the poloidal field in figure 1 a) for various differential rotation profiles a) rotation on spheres $\partial\omega/\partial r < 0$, b) rotation constant with latitude $\partial\omega/\partial\theta < 0$, c) general case with $\partial\omega/\partial r < 0$, $\partial\omega/\partial\theta > 0$, d) general case with $\partial\omega/\partial r < 0$, $\partial\omega/\partial\theta < 0$, e) general case with $\partial\omega/\partial r > 0$, $\partial\omega/\partial\theta > 0$ and f) rotation on cylinders $\partial\omega/\partial R < 0$. For details of the functional forms see Yoshimura et al (1984).

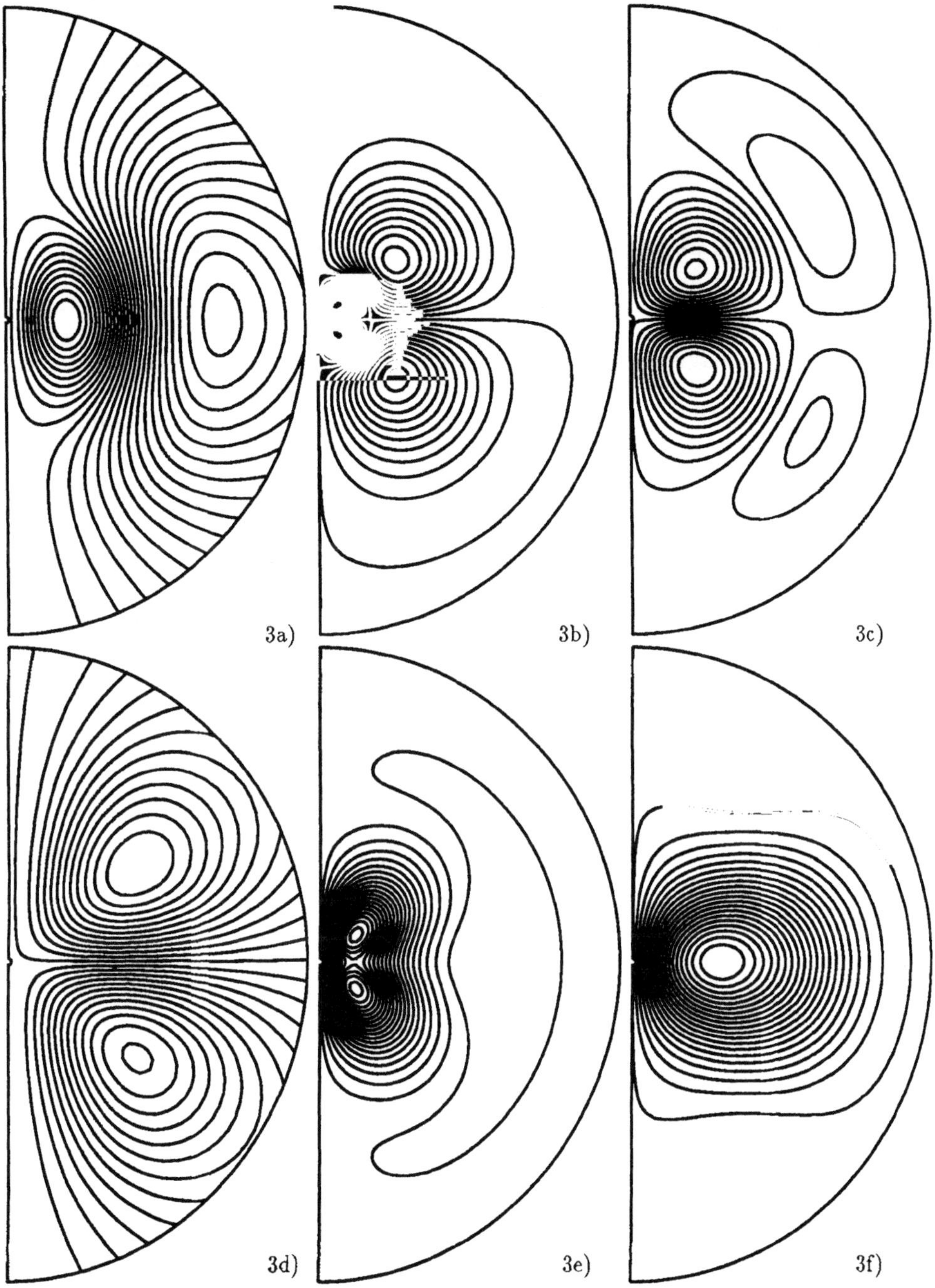

Figures 3 a) and d). Streamlines of poloidal magnetic field for a) ($\ell = 1, n = 1$) and d) ($\ell = 2, n = 0$) and vacuum boundaries. Figures 3 b) and c). Contours of toroidal magnetic field given the poloidal field from figure 3a) for two differential rotation profiles b) - as in figure 2 c) and c) - as in figure 2 f). Figures 3 e) and f). Contours of toroidal magnetic field given the poloidal field from figure 3d) for two differential rotation profiles e) - as in figure 2 c) and f) - as in figure 2 f).

Figure 3 shows the next higher order field configurations. These illustrate the differences between certain rotation profiles as the structure of the field becomes more complex. It also shows that for even order ℓ modes the toroidal fields cross the equator and this may not be compatible with the polarity rules for surface features.

4. CONCLUSIONS

The results of this preliminary investigation demonstrate that large toroidal fields may be generated from a small poloidal field and there is a marked sensitivity to the Sun's internal rotation. Obviously the differential rotation profiles used in this study may not be representative of the solar rotation, and, it is also known that even modest magnetic fields are enough to modify the differential rotation. Nevertheless the variation in field configuration due to ω is a point to be pursued.

The inclusion of a depth dependent diffusion coefficient has also indicated that large scale fields may exist in some regions. Boundary conditions seem to be most important in the outer layers. This makes the treatment of surface features difficult since the true condition is unknown, but is adequate for the inner regions.

The lower order components of the magnetic field exhibit decay times of the order of the solar age. More detailed calculations however are required before drawing any firm conclusions.

Further work on this type of problem is to include, other stars, coupling the calculations to the evolution of the rotating Sun (Endal and Sofia 1981, Sofia 1986), improved treatment of turbulent diffusion, incorporation of meridional circulation as a regeneration mechanism and estimating the back reaction on the flow.

This work was supported by NASA grant NAGW-777.

5. REFERENCES

DeVore, C.R., Sheeley Jr., N.R., Boris, J.P., Young Jr., T.R. and Harvey, K.L. 1985, *Austr. J. Phys.*,**38**,999.
Endal, A.S. and Sofia, S. 1981, *Ap. J.*,**243**,625.
Gilman, P.A. and Miller, J. 1981, *Ap. J. Suppl.*,**46**,211.
Gilman, P.A. 1983, *Ap. J. Suppl.*,**53**,243.
Glatzmaier, G.A. 1984, *J. Comp. Phys.*,**55**,461.
Glatzmaier, G.A. 1985a, *Ap. J.*,**291**,300.
Glatzmaier, G.A. 1985b, *Geophys. Astrophys. Fl. Dyn.*,**31**,137.
Krause, F. and Radler, K-H. 1980, *Mean-field Magnetohydrodynamics and Dynamo Theory* (Pergamon).
Kuklin, G.V. 1971, in *Solar Magnetic Fields* (Reidel), IAU Symp #43,737.
Nakagawa, Y. 1971, in *Solar Magnetic Fields* (Reidel), IAU Symp #43,725.
Nakano, T. and Umebayashi, T. 1986, *Mon. Not. R. astron. Soc.*, **221**,319.
Phillips, G.J. 1986, *Mon. Not. R. astron. Soc.*,**222**,111.
Ruzmaikin, A.A. 1985, *Solar Phys.*,**100**,125.
Sofia, S. 1986, this volume.
Stix, M. 1981, *Solar Phys.*,**74**,79.
Yoshimura, H. 1983, *Ap. J. Suppl.*,**52**,363.
Yoshimura, H., Wang, Z. and Wu, F. 1984, *Ap. J.*,**280**,865.

MASS TRANSPORT BY WAVE MOTION

E. Knobloch and J. B. Weiss
Department of Physics
University of California
Berkeley, California, 94720

ABSTRACT. Transport properties of waves require a Lagrangian description. Even in simple waves particle motions can be highly nonlinear, and these nonlinearities are essential for transport. One-dimensional kinematic waves are used to illustrate some of the possible phenomena that affect wave transport.

1. INTRODUCTION

Waves are ubiquitous in astrophysics. They are believed to play important roles in mixing of the solar interior, in the process of angular momentum extraction from the core of the Sun and from accretion disks, and in many processes of importance in astrophysics and geophysics. However, little is known about how waves might accomplish all the disparate tasks that are demanded of them. Waves in a stable medium can be excited externally by an appropriate mechanism. For example, internal gravity waves are excited in the overshoot region below the solar convection zone and propagate into the interior (Press 1981; Hurlburt *et.al.* 1986). In addition, many instabilities, such as the ABCD instability (Knobloch and Spruit 1983; Spruit, Knobloch, and Roxburgh 1983), are oscillatory in nature. Such instabilites may saturate at finite amplitude, and give rise to propagating finite amplitude waves. In either case one seeks to describe the effect of a sea of waves on the composition of, say, the solar interior, or the distribution of angular momentum. At present this remains an unsolved problem, although in the case of internal waves significant progress has been made (e.g. Müller 1976; McEwan 1983). An excellent review of mixing in the Sun is by Spruit (1985).

The purpose of this preliminary report is to indicate some of the ingredients that must be included in a credible theory of mixing. We do this by means of a tractable model problem. Specifically, we consider the transport properties of longitudinal compressive waves in one dimension, and focus attention on the motion of the particles caused by the passage of the wave. The wave produces oscillations in time at each point x, described by a velocity field $u(x,t)$. The motion of a test particle obeys the equation

$$\frac{dx}{dt} = u(x,t). \tag{1}$$

B. R. Durney and S. Sofia (eds.), The Internal Solar Angular Velocity, 221–228.

This equation provides a kinematic description of the particle motion; the velocity field u is assumed to be prescribed, so there is no dynamics in the problem. A parallel may be made with kinematic dynamo theories. Notice, however, that equation (1) is *nonlinear,* and hence even the simplest kinematic problem is nontrivial. To understand the transport properties of the wave we must follow the particles as they move with the wave, i.e., we wish to find a solution of (1),

$$x = X(t, x_0), \qquad X(0, x_0) = x_0, \tag{2}$$

in terms of the *initial* position x_0 of the particle. The description (2) is called Lagrangian and provides the correct framework for the problem.

In general it is not possible to obtain (2) explicitly. However, by a combination of analytical and numerical means one can learn a great deal about the motion of the particles. In the following we distinguish between monochromatic wave trains, nondispersive wave packets, and dispersive wave packets.

2. MONOCHROMATIC WAVE TRAINS

The motion of a particle in a monochromatic wave train is described by

$$\dot{x} = A\cos(kx - \omega t), \tag{3}$$

where $\omega = \omega(k)$ is the frequency, and $c = \omega/k$ is the phase speed. Traditionally, such equations are solved by a perturbation expansion in the amplitude A of the wave:

$$x = x_0 + Ax_1 + \cdots. \tag{4}$$

One obtains

$$\dot{x} = A\cos(kx_0 - \omega t) + \frac{A^2}{c}\sin^2(kx_0 - \omega t) + \cdots. \tag{5}$$

At leading order in A the particle oscillates about its initial position x_0. However, at second order one finds that the particle drifts in the direction of the phase velocity of the wave, with a speed

$$\langle \dot{x} \rangle = \frac{1}{2}\frac{A^2}{c} + O(A^3), \tag{6}$$

where $\langle\cdots\rangle \equiv \frac{2\pi}{\omega}\int_0^{2\pi/\omega}(\cdots)\,dt$. This is known as the Stokes' drift and is a purely *nonlinear* effect. But even this calculation fails to capture the possibility of *particle trapping* by the wave at sufficiently large amplitudes. To understand trapping we introduce the comoving coordinate $s = x - ct$, and write (3) in the form

$$\dot{s} = -\frac{dV}{ds}, \qquad V \equiv \frac{c}{k}\left(ks - \frac{A}{c}\sin ks\right). \tag{7}$$

The potential V has the form of a washboard, and is drawn in Fig. 1 for the two cases (a) $A < c$ and (b) $A > c$. In (a) there are no minima in V and the particle drifts backwards relative to the wave, with a speed that is controlled by

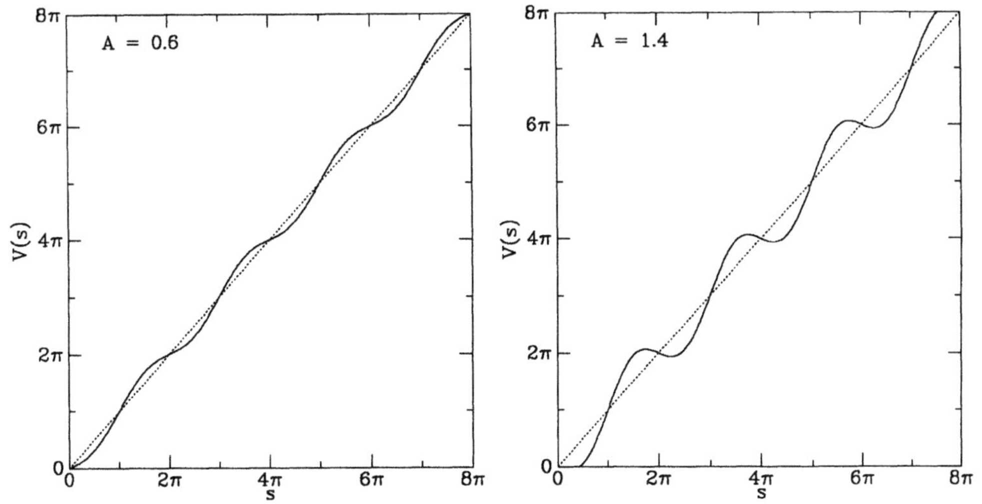

Figure 1. Potential $V(s)$ as a function of the comoving coordinate s for a particle in a monochromatic wave train. The wave trains have wavevector $k = 1$, speed $c = 1$, and amplitudes below and above the critical amplitude for trapping, $A = c$.

the amplitude of the corrugation of the washboard. As A increases through c, local minima and maxima appear in saddle-node bifurcations, and since V is locally a Liapunov function, the particle comes to rest at one of the minima. In the original frame the particle is carried along by the wave at the speed c. The exact expression for the drift speed,

$$\langle \dot{x} \rangle = \begin{cases} c\left(1 - \sqrt{1-(A/c)^2}\right), & A \leq c; \\ c, & A \geq c, \end{cases} \tag{8}$$

includes both the perturbation result (6) and the saturation at speed c due to trapping, and is shown in Fig. 2. Note that a standing wave produces no transport, since the particles cannot get past the nodes.

3. NONDISPERSIVE WAVE PACKETS

In astrophysical situations waves are rarely produced continuously. Instead they come in bursts, which are described by wave packets of the form

$$u(x,t) = \operatorname{Re} A \int f(k) e^{i(kx-\omega(k)t)}\, dk. \tag{9}$$

A wave packet is nondispersive if $\omega'(k) = \omega/k = c$, a constant. We choose $f(k) = \sqrt{\frac{a}{2\pi}} \exp\left[-\frac{1}{2}a(k-k_0)^2\right]$, and study the equation

$$\dot{x} = |A| \exp\left[-\frac{(x-ct)^2}{2a}\right] \cos\left[k_0(x-ct)+\alpha\right], \tag{10}$$

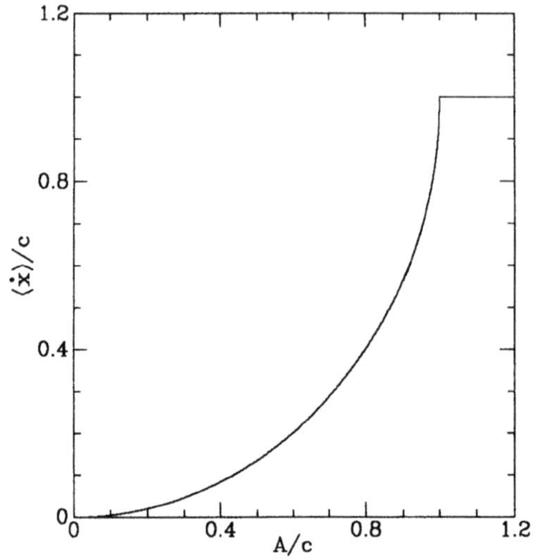

Figure 2. Drift speed for a particle in a monochromatic wave train as a function of the wave amplitude.

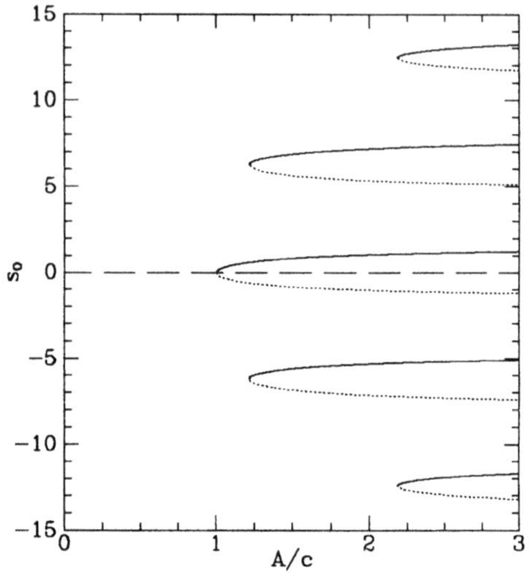

Figure 3. Bifurcation diagram for the position of maxima and minima of the potential for a particle in a nondispersive wave packet with $k_0 = 1$, $a = 100$, and $\alpha = 0$. Solid lines denote maxima, while dotted lines denote minima.

where $\alpha = \arg A$. In this case the fluid is undisturbed until the packet arrives. If the amplitude of the wave is not too large, the particles interact with it for only a finite time. When the wave departs, the particles are deposited at new positions $x(x_0)$. The net displacement is given by

$$\Delta x = \int_{-\infty}^{\infty} \dot{X}(t, x_0)\, dt, \tag{11}$$

where $x = X(t, x_0)$ solves (10). It is simple to show that

$$\Delta x = \int_{-\infty}^{\infty} \frac{|A| \cos(k_0 s + \alpha)\, ds}{c \exp(s^2/2a) - |A| \cos(k_0 s + \alpha)}\ , \tag{12}$$

and hence that all particles are transported by the same amount. In addition, one sees that $\Delta x = O(|A|)$, for small $|A|$, and that trapping requires that $u(s)$ exceeds c, for some s. The particles are then trapped in the wave packet, and carried at speed c. Evidently trapping can only occur if $|A| \geq c$. The locations s_0 of the minima and maxima of the comoving potential are shown as a function of $|A|$ in Fig. 3.

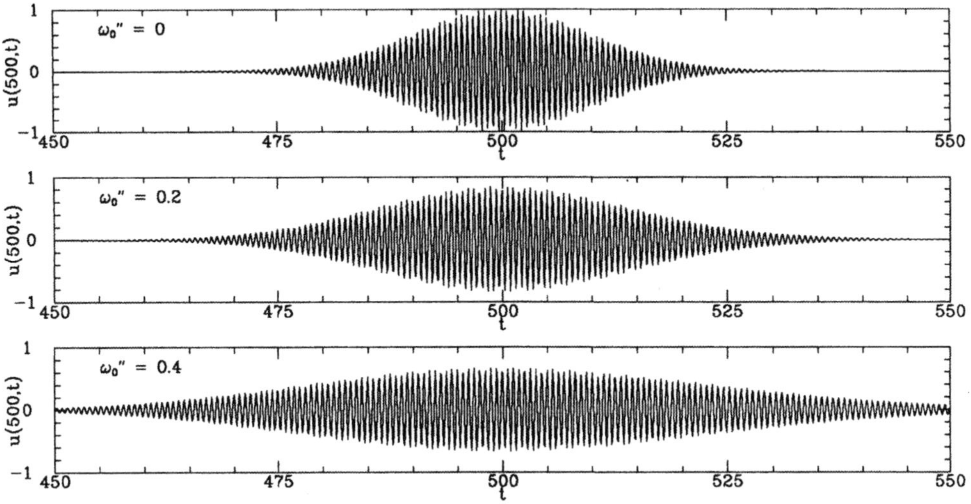

Figure 4. Amplitude at $x = 500$ of wave packets launched from $x = 0$ at $t = 0$. The parameters of the wave packets are $A = 1$, $a = 100$, $k_0 = 10$, $\omega_0 = 10$, $c_g = 1$, $\alpha = 0$, and $\omega_0'' = 0, 0.2$, and 0.4.

4. DISPERSIVE WAVE PACKETS

More generally, we write

$$\omega(k) = \omega(k_0) + (k - k_0)\omega'(k_0) + \frac{1}{2}(k - k_0)^2\omega''(k_0) + \cdots \tag{13}$$

and obtain in place of (10)

$$\dot{x} = \frac{\sqrt{a}\,|A|}{[a^2 + (\omega_0''t)^2]^{\frac{1}{4}}} \exp\left[-\frac{a}{2}\frac{(x - c_g t)^2}{a^2 + (\omega_0''t)^2}\right] \cos\left[k_0 x - \omega_0 t + \phi(x,t)\right], \tag{14}$$

where

$$\omega_0 = \omega(k_0), \qquad c_g = \left.\frac{d\omega}{dk}\right|_{k=k_0}, \qquad \omega_0'' = \left.\frac{d^2\omega}{dk^2}\right|_{k=k_0}, \tag{15}$$

and

$$\phi(x,t) = \frac{\omega_0''t}{2}\frac{(x - c_g t)^2}{a^2 + (\omega_0 t)^2} - \tan^{-1}\frac{\omega_0''t}{a} + \alpha. \tag{16}$$

Equation (14) describes the motion of a particle in a spreading wave packet travelling with the ***group*** velocity c_g (Fig. 4). There are two important differences between dispersive and nondispersive wave packets. For dispersive waves, the particle displacements $\Delta x(x_0)$ depend on x_0 because the wave changes as it propagates, and particles are trapped only temporarily, since as the wave disperses its amplitude diminishes. The particle distribution after the passage of the

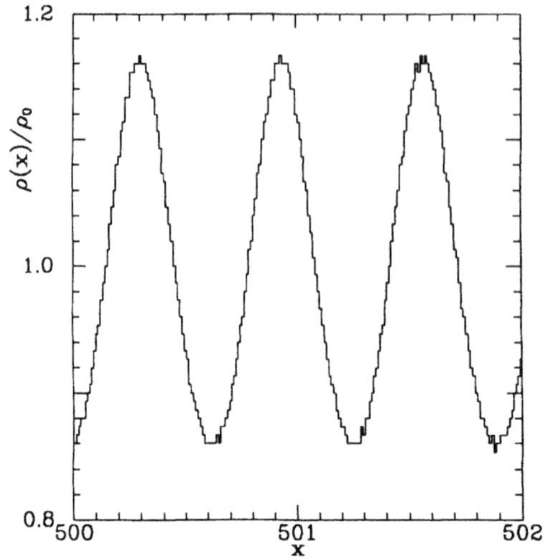

Figure 5. Density of particles after passage of a dispersive wave with k_0 near a root of the dispersion relation, $w(k) = 0$. The parameters of the wave packet are $A = 0.01$, $a = 100$, $k_0 = 10$, $\omega_0 = 1$, $c_g = 10$, $\omega_0'' = 1$, and $\alpha = 0$.

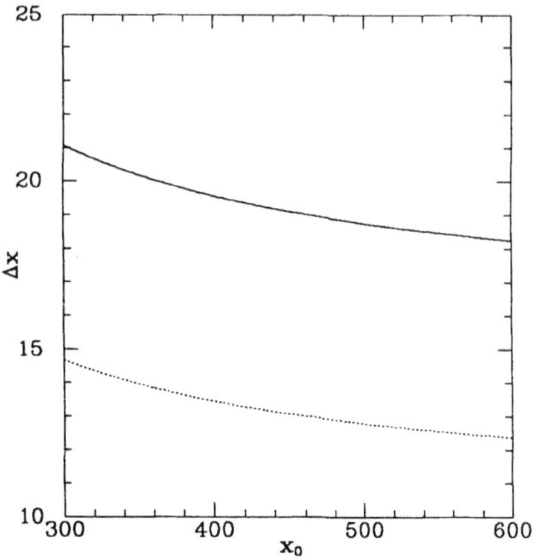

Figure 6. Particle displacement Δx as a function of initial position x_0 after passage of a dispersive wave packet. Solid and dotted lines result from the passage of wave packets with $c_g = 0.8$ and $c_g = 1.2$ respectively. Other parameters are $A = 1.2$, $a = 100$, $k_0 = 10$, $\omega_0 = 10$, $\omega_0'' = 1$, and $\alpha = 0$.

wave packet $\rho(x)$ is given in terms of an initially uniform distribution ρ_0 by the conservation law:

$$\rho(x)\, dx = \rho_0\, dx_0, \tag{17}$$

where $x = x_0 + \Delta x(x_0)$. If $|A|$ is small, we obtain

$$\begin{aligned} \rho(x) &= \rho_0 \left[1 - \left.\frac{\partial \Delta x}{\partial x_0}\right|_{x_0 = x} + O(|A|^2) \right] \\ &= \rho_0 \left[1 + |A| \sum_i k_i \frac{\sqrt{2\pi a}}{|\omega'(k_i)|} e^{-\frac{1}{2}a(k_i - k_0)^2} \sin(k_i x + \alpha) + O(|A|^2) \right], \end{aligned} \tag{18}$$

where k_i are those roots of $\omega(k) = 0$ that fall within the wave packet. The above procedure is equivalent to solving the conservation equation

$$\frac{\partial \rho}{\partial t} + \frac{\partial}{\partial x}[\rho u(x,t)] = 0 \tag{19}$$

with $u(x,t)$ given by (14). The density oscillations that occur if a root k_i falls where $f(k)$ is substantially different from zero are shown in Fig. 5. Otherwise, Δx, and hence $\rho(x)$, depend on x monotonically, as shown in Fig. 6.

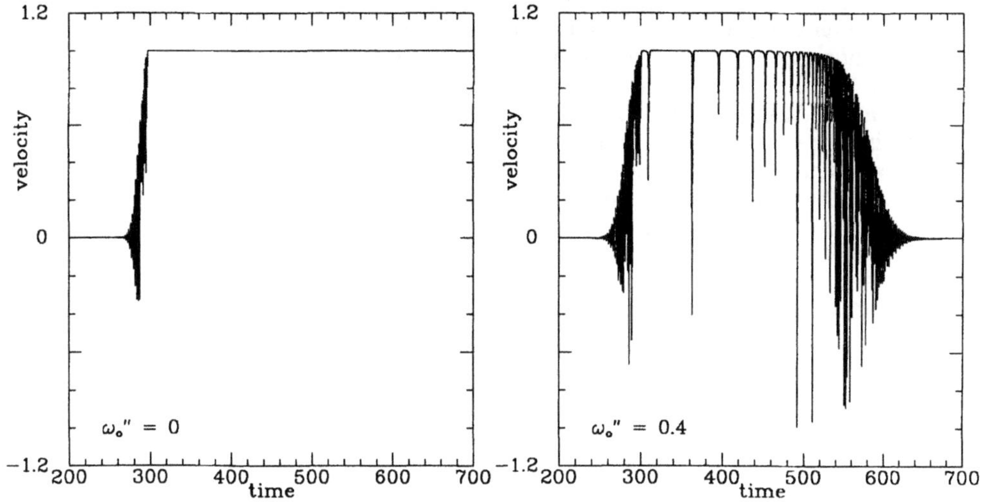

Figure 7. Velocity of a particle $\partial X(t, x_0)/\partial t$ in nondispersive ($\omega_0'' = 0$) and dispersive ($\omega_0'' = 0.4$) wave packets. Other parameters are $A = 1.5$, $a = 100$, $k_0 = 10$, $\omega_0 = 10$, $c_g = 1$, and $\alpha = 0$.

The velocity of a particle, $\partial X(t, x_0)/\partial t$, is shown in Fig. 7 as a function of time, for a particle starting at $x_0 = 300$. The wave packet is launched from $x = 0$ at $t = 0$ with an initial maximum amplitude large enough to give trapping. In the dispersive case trapping is temporary, as the particle leaks past the successive maxima of the wave packet. In the nondispersive case it is permanent.

5. DISCUSSION

The results presented above show that even simple one-dimensional waves can produce complicated particle motions. The amplitude of the wave determines the drift speed of the particles, or, in the case of a wave packet, the displacement of the particles from their initial position. These effects show that nonlinearities are necessary for transport. The distinction between wave trains and nondispersive and dispersive wave packets is also crucial; quite different transports obtain in the different cases. The situation is complicated by the fact that for most waves the phase speed depends on the amplitude. In addition, if a wave packet propagates in a slowly varying medium its amplitude varies according to the conservation of "wave action" (Whitham 1974). Thus particles that may be trapped can become untrapped (or vice versa) as the packet propagates. Of course, the kinematic assumption fails as soon as the wave has trapped a large enough number of particles to create pressure gradients that change the velocity field. This happens rapidly in the trapping regime, when the wave is in a "snowplow" phase. On the other hand, the description of the wave in terms of wave packets allows one to use methods of geometrical optics (Whitham 1974) to study the propagation of waves in inhomogeneous media, such as the stratified interior of the Sun.

In one dimension no mixing takes place. For this, overturning motions such as occur when a wave breaks are essential. However, wavebreaking too is a non-linear phenomenon, and its understanding is fundamental to the mixing process (McEwan 1983). Even when overturning occurs, mixing occurs only in a coarse-grained sense. Microscopic mixing requires the inclusion of particle diffusion. Dissipation is also one mechanism by which a wave can deposit its momentum (or angular momentum). Other possibilities include reflection and wavebreaking. These processes are undoubtedly important in determining the role played by waves in the internal rotation of the Sun.

6. ACKNOWLEDGEMENTS

This work was supported by the California Space Institue under grant CS 09–86, and the IBM DACE grant. We are grateful to J. D. Lerner for helpful discussions.

7. REFERENCES

Hurlburt, N. E., J. Toomre, and J. M. Massaguer, 1986, Astrophys. J., in press.

Knobloch, E., and H. C. Spruit, 1983, Astron. Astrophys., **125**, 59.

McEwan, A. D., 1983, J. Fluid Mech., **128**, 59.

Müller, P., 1976, J. Fluid Mech., **77**, 789.

Press, W. H., 1981, Astrophys. J., **245**, 286.

Spruit, H. C., 1984, *Hydromagnetics of the Sun*, (Proc. Fourth European Meeting on Solar Physics), European Space Agency, **ESA SP–220**, 21.

Spruit, H. C., E. Knobloch, and I. W. Roxburgh, 1983, Nature, **304**, 320.

Whitham, G. B., 1974, *Linear and Nonlinear Waves*, Wiley.

Differential Rotation in Stars

S. R. Sreenivasan and W. J. F. Wilson,

Department of Physics, The University of Calgary

Calgary, AB Canada T2N 1N4

I. Introduction

A considerable amount of observational material now exists from studies of stellar and solar pulsation as well as chemical peculiarities in upper main sequence stars and supergiants to enable us to understand the nature of rotation in stars. The nature of stellar rotation has a bearing on many problems ranging from the generation of magnetic fields to the occurrence and consequences of stellar winds. It is becoming increasingly evident that the consequences of stellar rotation to the structure and evolution of stars are amenable to critical observational scrutiny (eg. these proceedings). We shall concentrate on an aspect of rotation that has not been investigated in much detail so far. This concerns the effect of rotation on the size of the convective cores of stars. A great amount of effort has been invested recently to study carefully the question of meridional circulation in stars by the Tassouls (1982-6). Meridional circulation is one consequence of differential rotation in stars, and addresses the question of mixing of

B. R. Durney and S. Sofia (eds.), The Internal Solar Angular Velocity, 229–234.

material in the radiative envelope of stars. The core is treated as a rigid, impenetrable boundary and is taken to rotate as a solid body. Thus no mixing is permitted by caveat between the core and the radiative envelope. In the case of the Sun and solar-type stars, it is not clear what modifications would result when no convective cores are present. There is also some argument about the precise physical nature of the boundary layer between the core and the envelope when convective cores are present. The Tassouls maintain that it is a thermo-viscous boundary layer where as Sakurai (1986) argues that it is an Ekman boundary layer. Whatever type it is, one thing is clear and that is that it is a region of strong differential rotation and shear. (See also Endal and Sofia 1976, 1978). The resulting consequence is mixing between the convective core and the radiative envelope. The core size should, therefore, increase.

In stellar interior computations, one determines the core size by identifying the radius in a model at which the radiative gradient is equal to the adiabatic gradient. When one considers rotating models the structure variables are functions of the angular velocity, in addition. If the model is presumed to rotate as a solid body, one gets increased central densities and pressures, but reduced central temperatures and consequently, lower luminosity. The effective radius of the star as well as the size of the core are smaller. This is a well-known result and has been obtained by many people (see Sreenivasan and Wilson 1986). When the model is rotating as a gaseous body one needs to know the law of rotation: $\Omega = \Omega\ (\underline{r},t)$ to determine the structural changes in the star and to follow its evolution with time. This is not an easy proposition and so one either assumes that specific angular momentum is constant or some other perscribed rotation law. If, in addition, one assumes that the convective core rotates as a rigid body either because it is convenient and simple (as the Tassouls and others do), or because convection and differential rotation interact to bring about solid body rotation (as Kippenhahn - see discussion in Tassoul (1980) - did) one can compute the structural changes if one knows the angular velocity of the core-boundary. This involves an initial model (either rigidly rotating as in Endal and Sofia (1978) or a non-rotating model as in Sreenivasan and Narasimha (1986)) and some iteration. In this latter case, one also gets a smaller core because one is essentially forcing the core to rotate as a rigid body by knowing the core-size. It is not obvious, therefore, that the core will rotate as a rigid body at the same size as it would when convection interacting with differential rotation brought about rigid rotation.

We have recently shown that the radiative gradient for a rotating model is larger than that for a non-rotating model when differential rotation preserving specific angular momentum is taken into account but smaller if rigid rotation is imposed (Sreenivasan & Wilson 1986), other things being the same. It should be noted that the ratio of the centrifugal force to gravity decreases as you go from the surface to the centre of the star if the star is in rigid rotation. The opposite is the case when it rotates such that the specific angular mementum is constant. This means that the distortion from sphericity becomes worse in the inner regions of the star. As the star evolves, the convective core shrinks in size thus spinning faster and distorting more its sphericity if specific angular momentum remains constant. Angular momentum has to be transferred outwards. Narasimha and Sreenivasan (1986) have shown, in addition, that such models are overstable to non-radial pulsations and these pulsations effectively must transfer angular momentum outwards as well. One finds a situation analogous to the occurrence of semi-convection in massive star models in conservative mass evolution. The consequence should again be an enlarged core with lower angular momentum.

II. Consequences

What, if any, one might ask are the consequences of such core-growth? Classically, (rigidly) rotating stellar models resulted in stars appearing at a location shifted downward and to the right of the initial position occupied by non-rotating models of the same mass and composition (Maeder and Peytremann 1972). If core-size is enlarged due to differential rotation, one gets the opposite trend. Stars would be more luminous compared to the classical model predictions. They could occupy the same position as those of non-rotating models. If one studies meriodional circulation in such models, the circulating elements would have less distance to cover, the process would be more efficient, and if stellar winds are present, they would make the process even more efficient and would reveal altered chemical compositions at the surface. The chemically peculiar stars of the upper main sequence, and the abundance anomalies of the peculiar giants would find a natural explanation. In addition, the Blue Stragglers, The High Latitude F Supergiants, and the widening of the main sequence noted in massive stars would all be equally more straightforward to understand. We have discussed these aspects elsewhere. (Sreenivasan and Wilson: 1985 and 1986). Furthermore, they provide evidence that stars rotate like gaseous bodies. It is even possible to understand, combined with our results of pulsation studies, the phenomena of β cephei stars and Be and Shell stars, on a unified basis.

We, therefore, conclude that mixing between the convective core and the radiative envelope, a subject that has not been investigated carefully or in any detail, deserves more careful study and offers the possibility of understanding many diverse and apparently unconnected phenomena, in a relatively straightforward manner.

Our work is supported by a Canada NSERC research grant to SRS.

References

Endal, A.S. and Sofia 1976 Ap. J. 210, 184.

________________ 1978 Ap. J. 220, 279.

Maeder, A. and E. Peytremann 1972 Astron. Ap. 21, 279.

Narasimha, D. and Sreenivasan, S.R. 1986 in Stellar Pulsation, a memorial to J.P. Cox (ed. A. Cox) (in press).

Sakurai, T. 1986 Geophys. Ap. Fluid Dyn. 36, 257.

Sreenivasan, S.R. and Narasimha, D. 1986 BAAS 17, 895.

Sreenivasan, S.R. and Wilson, W.J.F. 1980 in Stellar Pulsation Instabilities ed. D. Fischel, J.W. Lesh and W.M. Sparks (Greenbelt, MD: NASA TM 80625) 363.

Sreenivasan, S.R. and Wilson, W.J.F. 1985a Ap. J. 292,506.

________________________________ 1985b BAAS. 17, 600.

Sreenivasan, S.R. and Wilson, W.J.F. 1985c in Mass-loss from Red Giants (Dordrecht: D. Reidel) ed. M. Morris and B. Zuckermann, 261.

Sreenivasan, S.R. and Wilson, W.J.F. 1986a in Cool Stars, Stellar Systems and the Sun (New York; Springer) ed. M. Zeilik and D.M. Gibson, 426.

Sreenivasan, S.R. and Wilson, W.J.F. 1986b in Circumstellar Matter (Dordrecht: D. Reidel) ed. I, Appenzeller and C. Jordan, (in press).

Sreenivasan, S.R. and W.J.F. Wilson 1986c in Luminous Stars and Associations (Dordrecht: D. Reidel) ed. C.W.H. deLoore et. al, 385.

Sreenivasan, S.R. and Wilson, W.J.F. 1986d Ap. J. submitted.

Tassoul, J.L. 1980 Theory of Rotating Stars (Princeton: Princeton University Press)

Tassoul, J.L. and Tassoul, M. 1982a Ap. J. Suppl. 42, 317.

1982b Ap. J. 261, 265.

1982c Ap. J. 261, 273.

1983a Ap. J. 264, 298.

1983b Ap. J. 267, 334.

1983c Ap. J. 271, 315.

1984a Ap. J. 279, 384.

1984b Ap. J. 286, 350.

1986 Geophys. Ap. Fluid. Dyn. 36, 303.

THE GENERALIZATION OF MIXING LENGTH THEORY TO ROTATING CONVECTION ZONES AND APPLICATIONS TO THE SUN

Bernard R. Durney

National Solar Observatory
National Optical Astronomy Observatories*
Tucson, Arizona 85726 USA

ABSTRACT

The consequences of a balance between the Coriolis forces, pressure gradients and buoyancy forces in a compressible medium are investigated (the Taylor-Proudman theorem). A simple proof is given that if this balance holds, then the latitudinally dependent part of the superadiabatic gradient ($\nabla\Delta T$) is determined by the angular velocity, Ω, and it is of the order of $2\Omega_o^2 T/7g$ for rotation laws other than Ω constant along cylinders (it vanishes in this case). Here Ω_o is the average angular velocity, T the temperature and g gravity. In the lower part of the solar convection zone (SCZ), $2\Omega_o^2 T/7g$ is of the order of $\nabla\Delta T_r$, itself, i.e., very large.

The generalization of the mixing length theory to rotating convection zones is developed with particular emphasis on the approximations involved. The arbitrary parameters are the dimensions of the turbulent eddies. For large values of Ω_o, these eddies (with dimensions, say, ℓ_{pa}, ℓ_{pe} along and perpendicular to the axis of rotation), align themselves along the axis of rotation. Of particular importance is the "elongtion law," i.e., the equation relating ℓ_{pa}/ℓ_{pe} to α ($= 4\Omega_o^2 T/g\nabla\Delta T_r$), the dimensionless parameter characterizing the strength of rotation: surprisingly, there is little freedom in the choice of physically realistic elongation laws. It is shown that the Reynolds stresses transport angular momentum towards the equator and towards the deeper layers of the SCZ. This angular momentum transport tends to generate angular velocities with $\omega_o(r)$ decreasing outwards and negative $\omega_2(r)$ (equatorial acceleration), in conflict with a constant Ω along cylinders ($\Omega = \omega_o(r) + \omega_2(r)P_2(\cos\theta) + \omega_4(r)P_4(\cos\theta) + \cdots$; θ is the polar angle and $P_{2n}(\cos\theta)$ are Legendre polynominals). Meridional motions should change this rotation law. Large latitudinal variations ($\nabla_\theta\Delta T$) in the superadiabatic gradient should, however, be present in the lower part of the SCZ. The numerical solution of the energy equation shows that large values of $\nabla_\theta\Delta T$ do not necessarily imply large latitudinal variation in the energy flux.

The behaviour of the angular velocity at the boundaries of the SCZ is investigated. In order to include the overshoot region, no assumption is made concerning the radial variation of the turbulent viscosity. It is shown that for values of r slightly larger then R_c (the lower boundary), $\omega_2(r)$ varies little with r and $\omega_o(r)$ decreases outwards as suggested by dynamo theory.

*Operated by the Assocation of Universities for Research in Astronomy, Inc. under contract with the National Science Foundation.

B. R. Durney and S. Sofia (eds.), The Internal Solar Angular Velocity, 235–262.

1. INTRODUCTION

This paper is concerned with the following main subjects:

a) The consequences of a Taylor - Proudman balance for compressible flows (§ 3).

The Taylor-Proudman balance in the radial and latitudinal momentum equations is a balance between Coriolis forces, pressure gradients and buoyancy forces. No assumption is made concerning the azimuthal equation of motion. For the sun, the relevance of such a balance comes from the large value of the Coriolis force: differential rotation is of the order of rotation itself, i.e., the Coriolis force is very large. It is instructive to compare the Coriolis term, $2U_\phi\Omega_o$, with the largest turbulent-viscous term, namely, $\rho^{-1}\frac{d}{dr}(\mu\frac{dU_\theta}{dr})\approx\nu_t U_\theta/\ell_\rho\ell_u$. Here $\mu=\rho\nu_t$ with $\nu_t=\tau \underset{\sim}{v}^2/3$, the turbulent viscosity; τ and $\underset{\sim}{v}^2$ are respectively the life-time and mean square velocity of the turbulent eddy, and ℓ_ρ, ℓ_u are the scale of variations for the density and U_θ. An estimate for ν_t can be obtained as follows: we assume isotropy ($\underset{\sim}{v}'^2=3v_r'^2$), then from the energy equation for the turbulent motions (cf equation 5D.4), from the expression for the convective flux ($F=\tau\rho c_p v_r'^2\nabla\Delta T$) and from $\tau=\ell/v_r'$ (ℓ is the mixing length) we obtain $\nu_t=(2Fg\ell^4/15p)^{1/3}\approx 3.8\ 10^{13}cm^2s^{-1}$ for the lower CZ; p is the pressure and $\ell\approx 8\ 10^9cm$. Now, $\ell_\rho\approx\ell$ and we also assume $\ell_u\approx\ell$; the viscous term (VT) is, therefore, of the order of $6\ 10^{-7}U_\theta s^{-1}$. For the Coriolis term (CT) we find $2\Omega_o U_\phi\approx 2\Omega_o^2 r/10=6.7\ 10^{-2}cms^{-2}$. Clearly $VT<<CT$ for all reasonable values of U_θ.

For equatorial symmetry, the expansions of the angular velocity and mean entropy in Legendre polynomials are $\Omega=\Omega_o(\omega_o(r)+\omega_2(r)P_2(\cos\theta);\ +\omega_4(r)P_4(\cos\theta)+..)$; $S=S_o(r)+S_2(r)P_2(\cos\theta)+S_4(r)P_4(\cos\theta)+\cdots$; where θ is the polar angle. It is shown that if the Taylor-Proudman balalnce holds then $S_2(r), S_4(r)\cdots$ are given by products of $2\Omega_o^2T/7g$ with dimensionless, linear functions of $\omega_2, \omega_4, ..., \omega_o', \omega_2', \omega_4'.....$ (the primes denote derivatives with respect to r). These functions vanish if Ω is constant along cylinders and then, $S_2=S_4\cdots=0$; *also* if $S_2, S_4\cdots$ do not vanish, Ω cannot be constant along cylinders. The numerical solution of the energy equation suggests that considerable values of $S_2, S_4\cdots$ *at* $r\approx R_c$ are compatible with uniform energy flux.

b) The generalization of the mixing length theory to rotating convection zones (§ 4, 5, 6, 7, 9).

Within the limits imposed by space, a generalization of the mixing length theory to rotating convection zones is presented which is as comprehensive as possible (see also Durney and Spruit 1979, Durney 1985). An approach of this type to the theory of rotating convection zones must be judged, partly, on the results. The main results here are the expressions for the velocity correlations in terms of the mean quantities (see § 5C). In each single case the properties of the velocity correlations are found to be physically realistic (see § 6, f.ex). In the calculation of the velocity correlations, the mean flow is neglected in the equation for the turbulent velocities (the turbulent and mean quantities are then coupled through the mean superadiabatic gradient). An estimate of the effect of the mean flow on the velocity correlations is presented in section 7.

The constraints imposed by the Taylor-Proudman theorem are discussed in § 9A in the light of the numerical solution of the energy equation.

In § 9B, the equation for Ω is solved numerically; only the Reynolds stresses are included, the meridional motions are neglected. The numerical results show that $\omega_o(r)$ decreases outward, $\omega_2(r)$ is negative (equatorial acceleration) and varies little with r. The remaining $\omega_{2n}(r)$ are small. According to present day observations it appears that the solar $\omega_o(r)$ differs considerably more from this calculated value than the solar $\omega_2(r)$ does. Assume that meridional circulations change this "Reynolds stress rotation law," mainly to alleviate the constraints imposed by the Taylor-Proudman balance. The question that needs to be answered is then: why is $\omega_o(r)$ mostly changed [and not $\omega_2(r)$] ?

c) The behaviour of the angular velocity at the boundaries of the solar convection zone (§ 8).

The lower boundary is important because it is the region where the solar dynamo is generated. The surface boundary is important because it is here that a comparison between theory and observations (present and future) will be easier to carry out. The "boundaries" of the SCZ are not well defined. They could be thought as the values of r at which the superadiabatic gradient changes sign or as the boundaries of the overshoot region. To include both cases, no assumption was made concerning the radial dependence of the turbulent viscosity, ν_t. It was, however, assumed that ν_t is isotropic; there can be little doubt that it is of considerable importance to abandon, eventually, this assumption. In principle this is feasible, but quite involved in particular for the lower boundary.

Other approaches to the theory of rotating convection zones are presented in the papers by Glatzmaier (1987), Stix (1987) and Rudiger and Tuominen (1987); see also Chan (1987).

2. THE MEAN EQUATIONS, CONSERVATION PROPERTIES

i) In a system of coordinates rotating with an angular velocity Ω_o the general non-axisymmetric momentum equation can be written

$$\rho \partial \underset{\sim}{u}/\partial t + \rho \underset{\sim}{u}\cdot\nabla\underset{\sim}{u} = 2\rho\underset{\sim}{u}\times\underset{\sim}{\Omega}_o - \nabla p_t + \underset{\sim}{g}\,\rho_t \tag{2.1a}$$

$$\frac{d}{dt} = \frac{\partial}{\partial t} + \Omega_o \frac{\partial}{\partial\phi} \tag{2.1b}$$

$$p_t = p + p'; \quad \rho_t = \rho + \rho'; \quad \underset{\sim}{u} = \underset{\sim}{U} + \underset{\sim}{u}' \tag{2.1c}$$

where $p, \rho, \underset{\sim}{U}$ are the mean - and $p', \rho', \underset{\sim}{u}'$ the turbulent - values of the pressure, density and velocity. In the terms containing $\underset{\sim}{u}$, ρ_t has been replaced by ρ. To obtain the equation for the mean flow $\underset{\sim}{U}$, we take the average of equation (2.1a) and assume that the mean flow is axisymmetric. The average of $\rho\underset{\sim}{u}\cdot\nabla\underset{\sim}{u}$ gives rise to $\rho\underset{\sim}{U}\cdot\nabla\underset{\sim}{U}$ and to the Reynolds stresses:

$$<\rho\,\underset{\sim}{u}\cdot\nabla\underset{\sim}{u}> = \rho\underset{\sim}{U}\cdot\nabla\underset{\sim}{U} + <\rho u'^i\partial_i\underset{\sim}{u}'> = \rho\underset{\sim}{U}\cdot\nabla\underset{\sim}{U} + \partial_i<\rho u'^i\underset{\sim}{u}'> \tag{2.2a}$$

$$\partial_i <(\rho u'^i\underset{\sim}{u})> = div\mathbf{T}; \quad T^{ij} = <\rho u'^i u'^j> \tag{2.2b}$$

In equation (2.2b) the bracket denotes an average, repeated indices are summed and ∂_i is in general a covariant derivative. The second equality in equation (2.2a) makes use of the continuity equation for the turbulent flow in the following form: div $(\rho\underset{\sim}{u}')=0$. With the help of equations (2.2), the mean momentum equation can be written.

$$\rho\partial\underset{\sim}{U}/\partial t + \rho\underset{\sim}{U}\cdot\nabla\underset{\sim}{U} + div\mathbf{T} = 2\rho\underset{\sim}{U}\times\underset{\sim}{\Omega}_o - \nabla p + \rho\underset{\sim}{g} \tag{2.3}$$

The conservation of angular momentum follows from equation (2.3). It is not difficult to show that

$$\partial J(r)/\partial t = -\int_R^r \frac{\partial}{\partial r}[\sin^2\theta \rho r^3 U_r(U_\phi + \Omega_o r \sin\theta)]drd\theta d\phi - \int_R^r \frac{\partial}{\partial r}[\sin^2\theta\ r^3\ T_{r\phi}]drd\theta\ d\phi \quad (2.4a)$$

where

$$J(r) = \int_R^r \rho r^3 \sin^2\theta\ (U_\phi + \Omega_o\ r \sin\theta)drd\ \theta\ d\phi \quad (2.4b)$$

is the angular momentum of the shell (R,r) in the inertial system of coordinates.

ii) The mean energy equation can be written

$$\rho T(\frac{\partial}{\partial t} + \underset{\sim}{U}\cdot\nabla)S + div\underset{\sim}{F} + T^{ij}\,\partial_j U_i = 0 \quad (2.5a)$$

where S is the mean entropy and

$$\underset{\sim}{F} = \underset{\sim}{F}_c + \underset{\sim}{F}_{rad} + \frac{1}{2}\,\rho\,\underset{\sim}{U}\langle\underset{\sim}{u}'^2\rangle \quad (2.5b)$$

is the sum of the energy fluxes due to convection, radiation and kinetic energy of the turbulent motions. The third term in equation (2.5a), namely, $T^{ij}\,\partial_j U_i$ is the turbulent dissipation function. It is clear (see Durney, 1981) that the derivation of equation (2.5a) is not rigurous. However, it is apparent, that this equation is a generalization, for turbulent motions, of the general equation for heat transfer (see Landau and Lifshitz 1959, p 185, Equation 49.4). In this paper it will be assumed that the energy flux is due, entirely, to convection:

$$F^i = \tau\,\rho\,c_p < u'^i u'^j > \nabla\Delta T_j \quad (2.6a)$$

$$\underset{\sim}{\nabla}\Delta T = (1-1/\gamma)T\nabla p/p - \nabla T \quad (2.6b)$$

In these equations, τ is the life-time of the turbulent eddy, c_p is the specific heat at constant pressure and γ ($=c_p/c_v$) is the ratio of specific heats. Equation (2.6a) is a generalization of the usual mixing length expression for the convective flux (see Durney and Spruit 1979, hereafter DS).

A useful relation is the following:

$$\underset{\sim}{\nabla}\,\Delta T = -\,T\nabla S\ /c_p \quad (2.7)$$

which expresses the superadiabatic gradient, $\underset{\sim}{\nabla}\,\Delta T$, in terms of the mean entropy S.

With the help of the momentum equation it is not difficult to show from equation (2.5a) that energy is conserved:

$$\frac{\partial}{\partial t}(c_v\,\rho\,T + \frac{1}{2}\,\rho\,\underset{\sim}{U}^2) + div\,[(c_v\,\rho\,T + \frac{1}{2}\,\rho\,\underset{\sim}{U}^2 + p)\,\underset{\sim}{U} + U_i\,T^{ij} + \underset{\sim}{F}\,] = \rho\,\underset{\sim}{g}\cdot\underset{\sim}{U} \quad (2.8)$$

In equation (2.8), $U_i\,T^{ij}$ is the vector obtained by contracting $\underset{\sim}{U}$ with the tensor **T**. Assume that $U_r = 0$ at $r = R_c, R_o$ (the lower and upper boundaries of the SCZ) and that $\partial / \partial t = 0$. Integrating equation (2.8) over the entire convection zone gives

$$\int_{R_o} F_r\,ds = \int_{R_c} F_r\,ds + \int_{R_o}(U_\theta\,T_{r\theta} + U_\phi\,T_{r\phi})\,ds - \int_{R_c}(U_\theta\,T_{r\theta} + U_\phi\,T_{r\phi})\,ds \quad (2.9)$$

where the integrals are surface integrals. Conservation of flux follows if $T_{r\theta} = T_{r\phi} = 0$ *at* $r = R_c, R_o$ It should be stressed that to obtain this intuitively valid result, the turbulent dissipation function, $T^{ij}\partial_j U_i$, must be included in equation (2.5a).

iii) The remaining equations for the large-scale quantities are the continuity equation, $div\ \rho \underset{\sim}{U} = 0$, ant the perfect gas law. It is clear that the equations for the mean quantities are not closed (a familiar difficulty in the theory of turbulence). In sections 4 and 5 we will derive (in the mixing length approximation) the expressions for the velocity correlations, $< u'^i u'^j >$, in terms of the mean flow.

In the next section we consider an important theorem in rotating flows, namely, the Taylor-Proudman theorem.

3. THE TAYLOR-PROUDMAN THEOREM

i) The Taylor-Proudman theorem explores the consequence of a balance between the Coriolis forces, pressure gradients and buoyancy forces in the momentum equation (2.3), namely: $2\,\underset{\sim}{U} \times \underset{\sim}{\Omega}_o - \frac{1}{\rho}\nabla p + \underset{\sim}{g} = 0$. Concerning the sun, the relevance of such a balance lies in the fact that differential rotation (U_ϕ) is large, of the order of rotation itself. The r-, and θ - components of the Coriolis force are, therefore, very large and reasonable estimates of $\underset{\sim}{U}\cdot\nabla\,\underset{\sim}{U}$ and div **T** in equation (2.3) invariably suggest that a balance given by

$$2\,\Omega_o\,U_\phi \sin\theta - \frac{1}{\rho}\frac{\partial p}{\partial r} - g(r) \equiv E_r = 0;\quad 2\,\Omega_o\,U_\phi \cos\theta - \frac{1}{r\rho}\frac{\partial p}{\partial \theta} \equiv E_\theta = 0 \qquad (3.1)$$

should be a good approximation for the r-, and θ -components of the mean momentum equation over most of the SCZ. The regions where equations (3.1) are most likely to fail are near the upper and lower boundaries of the SCZ: at $r = R_o, R_c$ the meridional and azimuthal flows must satisfy free surface boundary conditions, namely, the $r\theta$ and $r\phi$ components of the viscous stress must vanish. It is apparent that to satisfy these boundary conditions, the viscous terms must be retained in the momentum equation. To obtain equations (3.1), these terms, contained in div **T** (see section 4) were neglected.

The consequences of equations (3.1) for the case when the SCZ is strictly adiabatic ($p = f(\rho)$) are well known (see, f. ex., Pedloski, 1979, p. 43): from equations (3.1) it follows that $\partial E_r/\partial\theta - \partial(\,r\,E_\theta\,)/\partial r = 0$ or $\cos\theta\,\partial U_\phi/\partial r - r^{-1}\sin\theta\,\partial U_\phi/\partial\theta = 0$ which is a first order partial differential equation, admitting $U_\phi = f(\,r\sin\theta\,)$ as the general solution. *The angular velocity must be constant along cylinders.*

ii) The SCZ is, however, not strictly adiabatic. To understand, in this case, the implications of equations (3.1) we proceed as follows (see Durney 1976, 1985): let q_u be any physical quantity in the absence of rotation and δq the perturbation introduced by rotation, then, $q = q_u + \delta q$. With the help of the unperturbed momentum equation, $-\rho_u^{-1}\nabla p_u + \underset{\sim}{g} = 0$, it can be readily shown that

$$\underset{\sim}{RH} \equiv -\frac{1}{\rho}\nabla p + \underset{\sim}{g} = -\frac{1}{\rho_u}\nabla\delta p + \rho_u^{-2}\,\delta\rho\,\nabla p_u \qquad (3.2)$$

It follows from the definition of entropy ($S = c_v\ ln\ p\rho^{-\gamma}$) that $\delta S = -c_p\,[(1-1/\gamma)\delta p/p_u - \delta T/T_u]$ or, with the help of the gas law:

$$\delta p = \frac{\gamma p_u}{\gamma - 1} [\, \delta\, T/T_u - \delta\, S/c_p \,]\, ; \quad \delta \rho = \frac{\rho_u}{\gamma - 1} [\, \delta T/T_u - \gamma\, \delta\, S/c_p \,] \tag{3.3}$$

Equation (3.2) can then be written

$$\underset{\sim}{RH} = - \frac{\gamma R_g}{\mu(\gamma - 1)} [\nabla \delta T - T_u c_p^{-1} \nabla \delta S + (\delta T / T_u) \nabla \Delta T_u] \tag{3.4}$$

Here the last two terms in the bracket are, by virtue of equation (2.7), the perturbation in the superadiabatic gradient. Since the SCZ is adiabatic to a good approximation, the last term in equation (3.4) is negligible over most of the convection zone and

$$\delta\, \underset{\sim}{\nabla} \Delta T = -\, T_u\, c_p^{-1}\, \nabla\, \delta\, S \tag{3.5}$$

a useful reation. Replacing $-\rho^{-1} \nabla p + \underset{\sim}{g} = 0$ in Equation (3.1) by $\underset{\sim}{RH}$ and forming $\partial E_r / \partial \theta - \partial (r E_\theta) / \partial r$ we now obtain

$$2\, \Omega_o\, (\cos\theta\, U_\phi \partial U_\phi / \partial r - r^{-1} \sin\theta\, \partial\, U_\phi / \partial\, \theta\,) = - \frac{1}{r} \frac{dT_u}{dr} \frac{\partial \delta S}{\partial \theta} \tag{3.6}$$

where $dT_u/dr \approx -\, g \,/\, c_p$ due to the adiabatic character of the CZ. We expand, next, Ω, the angular velocity in the rotating system of coordinates ($U_\phi = r \sin\theta\, \Omega$) in Legendre polynomials

$$\Omega = \Omega_o\, (\, \omega_o\, (r) + \omega_2\, (r)\, P_2\, (\cos\theta\,) + ..\,) \tag{3.7}$$

and neglect $\omega_4, \omega_6 \cdots$; Equation (3.6) can then be written

$$\frac{2\Omega_o^2 T_u}{7g} \frac{1}{r} \frac{\partial A}{\partial \theta} = - \frac{T_u}{c_p} \frac{1}{r} \frac{\partial \delta S}{\partial \theta} \tag{3.8a}$$

$$A = r\, B_1\, P_2\, (\cos\theta\,) \,/\, 3 + 3\, r\, B_2\, P_4\, (\cos\theta\,) \,/\, 5 \tag{3.8b}$$

$$B_1 = 12\, \omega_2 + r\, \omega'_2 + 7\, r\, \omega'_o; \quad B_2 = r\, \omega'_2 - 2\, \omega_2 \tag{3.8c}$$

In virtue of equation (3.5), the right hand side of equation (3.8a) is the θ -component of the perturbation in the superadiabatic gradient. Equation (3.8a) is identical to equation (24b) of Durney (1976). In that paper this equation was obtained by solving explicitly for the perturbed pressure and temperature in terms of $\omega_o\, (r)$ and $\omega_2\, (r)$ [Eq. (16), (17) of Durney (1976)] This is clearly feasible since there are three equations (equations (3.1) and the gas law) for the three unknown, $\delta\, p$, $\delta\, \rho$, and $\delta\, T$. If the dependence of g on r is neglected it follows from equation (3.8a) that (f (r) is an arbitrary function of r)

$$\frac{2\Omega_o^2 T_u}{7g} \frac{\partial A}{\partial r} + f(r) = - \frac{T_u}{c_p} \frac{\partial \delta S}{\partial r} \tag{3.8d}$$

The expressions for ω_o, ω_2 such that $B_1 = B_2 = 0$ can be easily derived from equation (3.8c). It is found: $\omega_2 = c_1 r^2$ and $\omega_o + \omega_2 = c_2$ where c_1 and c_2 are constants. Therefore, $\Omega = \Omega_o [c_2 - (3/2) c_1 (r \sin\theta\,)^2\,]$, i.e., rotation is constant along cylinders. Other rotation laws generate perturbations in the superadiabatic gradient which are very roughly of the order of $2\Omega_o^2 T_u / 7g$ according to equations (3.8a) and (3.8d). In Fig. 1, $2\Omega_o^2 T_u / 7g \nabla \Delta T_u$ is plotted against r for the SCZ. It is clear that if Ω is not constant along cylinders large (θ -dependent!) perturbations in the superadiabatic gradient (likely associated with large θ -dependent perturbations in the flux; see, however, section 9A) must be present in the lower part of the SCZ.

It should be kept in mind, however, that due to the efficiency of the turbulent diffusion of heat, large latitudinal variations in the convective flux, in the deeper layers of the SCZ, are very much reduced at the surface (see Spruit 1977 a, b). In relation to dynamo theory one of the key questions that needs to be answered is where departures from the $\Omega = f(r\sin\theta)$ - law are localized and what is their origin.

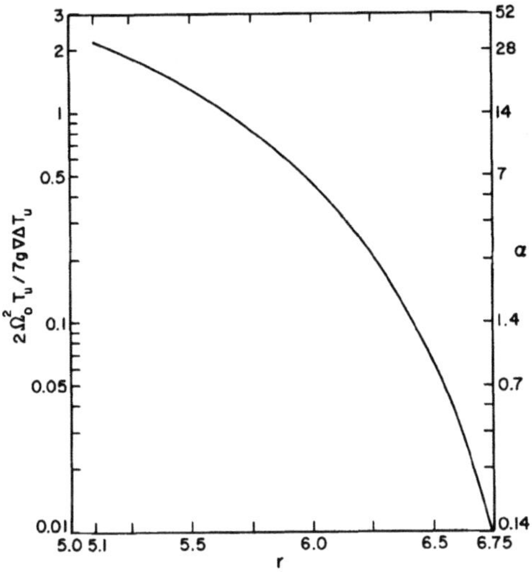

Fig. I

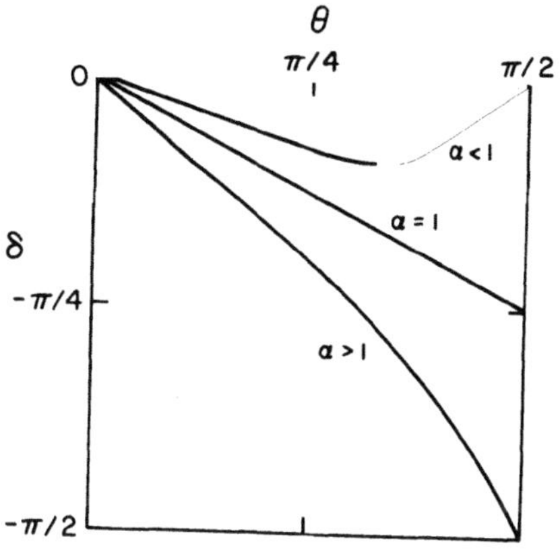

Fig. 2

Fig. 1 $4\Omega_o^2 T_u/g\nabla\Delta T_u$ $(=\alpha)$ and $\alpha/14$ vs radial distance r in the solar convection zone. Units of r: $10^{10}cm$

Fig. 2 Schematic plot of the tilt angle, δ, vs. θ for $\alpha < 1$, $\alpha = 1$, $\alpha > 1$

4. EQUATIONS FOR THE TURBULENT MOTIONS

i) In a system of coordinates rotating with an angular velocity Ω_o we write the momentum equation for the turbulent motions as follows:

$$\rho D\underset{\sim}{u}'/Dt + \rho\underset{\sim}{u}'\cdot\nabla\underset{\sim}{u}' + \rho\underset{\sim}{u}'\cdot\nabla\underset{\sim}{U} = 2\rho\underset{\sim}{u}' \times \underset{\sim}{\Omega}_o - \nabla p' + \underset{\sim}{g}\rho' \tag{4.1a}$$

$$\frac{D}{Dt} = \frac{\partial}{\partial t} + \Omega_o\frac{\partial}{\partial\phi} + \underset{\sim}{U}\cdot\nabla \tag{4.1b}$$

which is obtained by subtracting the equation for the mean flow from equation (2.1a) (strictly speaking we should have written $\rho\underset{\sim}{u}'\cdot\nabla\underset{\sim}{u}' - <\rho\underset{\sim}{u}'\cdot\nabla\underset{\sim}{u}'>$ in equation 4.1a). The operator D/Dt denotes time differentiation following rotation and the mean flow. We introduce the life-time of the eddy:

$$D\underset{\sim}{u}'/DT + \underset{\sim}{u}'\cdot\nabla\underset{\sim}{u}' \approx \underset{\sim}{u}'/\tau \tag{4.2}$$

Equation (4.1a) simplifies then to

$$\rho\underset{\sim}{u}'/\tau = -\rho\underset{\sim}{u}'\cdot\nabla\underset{\sim}{U} + 2\rho\underset{\sim}{u}' \times \underset{\sim}{\Omega}_o - \nabla p' + \underset{\sim}{g}\rho' \tag{4.3}$$

ii) The approximations involved in the energy equation for the turbulent flow will be clear from the derivation below. In general, motions are adiabatic if $ds/dt + \underset{\sim}{u}\cdot\nabla s = 0$, where d/dt is defined by equation (2.1b); s is the total entropy, i.e., $s = S + s'$ where S, s' are, respectively, the mean and fluctuating components of the entropy. We assume that the turbulent motions are adiabatic and write the fluctuating component of the above equation as follows $ds'/dt + \underset{\sim}{u}'\cdot\nabla S + \underset{\sim}{U}\cdot\nabla s' + \underset{\sim}{u}'\cdot\nabla s' = 0$ Now from $s = c_v \ell n\, (p_t \rho_t^{-\gamma})$ it follows that (see Equation (2.1c) for the definitions of $p_t, p, p' \cdots$) $\nabla s = c_v(\nabla p_t/p_t - \gamma\nabla\rho_t/\rho_t) \approx c_v(\nabla p_t/p - \gamma\,\nabla\rho_t/\rho) = c_v\,(\nabla p'/p - \gamma\nabla\rho'/\rho) + c_v\,(\nabla p/p - \gamma\nabla\rho/\rho) = \nabla s' + \nabla\, S$ Analogously $\partial s'/\partial t = c_v[(\partial p'/\partial t)/p - \gamma(\partial\rho'/\partial t)/\rho]$. The adiabatic equation for the turbulent flow, written above, simplifies then to

$$\frac{1}{p}\left[\frac{Dp'}{DT} + \underset{\sim}{u}'\cdot\nabla(p + p')\right] = \frac{\gamma}{\rho}\left[\frac{D\rho'}{Dt} + \underset{\sim}{u}'\cdot\nabla(\rho + \rho')\right] \tag{4.4}$$

Since the turbulent pressure p' tends rapidly to equilibrium with the undisturbed pressure, p, we neglect $Dp'/Dt + \underset{\sim}{u}'\cdot\nabla p'$ in equation (4.4), and introduce again the lifetime of the eddy:

$$D\rho'/Dt + \underset{\sim}{u}'\cdot\nabla\rho' \approx \rho'/\tau \tag{4.5}$$

With the help of these approximations equations (4.4) reduces to

$$\rho'/\tau = -\,\rho\underset{\sim}{u}'\cdot\underset{\sim}{\nabla}\Delta T/T \tag{4.6}$$

where the superadiabatic gradient has been defined in equation (2.6b).

iii) The continuity equation for the turbulent flow will be approximated to

$$div\ \underset{\sim}{u}' = 0 \tag{4.7}$$

Equations (4.3), (4.6) and (4.7) are the momentum, energy and continuity equations for the turbulent flow in the mixing length approximation. It is clear that the present derivation of these equations closely follows Cowling's analysis on the condition for turbulence in rotation stars (Cowling 1951; Durney and Spruit 1979). As Equation (4.7) shows, the mixing length approximation is Boussinesq, that is, the convected elements are assumed to have small dimensions compared with the scale of variation of macroscopic variables like pressure, density, etc. The Boussinesq approximation can, however, be valid for strongly stratified convection zones.

iv) The mixing length approximation of the momentum equation for the turbulent flow will be approximated further by neglecting the term $-\rho\underset{\sim}{u}'\cdot\nabla\underset{\sim}{U}$ in the righthand side of equation (4.3) *We also assume that* $\underset{\sim}{g}$ *and* $\underset{\sim}{\nabla}\,\Delta T$ *are in the radial direction* and designate by $\nabla\Delta T$ the radial component of the mean superadiabatic gradient. To summarize: In this paper the Reynolds stresses will be evaluated by solving (for specified dimensions of the convective elements) the following equations for the turbulent motions (see Paper DS; we use $\underset{\sim}{v}'$ to designate the turbulent velocities when $-\rho\underset{\sim}{u}'\cdot\nabla\underset{\sim}{U}$ is neglected in equation (4.3)

$$\rho\underset{\sim}{v}'/\tau = 2\rho\underset{\sim}{v}' \times \underset{\sim}{\Omega}_o - \nabla p' + \underset{\sim}{g}\rho' \tag{4.8a}$$

$$\rho'/\tau = -\rho v_r'\nabla\Delta T/T \tag{4.8b}$$

$$div\ \underset{\sim}{v}' = 0 \tag{4.8c}$$

The following important fact is apparent from equations (4.8): *the large-scale quantities can influence the turbulent quantities through* $\nabla\Delta T$, *the mean superadiabatic gradient.*

v) Assume that with the help of equations (4.8), the velocity correlations $<v'^i v'^j>$ have been calculated in terms of the mean quantities (see Section 5). The effect of the mean flow $\underset{\sim}{U}$, can then be incorporated approximately with the help of a perturbation procedure as follows: assume that the mean flow does not significantly alter

τ, ρ' and p'. Subtracting equations (4.8a) from equation (4.3) gives

$$(\underset{\sim}{u}' - \underset{\sim}{v}')/\tau - 2(\underset{\sim}{u}' - \underset{\sim}{v}') \times \underset{\sim}{\Omega}_o = - \underset{\sim}{v}' \cdot \nabla \underset{\sim}{U} \tag{4.9}$$

where, consistent with a perturbation procedure, we have replace u' by $\underset{\sim}{v}'$ in the term containing the mean flow. As an illustration we assume that $\Omega_o = \tilde{0}$. Neglecting terms quadratic in τ we obtain

$$< u'^i u'^j > = < v'^i v'^j > - \tau[< v'^i v'^k > \partial_k U^j + < v'^j u'^k > \partial_k U^i] \tag{4.10}$$

where for cartesian coordinates $\partial_k U^j = \partial U^j / \partial x^k$; for more general system of coordinates $\partial_k U^j$ are coveriant derivatives. In equation (4.10) repeated indices are summed. *In this approximation the effect of the mean flow reduces to a turbulent viscosity.*

5. CALCULATION OF THE REYNOLDS STRESSES

As stated in the previous section, the Reynolds stresses will be evaluated by solving equations (4.8) for specified elimensions of the convective eddies. It should be noticed that the method of solution remains viable if $\underset{\sim}{U}$ and the θ component of $\underset{\sim}{\nabla} \Delta T$ are retained

Equations (4.8) are linear and homogenous in the unknown quantities $\underset{\sim}{v}', p'$, and ρ'. The solution of these equations expresses $\underset{\sim}{v}'$, p' and ρ' in terms of an unknown amplitude, w and the eddy's dimension. The calculation of the Reynolds stresses involves the following steps:

a) Solution of equations (4.8) for $p', \rho', \underset{\sim}{v}'$ proportional to exp $(i\underset{\sim}{k} \cdot \underset{\sim}{r})$ (basic waves or rolls)

b) Combination of several values of $\underset{\sim}{k}$ to form eddies of specified dimensions

c) Calculation of the amplitude, w, undetermined in a) and b).

d) Calculation of the spatial averages $< v'^i v'^j >$

We discuss these points in turn:

5A. Solution of equations (4.8) for basic waves or rolls.

We take $p', \rho', \underset{\sim}{v}'$ proportional to exp $(i\underset{\sim}{k} \cdot \underset{\sim}{r})$ ($\underset{\sim}{k}$ is the eddy wavevector):

$$p', \rho' \underset{\sim}{v}' \sim \exp(ik_r r + ik_\theta r\theta + ik_\phi r \sin\theta \, \phi) \tag{5A.1}$$

where r, $r\theta$, $r\sin\theta\phi$ are the radial, meridional and azimuthal coordinates, respectively. Equations (4.8) can then be written

$$\rho \underset{\sim}{v}'/\tau = 2\rho \underset{\sim}{v}' \times \underset{\sim}{\Omega}_o - i\underset{\sim}{k}p' + \underset{\sim}{g}\rho' \tag{5A.2a}$$

$$\rho'/\tau = - \rho v_r' \nabla \Delta T / T \tag{5A.2b}$$

$$\underset{\sim}{k} \cdot \underset{\sim}{v}' = 0 \tag{5A.2c}$$

Since equations (5A.2) are linear and homogeneous in $p', \rho', \underset{\sim}{v}'$, it is possible to show the following two important points [Durney and Spruit 1979, Durney 1983].

i) Equations (5A.2) admit only a solution if

$$1/\tau^2 = (gK^2/Tk^2)\nabla\Delta T(1 - \alpha(\underset{\sim}{\omega}_o \cdot \underset{\sim}{k})^2/K^2) \tag{5A.3a}$$

$$K^2 = k_\theta^2 + k_\phi^2;\ \ k^2 = K^2 + k_r^2;\ \ \alpha = 4\Omega_o^2 T/g\nabla\Delta T \tag{5A.3b}$$

$$\underset{\sim}{\omega}_o = \underset{\sim}{\Omega}_o/\Omega_o \ \ (unit\ vector) \tag{5A.3c}$$

ii) If $1/\tau^2$ is given by equation (5A.3a), then

$$\underset{\sim}{v}' = w[\underset{\sim}{P} + \frac{2\Omega_o\tau\,(\underset{\sim}{\omega}_o\cdot\underset{\sim}{k})}{k}\underset{\sim}{T}] \tag{5A.4a}$$

$$\underset{\sim}{P} = \underset{\sim}{k}\times(\underset{\sim}{n}\times\underset{\sim}{k})/|\underset{\sim}{k}\times(\underset{\sim}{n}\times\underset{\sim}{k})|;\ \ \underset{\sim}{T} = \underset{\sim}{n}\times\underset{\sim}{k}/|\underset{\sim}{n}\times\underset{\sim}{k}| \tag{5A.4b}$$

$$\underset{\sim}{n} = (\ unit\ vector\ in\ the\ radial\ direction\ ) \tag{5A.4c}$$

$$\underset{\sim}{P}^2 = \underset{\sim}{T}^2 = 1;\ \ \underset{\sim}{k}\cdot\underset{\sim}{P} = \underset{\sim}{k}\cdot\underset{\sim}{T} = 0;\ \ \underset{\sim}{P}\cdot\underset{\sim}{T} = 0 \tag{5A.4d}$$

$$\underset{\sim}{P} = (K^2, -k_rk_\theta, -k_rk_\phi)/Kk;\ \ \underset{\sim}{T} = (0, -k_\phi, k_\theta)/K \tag{5A.4e}$$

In relation to these equations the following comments are in order: Equations (5A.3a) is an expression for the life-time of the convective eddy. It is clear that if $\alpha > 1$, then convection can influence the convective instability (Tayler 1973). This stabilizing effect of rotation (τ tends to increase) can be reduced if the eddy elongates along the axis of rotation, i.e., if $(\underset{\sim}{\omega}_o\cdot \underset{\sim}{k}^2)/k^2$ becomes smaller than unity. Equation (5A.4a) expresses the turbulent velocity $\underset{\sim}{v}'$, in terms of polaidal, $\underset{\sim}{P}$, and toroidal, $\underset{\sim}{T}$, vectors. The vectors $\underset{\sim}{P}$ and $\underset{\sim}{T}$ (with components given by equations (5A.4e) in a spherical system of coordinates) are local definitions of the well-known poloidal and toroidal vectors (see Chandrasekhar, 1961). Rotation introduces a toroidal component of the turbulent flow of relative amplitude $2\Omega_o\tau(\underset{\sim}{\omega}_o\cdot\underset{\sim}{k})/k$. It should be noticed that $(2\Omega_o\tau)^2 \sim 4\Omega_o^2 T/g\nabla\Delta T$ (see Tayler, 1973). Therefore, if rotation is sufficiently strong to influence the convective instability, then Ω_o will be able also to influence the dynamics of the motion. Again it is clear from equation (5A.4a) that the toroidal component of $\underset{\sim}{v}'$ will be reduced if $\underset{\sim}{\omega}_o\cdot\underset{\sim}{k}/k << 1$. Finally it should be stressed that equations (5A.3a) and (5A.4a) are valid for arbitrary large values of $\Omega_o\tau$ (see Durney 1983).

Since equations (5A.2) are homogenous, the amplitude w in equation (5A.4a) is undetermined for the moment.

5B. Solution of equations (4.8) for eddies of specified dimensions

Consider the wave vector $\underset{\sim}{k} = k_r, k_\theta, k_\phi$, assume that $\Omega_o = 0$ and define the eight wave-vectors $\underset{\sim}{k}_i$ by taking for $k_{ir}, k_{i\theta}\, k_{i\phi}$ all the possible sign combinations $(\pm k_r, \pm k_\theta \pm k_\phi)$. Therefore;

$$\underset{\sim}{k}_i = k_{ir}, k_{i\theta}, k_{i\phi};\qquad |k_{ir}|, |k_{i\theta}|, |k_{i\phi}|\ independent\ of\ i\ (i = 1, 2, \cdots 8). \tag{5B.1}$$

Then if w and k_r, k_θ, k_ϕ are slowly varying functions of r it is clear that

$$\underset{\sim}{v}' = \frac{w}{8}\sum_{i=1}^{8} \underset{\sim}{P}(\underset{\sim}{k}_i)e^{i\underset{\sim}{k}_i\cdot\underset{\sim}{r}} \tag{5B.2}$$

is a solution of equations (4.8): these equations are linear and the main contribution to the derivatives will come from exp $(i\underset{\sim}{k}_i\cdot\underset{\sim}{r})$, varying in an eddy-scale. The contributions to the derivatives due to the spatial dependence of $\underset{\sim}{k}, w, \rho....$ varying on a mean-scale is small. In equation (5B.2), $\underset{\sim}{P}(\underset{\sim}{k}_i)$ is the poloidal vector $\underset{\sim}{P}$ evaluated with $\underset{\sim}{k} = \underset{\sim}{k}_i$. Explicitly it is found

$$v'_r = \frac{Kw}{k}\cos(k_r r)\cos(k_\theta r\,\theta)\cos(k_\phi r\sin\theta\,\phi) \quad (5B.3a)$$

$$v'_\theta = \frac{k_r k_\theta w}{Kk}\sin(k_r r)\sin(k_\theta r\,\theta)\cos(k_\phi r\sin\theta\,\phi) \quad (5B.3b)$$

$$v'_\phi = \frac{k_r k_\phi w}{Kk}\sin(k_r r)\cos(k_\theta r\,\theta)\sin(k_\phi r\sin\theta\,\phi) \quad (5B.3c)$$

$$p' = \frac{wk_r\rho}{Kk\tau}\sin(k_r r)\cos(k_\theta r\,\theta)\cos(k_\phi r\sin\theta\,\phi) \quad (5B.3d)$$

Clearly div $\underset{\sim}{v}' = 0$ (it should be kept in mind that the coordinates are $R = r$, $\Theta = r\theta$, $\Phi = r\sin\theta\,\phi$) and the eddies have dimensions π/k_r, π/k_θ and π/k_ϕ along R,Θ and Φ .

We consider now the case with rotation. There is an essential difference between the cases with $\Omega_o = 0$ and $\Omega_o \neq 0$. If $\Omega_o = 0$, τ given by equation (5A.3a), is the same for all k_i's defined by equation (5B.1). If $\Omega_o \neq 0$, τ depends on the particular value of $\underset{\sim}{k}_i$ ($\underset{\sim}{\omega}_o \cdot \underset{\sim}{k}_i$ depends on i). Due to this dependence of τ on $\underset{\sim}{k}_i$ it is clear that it is inconsistent to define $\underset{\sim}{v}' = \sum_{i=1}^{8}[\underset{\sim}{P}(\underset{\sim}{k}_i) + \frac{2\Omega_o\tau_i(\underset{\sim}{\omega}_o \cdot \underset{\sim}{k}_i)}{k}\underset{\sim}{T}(\underset{\sim}{k}_i)]\, e^{i\underset{\sim}{k}_i \cdot \underset{\sim}{r}}$ as the velocity field satisfying equations (4.8). The way out of this dichotomy, chosen in paper DS, was to replace $(\underset{\sim}{\omega}_o \cdot \underset{\sim}{k})^2$ in equation (5A.3a) by its average over the values of $\underset{\sim}{k}_i$, which obviously is not a satisfactory procedure. The definitive solution of this problem, which has its roots in interesting physics is the following: from a study of the linearized equations describing convective motions in a rapidly rotating thin spherical shell, Flasar and Gierasch (1978) found that the solutions to these equations were proportional to (with our notation) $W(r)\exp[\sigma t + i(k_\theta r\theta + k_\phi r\sin\theta\phi + \lambda_1 k_\theta r)] \equiv FG$, where σ the growth rate $(\sim 1/\tau)$. This result was obtained by taking the radial velocities to vanish at the boundaries of the spherical shell. As noticed by Hathaway (1984) there is a non-trivial difference between equation (5A.1) and expression FG above: Equation (5A.1) does not contain the term, exp $(i\lambda_1 k_\theta r)$, which represents a poleward tilt of the eddy. Furthermore, Hathaway obtained the following result: if equation (5A.1) is replaced by $\exp[\sigma t + ir(k_r + \lambda_1 k_\theta) + ik_\theta r\theta + ik_\phi r\sin\theta\,\phi] \equiv H$ then (for the appropriate value of λ_1) σ is the same for all $\underset{\sim}{k}_i$. Hathaway's significant step forward is, however, not the definitive answer to the "tilt problem". It is apparent that in the expression, H, above, k_r has received a preferential treatment: H can essentially be obtained from equation (5A.1) by replacing $k_r \rightarrow k_r + \lambda_1 k_\theta$, whereas k_θ is left unchanged. The origin of this dissimilarity between k_r and k_θ lies in the rigid boundary conditions imposed by Flasar and Gierasch on the eddy's velocity. It is clear that in the present context, no such boundary conditions should be imposed. Keeping this in mind, the introduction of the tilt proceeds then along the following lines (see Durney 1983): the eddy's dimensions cannot be defined in the $\underset{\sim}{n}_r$, $\underset{\sim}{n}_\theta$ system of coordinates (unit vectors in the radial and latitudinal directions) but must be defined in a system of coordinates (say $\underset{\sim}{n}_1$, $\underset{\sim}{n}_2$) obtained from $\underset{\sim}{n}_r$, $\underset{\sim}{n}_\theta$ by a rotation:

$$\underset{\sim}{n}_1 = \underset{\sim}{n}_r\cos\delta + \underset{\sim}{n}_\theta\sin\delta; \quad \underset{\sim}{n}_2 = -\underset{\sim}{n}_r\sin\delta + \underset{\sim}{n}_\theta\cos\delta \quad (5B.4)$$

The angle δ will be determined be requiring that the eight basic waves generating the eddy have all the same life-time. Let π/p_1, π/p_2, π/k_ϕ be the dimensions of the eddy in the $\underset{\sim}{n}_1$, $\underset{\sim}{n}_2$ system and consider the wave vector $\underset{\sim}{k} = p_1\underset{\sim}{n}_1 + p_2\underset{\sim}{n}_2 + k_\phi\underset{\sim}{n}_\phi$ with components p_1, p_2, k_ϕ in this system. By virtue of equation (5B.4) the components of $\underset{\sim}{k}$ in the $\underset{\sim}{n}_r$ $\underset{\sim}{n}_\theta$ system are given by

$$k_r = p_1\cos\delta - p_2\sin\delta;\quad k_\theta = p_1\sin\delta + p_2\cos\delta;\quad k_\phi \tag{5B.5}$$

For $k_{ir}, k_{i\theta}, k_{i\phi}$ (the components of the eight basic $\underset{\sim}{k}_i$'s generating the eddy) we take the values of k_r, k_θ, k_ϕ obtained from equation (5B.5) by assigning to p_1, p_2, k_ϕ the eight possible sign combinations, $\pm p_1, \pm p_2, \pm k_\phi$. We discuss now the determination of δ. From the equation for τ, equation (5A.3a), it is clear that the life-time of the eight basic waves will be the same if $K_i^2 - \alpha(\underset{\sim}{\omega}_o\cdot\underset{\sim}{k}_i)^2$ is independent of i, i.e., if in this expression the term in p_1p_2 vanishes. For δ we obtain

$$tg2\delta = -\alpha\sin2\theta/(1+\alpha\cos2\theta) \tag{5B.6}$$

It can be readily seen that the common value of τ is given by

$$1/\tau^2 = (g<K^2>/Tk^2)\nabla\Delta T(1-\alpha<k_y^2>/<K^2>) \tag{5B.7a}$$

where

$$<K^2> = 8^{-1}\sum_{i=1}^{8}(k_{i\theta}^2 + k_{i\phi}^2);\quad <k_y^2> = 8^{-1}\sum_{i=1}^{8}(\underset{\sim}{\omega}_o\cdot\underset{\sim}{k}_i)^2 \tag{5B.7b}$$

Therefore, $<K^2>$ and $<k_y^2>$ are average values over the eight values of i. Since τ is now the same for the eight basic waves or rolls it is clear that

$$\underset{\sim}{v}' = \frac{w}{8}\sum_{i=1}^{8}\left[\underset{\sim}{P}(\underset{\sim}{k}_i) + \frac{2\Omega_o\tau(\underset{\sim}{\omega}_o\cdot\underset{\sim}{k}_i)}{k}\underset{\sim}{T}(\underset{\sim}{k}_i)\right]e^{\underset{\sim}{k}_i\cdot\underset{\sim}{r}} \tag{5B.8}$$

is a solution of equations (4.8) with the desired properties: It is a real periodic velocity field with a $\pi/p_1, \pi/p_2, \pi/k_\phi$ periodicity in the $\underset{\sim}{n}_1, \underset{\sim}{n}_2, \underset{\sim}{n}_\phi$ system.

5C. Calculation of the spatial averages of the velocity correlation

The velocity field $\underset{\sim}{v}'$ given by equation (5B.8) contains two classes of terms, namely, rapidly varying functions of $\underset{\sim}{r}$, (exp $i\underset{\sim}{k}_i\cdot\underset{\sim}{r}$), and quantites like $\delta, \tau, \underset{\sim}{\omega}_o$ which vary with $\underset{\sim}{r}$, only slowly, on a global scale or on scale comparable with that of the mean quantities. Therefore, the velocity correlations, $v_\alpha'v_\beta'$ ($\alpha, \beta = r, \theta, \phi$), which can be calculated from Equation (5B.8), will contain, aside from these slowly varying quantities, rapidly varying functions of $\underset{\sim}{r}$ of the form exp $i(\underset{\sim}{k}_i + \underset{\sim}{k}_j)\cdot\underset{\sim}{r}$ (or exp $i(\underset{\sim}{p}_j + \underset{\sim}{p}_j)\cdot\underset{\sim}{x}$ in the $\underset{\sim}{n}_1, \underset{\sim}{n}_2$ system; the notation is obvious). It is clear that these oscillatory terms will not contribute to the spatial average of $v_\alpha'v_\beta'$; the only contribution, to $<v_\alpha', v_\beta'>$ will come from those terms of v_α', v_β' such that in exp $i(\underset{\sim}{k}_i + \underset{\sim}{k}_j)\cdot\underset{\sim}{r}$ the sum of the wave-vectors $\underset{\sim}{k}_i + \underset{\sim}{k}_j$ (or equivalently $\underset{\sim}{p}_i + \underset{\sim}{p}_j$) vanishes. With the help of these relations the calculation of $<v_\alpha'v_\beta'>$ reduces to some tedious algebra. These velocity correlations contain averages over the eight basic waves or rolls: let F by any function of $\underset{\sim}{p}_j$ (i=1,2,..8) We define this average as follows

$$<F> = \frac{1}{8}\sum_{i=1}^{8}F(\underset{\sim}{p}_j) \tag{5C.1}$$

We furthermore define

$$\beta = \theta + \delta;\quad \gamma_1 = p_1p_2\sin2\delta;\quad \gamma_2 = p_1p_2\sin2\beta;$$

$$\gamma_3 = p_1p_2\sin(\delta+\beta);\quad k_y = k_r\cos\theta - k_\theta\sin\theta \tag{5C.2}$$

$$t_{jk} = <v_j'v_k'> = w^2(t_{jk}^{(0)} + 2\Omega_o\tau t_{jk}^{(1)} + 4\Omega_o^2\tau^2 t_{jk}^{(2)}) \tag{5C.3}$$

In equation (5C.2), δ is given by equation (5B.6), and k_r and k_θ are given, for each value of i, by equation (5B.5). In the definition of $\gamma_1, \gamma_2, \gamma_3$ above, p_1 and p_2 stand for any p_{1i}, p_{2i}: the value of i is irrelevant since the expressions for the Reynolds stresses below contain only γ_1^2, $\gamma_1\gamma_2$, $\gamma_1\gamma_3$ which are independent of i. It can, then, be shown.

$$<k_r^2> = p_1^2\cos^2\delta + p_2^2\sin^2\delta \quad (5C.4a)$$

$$<k_\theta^2> = p_1^2\sin^2\delta + p_2^2\cos^2\delta \quad (5C.4b)$$

$$<k_y^2> = p_1^2\cos^2\beta + p_2^2\sin^2\beta \quad (5C.4c)$$

$$<k_y k_\theta> = p_1^2\sin\delta\cos\beta - p_2^2\sin\beta\cos\delta \quad (5C.4d)$$

$$<k_y k_r> = p_1^2\cos\delta\cos\beta + p_2^2\sin\delta\sin\beta \quad (5C.4e)$$

$$<k_r k_\theta> = (p_1^2 - p_2^2)\sin\delta\cos\delta \quad (5C.4f)$$

$$<K^2> = <k_\theta^2> + k_\phi^2 \quad (5C.4g)$$

$$<k_r^2 k_\theta^2/K^2> = \left[<k_r^2><k_\theta^2><K^2> - \gamma_1^2(k_\phi^2 + <k_r^2>)\right]/D;$$

$$D = <K^2>^2 - \gamma_1^2 \quad (5C.4h)$$

$$<k_r^2 k_\phi^2/K^2> = k_\phi^2(<k_r^2><K^2> + \gamma_1^2)/D \quad (5C.4i)$$

$$<k_r k_y k_\theta^2/K^2> = (<k_r k_y><k_\theta^2><K^2> - \gamma_1^2<k_r k_y> - \gamma_1\gamma_3 k_\phi^2)/D \quad (5C.4j)$$

$$<k_r k_y k_\phi^2/K^2> = k_\phi^2(<k_r k_y><K^2> + \gamma_1\gamma_3)/D \quad (5C.4k)$$

$$<k_y^2 k_\theta^2/K^2> = (<k_y^2><k_\theta^2><K^2> - \gamma_1^2<k_y^2> - \gamma_1\gamma_2 k_\phi^2)/D \quad (5C.4l)$$

$$<k_y^2 k_\phi^2/K^2> = k_\phi^2(<k_y^2><K^2> + \gamma_1\gamma_2)/D \quad (5C.4m)$$

and clearly $<\underset{\sim}{p}^2> = p_1^2 + p_2^2 + k_\phi^2 = k_r^2 + k_\theta^2 + k_\phi^2$. The non- vanishing components of $t_{jk}^{(N)}$ [see equation (5C.3)] are then given by

$$t_{rr}^{(0)} = <K^2>/k^2;\quad t_{r\theta}^{(0)} = -<k_r k_\theta>/k^2;$$

$$t_{\theta\theta}^{(0)} = <k_r^2 k_\theta^2/K^2>/k^2;\quad t_{\phi\phi}^{(0)} = <k_r^2 k_\phi^2/K^2>/k^2 \quad (5C.5a)$$

$$t_{r\phi}^{(1)} = <k_y k_\theta>/k^2;\quad t_{\phi\phi}^{(1)} = (<k_y k_r k_\phi^2/K^2> - <k_y k_r k_\theta^2/K^2>)/k^2 \quad (5C.5b)$$

$$t_{\theta\theta}^{(2)} = <k_y^2 k_\phi^2/K^2>/k^2;\quad t_{\phi\phi}^{(2)} = <k_y^2 k_\theta^2/K^2>/k^2 \quad (5C.5c)$$

From equation (5C.3) and (5C.5a) the following useful relation follows

$$w^2 = <v_r'^2>k^2/<K^2> \quad (5C.6)$$

5D. Calculation of the amplitude

We define

$$<\underset{\sim}{v}'^2> = <v_r'^2> + <v_\theta'^2> + <v_\phi'^2> \quad (5D.1)$$

with the help of equations (5C.3), (5C.4) and (5C.5) it can be readily shown that

$$<\underset{\sim}{v}'^2> = w^2(1 + 4\Omega_o^2\tau^2<k_y^2>/k^2) \quad (5D.2)$$

From the equation for the eddy's life-time (equation 5B.7a) it follows that

$$1 + 4\Omega_o^2\tau^2<k_y^2>/k^2 = g\tau^2<K^2>\nabla\Delta T/Tk^2 \quad (5D.3)$$

With the help of equations (5C.6) and (5D.3) we can write equation (5D.2) as follows

$$\frac{1}{2}\rho<\underset{\sim}{v}'^2> = \frac{1}{2}\tau^2\rho\frac{g}{T}\nabla\Delta T<v_r'^2> \quad (5D.4)$$

This equation closely resembles the well known expression relating the eddy kinetic energy with the work done by the buoyancy forces [see, f. ex., equation (7.5) of Schwarzschild 1958]. However, since the amplitude w, or $<v_r'^2>$, are undetermined, there is, as yet, no relation between $\tau^2 <v_r'^2>$ and the radial dimension of the eddy, namely the mixing length, ℓ ($\ell^2 = \pi^2/<k_r^2>$). We, therefore, define the amplitude by

$$\tau^2 <v_r'^2> = \ell^2; \quad w^2 = \ell^2 k^2/<K^2>\tau^2 \tag{5D.5}$$

The calculation of the Reynolds stresses is now complete. The arbitrary parameters are the mixing length ℓ (the radial dimension of the eddy) and p_2^2/p_1^2, k_ϕ^2/p_1^2 : from the equation for τ (Equation 5B.7a), and equation (5C.4) and (5C.5) it is clear that w^2 and the velocity correlations, defined in equation (5C.3), are dimensionless in $\underset{\sim}{k}$, and functions of p_2^2/p_1^2, k_ϕ^2/p_1^2.

6. SOME PROPERTIES OF THE TILT ANGLE AND VELOCITY CORRELATIONS

i) It is clear from equation (5B.6) that if $\alpha << 1$ also $\delta << 1$, an intuitively correct result. If $\alpha >> 1$, then $tg 2\delta \approx - tg 2\theta$ or $\delta \approx -\theta$, i.e., the eddy is aligned along the axis of rotation, again a very reasonable result. It is important to notice that with our convention for the sign of the angle δ, the eddy tilt polewards if $\delta < 0$. If $\alpha >> 1$, the denominator in equation (5B.6) vanishes for a certain value of $\theta = (\theta_o$, say; $\theta_o > \pi/4$) and becomes negative for $\theta > \theta_o$ Since the tilt must be poleward, i.e., $\delta < 0$, the continuous branch for δ must be chosen at $\theta = \theta_o$. Finally if $\alpha = 1$, then $\delta = -\theta/2$. Fig. 2 gives a schematic representation of the behaviour of δ for $\alpha < 1$, $\alpha = 1$, $\alpha > 1$.

The mean equations contain derivatives of the Reynolds stresses which are functions of δ. It is, therefore, of interest to consider the derivatives of δ. From equation (5B.6) we obtain.

$$\delta_\alpha' = \frac{\partial \delta}{\partial \alpha} = -\frac{1}{2}\sin 2\theta \cos^2 2\delta/(1 + \alpha \cos 2\theta)^2 = -\frac{1}{2}\sin 2\theta/[(1-\alpha)^2 + 4\alpha \cos^2\theta] \tag{6.1}$$

For $\alpha = 1$, $\delta_\alpha' = \dfrac{1}{4} \cot\theta$, which is infinite at the equator. This behaviour can be understood with the help of Fig. 2 which shows a discontinuity in the value of δ at the equator for $\alpha = 1$: at the equator, $\delta = -\pi/2, -\pi/4, 0$ for $\alpha >> 1$, $\alpha = 1$, $\alpha << 1$ respectively; it should be noticed that δ_α' remains finite at the equator for $\alpha = 1$ if $p_2 = p_1$ since in this case a tilt by $-\pi/2$ is undistinguishable from no tilt.

ii) We proceed now to a discussion of three important velocity correlations, namely, $<v_r' v_\phi'>$, $<v_\theta' v_\phi'>$, and $<v_\theta' v_r'>$. The first two are the only correlations which appear in the equation for U_ϕ and the third one, plays an important role in the expression for the convective flux (see section 9A).

From the above equations it is found:

$$<v_r' v_\phi'> = 2\Omega_o \tau <v_r'^2> (p_1^2 \sin\delta \cos\beta - p_2^2 \sin\beta \cos\delta)/<K^2> \tag{6.2}$$

Since $\delta < 0$ and $|\delta| \underset{\sim}{<} \theta$ (i.e., $\beta = \theta + \delta > 0$), $<v_r' v_\phi'>$ is always smaller than zero: as a particle goes up it must rotate slower due to conservation of angular momentum. For small values of $\Omega_o \tau$, $<v_\phi' v_r'>$ simplifies to $<v_\phi' v_r'> \sim -\Omega_o \tau <v_r'^2> \sin\theta$ in agreement with equation (3.2e) of paper DS.

The expression for $<v_\theta' v_\phi'>$ is given by

$$<v_\theta' v_\phi'> = \frac{2\Omega_o \tau <v_r'^2>}{D <K^2>} \left\{ [(k_\phi^2 - <k_\theta^2>) <K^2> + \gamma_1^2] <k_r k_y> + 2\gamma_1 \gamma_3 k_\phi^2 \right\} \tag{6.3}$$

The sign of $< v_\theta' v_\phi' >$ is important, assume, f. ex., that $< v_\theta' v_\phi' > 0$, then if $v_\phi' > 0$ also $v_\theta' > 0$: rapidly rotating particles are transported towards the equator. We will study the sign of $< v_\theta' v_\phi' >$ for $\alpha >> 1$. It can be readily shown that D > 0. From Equation (5C.4b) it is apparent that: $k_\phi^2 - < k_\theta^2 > = k_\phi^2 - p_2^2\cos^2\delta - p_1^2\sin^2\delta$. For $\alpha >> 1$, π/k_ϕ and π/p_2 are the eddy's dimensions perpendicular to the axis of rotation and π/p_1 is the dimension parallel to $\underset{\sim}{\Omega}_o$. Due to the elongation of the cell along $\underset{\sim}{\Omega}_o$ we expect $p_{pa} \equiv p_1 << k_\phi \approx p_2 \equiv p_{pe}$. The above expression simplifies then to $(p_{pe}^2 - p_{pa}^2)\sin^2\delta$, always positive. It can, furthermore, be shown that $\gamma_1\gamma_3 = (\alpha - 1)p_1^2 p_2^2 \sin^2 2\delta/(2\alpha\cos\theta)$ which is positive for $\alpha > 1$. It remains to study the sign of $< k_r k_y >$ (cf. equation 5C.4e). According to this equation, $< k_r k_y > < 0$, at first sight, since $p_2 >> p_1$, $\delta < 0$, $\beta > 0$. However, this is not the case: for large values of α, $\delta \approx -\theta$ and $\beta = \theta + \delta$ becomes small; in fact $\sin\beta \approx \sin\theta \cos\delta/\alpha$ for $\alpha >> 1$ and, in consequence, $< k_y k_r > = p_1^2[\cos\delta \cos\beta + (p_2^2/p_1^2)\sin\beta \sin\delta] \approx p_1^2\cos\delta [\cos\beta + (p_2^2/p_1^2\alpha)\sin\delta \sin\theta]$. For $p_2^2/p_1^2 \approx \alpha$, which on physical grounds is the most reasonable assumption concerning the elongation of the cell, it follows that $< k_y k_r > \approx p_1^2\cos\delta \cos^2\theta$. Therefore, $< v_\theta' v_\phi' >$ is larger than zero, i.e., *the formalism correctly predicts the existence of transport of angular momentum towards the equator.* For small values of α, $< v_\theta' v_\phi' >$ reduces to $2\Omega_o\tau < v_r'^2 > k_r^2\cos\theta(k_\phi^2 - k_\theta^2)/K^4$ in agreement with equation (32f) of paper DS. Since $k_\phi \approx k_\theta$ for small rotations, $< v_r' v_\phi' >$ becomes the dominant Reynolds stress in the equation for U_ϕ for slowly rotating stars.

The expression for $< v_r' v_\theta' >$ is found to be

$$< v_r' v_\theta' > = - < v_r'^2 > (p_1^2 - p_2^2)\sin\delta \cos\delta / < K^2 > \tag{6.4}$$

which is negative since $\delta < 0$ and $p_1 < p_2$

7. EFFECT OF THE MEAN FLOW ON THE VELOCITY CORRELATIONS

We assume that in equation (4.9) the main contribution to $\underset{\sim}{u}'$ comes from the U_ϕ component of $\underset{\sim}{U}$. We obtain

$$u_\phi' = v_\phi' - \tau \underset{\sim}{v}' \cdot \nabla U_\phi / E; \quad E = 1 + 4\Omega_o^2\tau^2 \tag{7.1a}$$

$$u_r' = v_r' - 2\Omega_o\tau^2\sin\theta \, \underset{\sim}{v}' \cdot \nabla U_\phi / E \tag{7.1b}$$

$$u_\theta' = v_\theta' - 2\Omega_o\tau^2\cos\theta \, \underset{\sim}{v}' \cdot \nabla U_\phi / E \tag{7.1c}$$

where we have approximated the covariant derivatives by ∂/∂_r, $r^{-1}\partial/\partial\theta$, $(r\sin\theta)^{-1}\partial/\partial\phi$. For U_ϕ we take $U_\phi = \Omega_o r \sin\theta \, \omega_2(R_c)(1 - \frac{3r^2}{2R_c^2}\sin^2\theta)$ corresponding to a rotation in cylinders $(\omega_o(R_c) = 0$; for a rotation in cylinders $\omega_o(r) + \omega_2(r) = \omega_2(R_c) =$ constant and $\omega_2(r) = r^2\omega_2(R_c)/R_c^2)$. We spproximate $u_\phi' \approx v_\phi'$ since the contribution of U_ϕ to u_ϕ' is only of the order $\Omega_o\tau/D$. We obtain

$$< u_r' u_\phi' > = < v_r' v_\phi' > - 2(\Omega_o\tau)^2\sin\theta \, W/E \tag{7.2a}$$

$$< u_\theta' u_\phi' > = < v_\theta' v_\phi' > - 2(\Omega_o\tau)^2\cos\theta \, W/E \tag{7.2b}$$

$$W = \omega_2(R_c)\,(1 - \frac{9r^2}{2R_c^2}\sin^2\theta)\,(\sin\theta < v_r' v_\phi' > + \cos\theta < v_\theta' v_\phi' >) \tag{7.2c}$$

With the help of the expressions for $< v_r' v_\phi' >$ and $< v_\theta' v_\phi' >$ valid for $\alpha >> 1$ $(k_\phi^2 = p_2^2 = \alpha p_1^2)$, equations (7.2) can be written

$$< u_r' u_\phi' > = < v_r' v_\phi' > (1 - Z\sin^2\theta / (1 + \cos^2\theta)^3) \tag{7.3a}$$

$$< u_\theta' u_\phi' > \; = \; < v_\theta' v_\phi' > (1 + Z/4) \tag{7.3b}$$

$$Z = 2(\Omega_o \tau)^2 \omega_2(R_c) \, (1 - \frac{9r^2}{2R_c^2} \sin^2\theta) \, (\sin^2\theta + 3\cos^4\theta)/E \tag{7.3c}$$

for $\Omega_o \tau >> 1$ and a rotation in cylinders with the observed differential rotation at the surface $(\omega_2(R_c) \approx -0.1)$ Z is of the order of 0.175 for sufficiently large values of θ: the mean flow tends to increase the angular momentum transport towards the equator and decrease the angular momentum transport towards the inner layers. The opposite is true for $\omega_2(R_c) \approx 0.1$. *It is essential to notice* that this result, already apparent from Eq (7.2c), differs from that obtained by Gilman and Glatzmaier. In their models the sign of $< u_\phi' \; u_r' >$ reverses and *becomes positive*, resulting in a net upward radial flux of angular momentum (see Glatzmaier, this volume). It appears, therefore, that the effect of U_ϕ on the global convective motions is considerably larger than its effect on the turbulent convective motions considered here. This is, clearly, not surprising. It is also clear that a negative value for $< u_\phi' \; u_r' >$ would help to reconcile theories of differential rotation and dynamo theory since it would tend to generate an inwards increasing angular velocity [see, f. ex., Eq (9B.1b)].

8. BEHAVIOUR OF Ω AT THE BOUNDARIES OF THE CONVECTION ZONE

In the *inertial* system of coordinates and due to axial symmetry, the equation for U_ϕ can be written

$$\underset{\sim}{U} \cdot \nabla U_\phi = \frac{1}{r^3} \frac{\partial}{\partial r} (r^3 T_{r\phi}) + \frac{1}{r \sin^2\theta} \frac{\partial}{\partial \theta} (\sin^2\theta \, T_{\theta\phi}) \tag{8.1}$$

where $T_{r\phi}$, $T_{\theta\phi}$ are the sum of the viscous and Reynolds stresses:

$$T_{r\phi} = \nu_t \rho r \sin\theta \; \partial\Omega/\partial r - \rho < v_r' v_\phi' >; \quad T_{\theta\phi} = \nu_t \rho \; \sin\theta \; \partial\Omega/\partial\theta \; - \; \rho < v_\theta' v_\phi' > \tag{8.2a}$$

$$\Omega = \Omega_o(\omega_o(r) + \omega_2(r) P_2(\cos\theta)) \tag{8.2b}$$

We have assumed that the turbulent viscosity is isotropic and since solid body rotation does not contribute to the viscous stresses, Ω , in equation (8.2b) is the angular velocity in the rotating system of coordinates. Legendre polynomials of higher order than two will be neglected in the expansion for Ω . Equation (4.10) shows that with the usual convention for the sign of the viscosity tensor, the Reynolds stresses should appear with a minus sign in equation (8.2a).

In the left hand side of equation (8.1) we approximate U_ϕ by $\Omega_o r \sin\theta$ (solid body rotation) and for the meridional motions we take

$$U_r = \sum_{L=2,4} \frac{\psi_L(r)}{r^2 \rho} P_L(\cos\theta); \quad U_\theta = \sum_{L=2,4} \frac{\psi_L'(r)}{L(L+1) r \rho} \frac{\partial P_L}{\partial \theta} \tag{8.3}$$

where $\psi_2(r)$, $\psi_4(r)$ are the stream functions and then the continuity equation is automatically satisfied.

The equations for ω_o and ω_2 are found by multiplying equation (8.1) by P_0, $P_2(\cos\theta)$ and integrating over θ :

$$d(\nu_t \rho r^4 \omega_o')/dr - 2\rho\nu_t r^2 \omega_2 + r^2(\psi_2'/3 + \psi_4'/10)$$
$$- \frac{1}{\Omega_o} \int_0^{\pi/2} d\theta [d(\rho r^3 < v_\phi' v_r' >)/dr + 2\rho r^2 \cot\theta < v_\theta' v_\phi' >] = 0 \tag{8.4a}$$

$$d(\nu_t \rho r^4 \omega_2') dr - 10 r^2 \nu_t \omega_2 - 2\psi_2 r + r^2(2\psi_2'/3 + \psi_4'/2)$$

$$-\frac{5}{\Omega_o}\int_0^{\pi/2} d\theta[P_2(\cos\theta)d(\rho r^3<v_\phi' v_r'>)/dr + 2r^2\cot\theta<v_\theta' v_\phi'>] = 0 \qquad (8.4b)$$

In equations (8.4), ω_o', ω_2', ψ_2', ψ_4' denote derivatives with respect to r and we stress that *we have not assumed that* ν_t *is constant.* Equations (8.4) must be solved subject to the following boundary conditions

$$\psi_2 = \psi_4 = r\nu_t\Omega_o\omega_o' - \int_0^{\pi/2} <v_r' v_\phi'> d\theta = r\nu_t\Omega_o\omega_2' -$$
$$5\int_0^{\pi/2} P_2(\cos\theta)<v_r' v_\phi'> d\theta = 0; \quad r = R_c, R_o \qquad (8.5)$$

which express that the flow is confined to the SCZ and that $\int_0^\pi T_{r\phi}P_0 d\theta = \int_0^\pi T_{r\phi}P_2 d\theta = 0$ at $r = R_c, R_o$ (free surface boundary conditions). Multiplying equation (8.4a) by 5, substracting equation (8.4b) and integrating this difference from R_b to r ($R_b = R_c$ *or* R_o, the lower or upper boundaries of the SCZ) we obtain, with the help of equation (8.5), the expression for the conservation of angular momentum:

$$\omega_o' = \omega_2'/5 - \psi_2/5\nu_t\rho r^2 + (\nu_t\Omega_o r)^{-1}\int_0^{\pi/2} d\theta[1 - P_2(\cos\theta)]<v_r' v_\phi'> \qquad (8.6a)$$

We notice that there are no contributions from the ψ_4 -component of the meridional flow nor from the $<v_\theta' v_\phi'>$ component of the stress tensor.

To obtain an equation for ω_2 we integrate equation (8.4b) from R_b to r, use the boundary conditions (8.5) and integrate $r^2(2\psi_2'/3 + \psi_4'/2)$ by parts. We obtain

$$\omega_2' = \frac{10}{\nu_t\rho r^4}[\int_{R_b}^r \nu_t\rho r^2\omega_2 dr + \frac{1}{\Omega_o}\int_{R_B}^r \rho r^2 dr\int_0^{\pi/2} \cot\theta<v_\theta' v_\phi'> d\theta]$$
$$+\frac{5}{\nu_r r\Omega_o}\int_0^{\pi/2} <v_\phi' v_r'> P_2(\cos\theta)d\theta - (2\psi_2/3 + \psi_4/2)/\nu_t\rho r^2 +$$
$$\frac{1}{\nu_t\rho r^4}\int_{R_b}^r r(10\psi_2/3 + \psi_4)dr \qquad (8.6b)$$

The P_2 -component of $<v_\phi' v_r'>$ in equations (8.6) is small: it is strictly zero [see equation 32.e) of paper DS] for slow rotation and the numerical calculations (see section 9) show that, even for large values of Ω_{oT} ($\approx$5, corresponding to the lower boundary of the SCZ), $\int_0^{\pi/2} <v_\phi' v_r'> P_2(\cos\theta)d\theta \approx 2\ 10^{-2}\int_0^{\pi/2} <v'_\phi v_r'> d\theta$. For the discussion that follows we neglect, therefore, the terms in $<v_\phi' v_r'> P_2(\cos\theta)$ in equations (8.6). At $r = R_c$ (we discuss the lower boundary which is the more important), the turbulent velocities and ψ_2 vanish (or are very small) and in consequence $\omega_o' \approx \omega_2' \approx 0$. Due to the rapid variation of ν_t with r, the Taylor-Proudman balance [equation (3.1)], does not need to be satisfied for $r < R_c + l_{TP}$, which defines l_{TP} (it is this rapid variation of ν_t with r, which leads to the free surface boundary conditions for the meridional flow). We next define l_v as the scale of variation of the turbulent velocities at the lower boundary (clearly $l_v \approx l_{TP}$) and proceed to discuss the rate of rise of the different terms in the right hand side of equations (8.6). As shown in section 3, if the Taylor-Proudman balance is satisfied, rotation laws that are not of the type $\Omega = f(r\sin\theta)$ (as will be shown to be the case near the boundaries) must be accompanied in the lower SCZ by large latitudinal variations in $\nabla\Delta T_r$, implying large latitudinal variations (ΔF) in the convective flux (for a qualification to this last statement see section 9). Very naively we assume that for $r \gtrsim R_c + \ell_{TP}$, the Taylor-Proudman balance, results in a tendency to generate these large values of ΔF with the consequence that for $r > R_c + \ell_{TP}$, the terms $-\psi_2/5\nu_t\rho r^2$ and $-(2\psi_2/3 + \psi_4/2)/\nu_t\rho r^2$ in equations (8.6) could become large. (It should be stressed that to be able to violate equation (3.1), the meridional flow needs, not only to be large, in the above sense, but truly gigantic) The terms containing integrals with respect to r in equation (8.6b) rise only slowly: f. ex., since $\ell_v << \ell_\rho$ (density scale height) $(10/\nu_t\rho r^4)\int_{R_c}^{R_c+\ell_v} \omega_2\nu_t\rho r^2 dr \approx 10\omega_2\ell_v/R_c^2$.

We discuss now the contributions of the Reynolds stresses to ω_o' and ω_2' . *It is essential to notice that these contributions behave very differently near the boundaries*: in equation (8.6b), due to the integral, the term in $<v_\theta' v_\phi'>$ will rise slowly, whereas even if $<v_r' v_\phi'>/\nu_t \approx 0$ at $r = R_c$, the term in $<v_r' v_\phi'>$ in equation (8.6a) will rapidly become negative (since $<v_\phi' v_r'>$ is smaller than 0) with the scale of variation of the turbulent velocities. To summarize, we expect

$$\omega_o' < 0; \quad \omega_2' \approx 0; \quad r - R_c \lesssim \ell_{TP} \approx \ell_v \tag{8.7}$$

Since in all likelyhood the solar magnetic field is generated near the lower boundary of the SCZ (Parker 1975), the above r- dependence of Ω is in agreement with the observed properties of the propagation of the dynamo wave (Parker 1955, Yoshimura 1975, Schussler 1987). An angular velocity increasing inwards has also been obtained by Schussler (1984), by the mechanism of vorticity expulsion.

For $r \sim R_c + \ell_{TP}$ the terms containing integrals in equation (8.6b) are still small (we recall that we neglected the term in $<v_\phi' v_r'> P_2(\cos\theta)$, however, as argued above, as a consequence of the Taylor-Proudman constraints, $-(2\psi_2/3 + \psi_4/2)/\nu_t\rho r^2$ could become large. Therefore

$$\omega_2' \approx -(2\psi_2/3 + \psi_4/2)/\nu_t\rho r^2; \quad r \approx R_c + \ell_{TP} + \delta r \tag{8.8}$$

where the magnitude of δr is unknown. Assume that in the region where equation (8.6a) and (8.8) are valid a transition to a rotation law with Ω approximately constant along cylinders takes place. Then since $\omega_o' + \omega_2'$ must vanish:

$$\psi_4 = -5\psi_2/3 + (5\rho r/3\Omega_o)\int_o^{\pi/2} d\theta <v_\phi' v_r'> \tag{8.9a}$$

$$\omega_o' = -\psi_2/6\nu_t\rho r^2 + (5/6\nu_t r\Omega_o)\int_o^{\pi/2} d\theta <v_\phi' v_r'> \tag{8.9b}$$

If Ω is a function only of $r\sin\theta$, it is furthermore necessary that the dependence of ω_2 on r be of the form $\omega_2 = cr^2$, with c a negative constant if the equator rotates faster than the poles. Therefore $\omega_o' = -\omega_2' > 0$ and from Equations (8.9) we obtain $\psi_2/\nu_t\rho r^2 < (5/\nu_t r\Omega_o)\int_o^{\pi/2} d\theta <v_\phi' v_r'>$, $\psi_4 \sim -5\psi_2/3$. In consequence $\psi_2 < 0$, $\psi_4 > 0$ and a large P_4 -component of the meridional motions, with the flow rising at the poles must be present. The flow sinks at the poles for the P_2 -component. It is tempting to speculate that this P_4 -component is at the origin of Kuhn's observations (Kuhn, 1986) of temperature differences at the solar limb and of Sporer's and Tuominen's observations concerning the latitudinal motions of sunspots (for $\theta > \approx 70^o$, the sunspots drift towards the equator and towards the poles for $\theta < \approx 70^o$). For ST's observtions it is necessary that $\psi_4 < 0$, but this could be a surface-relaxation effect to the interior dynamics. Whatever the case it should be noticed that if $\Omega = f(r\sin\theta)$, ψ_4 cannot vanish, since if $\psi_4 = 0$, the equation $\omega_o' = -2cr$ cannot be satisfied.

Will, the Taylor-Proudman constraint, tend to generate flows such that $\psi_2 < 0$ and $\psi_4 > 0$, as above? Equation (3.8d) shows that the latitudinal dependence of the superadiabatic gradient is given by $(rB_1)'P_2/3 + 3(rB_2)'P_4/5$ suggesting a similar dependence for the convective flux (see, hower, section 9). Here B_1 and B_2 are the functions of ω_o, ω_2 defined in Equation (3.8c) and it is apparent that $(r\,B_1)'<0$ and $(r\,B_2)'>0$ for the rotation law defined in Equation (8.7) (we neglect ω''_o, ω''_2 and take $\omega_2 < 0$). If meridional motions contribute to a θ -dependence of the flux of the above type then it is in fact necessary that $\psi_2 < 0$ and $\psi_4 > 0$.

At the upper boundary of the SCZ, Equations (3.1) do not impose a constraint on the superadiabatic gradient and Equation (8.7) should hold until the surface values of the meridional velocities, (generated in the lower solar convection zone) became large. Observations of high degree modes (Hill et al. 1986, Hill, 1987) show that the solar angular velocity indeed increases inward up to a depth of $\approx$ 1000 km. In the surface layers, $<v_r'v_\phi'> \approx -\Omega_o\tau<v_r'^2> \approx -\Omega_o\nu_t\sin\theta$ [see Equation (32.e) of paper DS] and equations (8.6a) and (8.6b) suggest that here (we neglect the terms containing integrals in equation 8.6b)

$$\omega_2' = -(2\psi_2/3 + \psi_4/2)/\nu_t\rho r^2; \quad \omega_o' = -(\psi_2/3 + \psi_4/10)/r^2\rho\nu_t - 1/r \tag{8.10}$$

Assume that for depths larger than 1000 km ω_o' ceases to be negative due to the presence of a negative ψ_4 in equation (8.10), the magnitude of this flow can be easily estimated from equation (8.10): $U_r \sim 10\nu_t/R_o \sim 10ms^{-1}$.

A final important remark concerning equation (8.6b). Since $\omega_2' = \psi_2 = \psi_4 = 0$ at $r = R_o$, we have

$$\int_{R_c}^{R_o}\nu_t\rho r^2\omega_2 dr \equiv \bar{\omega}_2\int_{R_c}^{R_o}\nu_t\rho r^2 dr = -\frac{1}{\Omega_o}\int_{R_c}^{R_o}\rho r^2 dr\int_0^{\pi/2}\cot\theta<v_\phi'v_\theta'>d\theta - \frac{1}{10}\int_{R_c}^{R_o}r(10\psi_2/3 + \psi_4)dr \tag{8.11}$$

where $\bar{\omega}_2$ is an average, heavily weighted for the regions of large density. *Clearly* $<v_\phi'v_\theta'>$ *greater than zero and flows rising at the poles generate equatorial acceleration.*

It has been suggested by Schmitt and Rossner (1983) and Weiss (1986) that the solar magnetic field is generated in the overshoot region. Since no assumption was made concerning the radial variation of ν_t, equations (8.6) remain valid in this region. The main property of the velocity correlations needed to obtain $\omega_o'(r)<0$, is $<v_r'v_\phi'> <0$ which is expected to be satisfied. Furthermore, the Taylor-Proudman balance is not expected to hold due to the rapid vairation of ν_t with r and (presumably) due to the weaker differential rotation. In summary, the difficulties associated with the large value of ω_2 in equation (3.8c) and with the extreme sensitivity of the convective flux to the magnitude of the superadiabatic gradient, disappear in the overshoot region.

9. NUMERICAL CALCULATIONS

The mean equations to be solved numerically are the following

$$\underset{\sim}{U}\cdot\nabla\underset{\sim}{U} + div\mathbf{T} = 2\rho\underset{\sim}{U}\times\underset{\sim}{\Omega}_o - \nabla p + \rho\underset{\sim}{g} \tag{9.1a}$$

$$T^{ij} = \rho<v'^iv'^j> - V^{ij} \tag{9.1b}$$

$$\rho T\underset{\sim}{U}\cdot\nabla S + div\underset{\sim}{F} = 0 \tag{9.1c}$$

$$F^i = \tau\rho c_p<v'^iv'^j>\nabla\Delta T_j = -\tau\rho T<v'^iv'^j>\partial S/\partial x^j \tag{9.1d}$$

$$div\rho\underset{\sim}{U} = 0; \quad p = \frac{Rg}{\mu}\rho T \tag{9.1e}$$

In equation (9.1b), V^{ij} is the usual isotropic viscous stress tensor with a viscosity given by $\mu = \rho\nu_t$ ($\nu_t = 10^{13} cm^2 s^{-1}$) . The turbulent viscous stress tensor [see equation (4.10) and notice that $\Omega_o \neq 0$] reduces to this form only when the non-diagonal components of $<v'^i v'^j>$ vanish and the diagonal components are equal, which is clearly not the case specially in the lower part of the SCZ. Even if the exact form of V^{ij} in equation (9.1b) is not too critical it should be kept in mind that the $V_{r\theta}$ and $V_{r\phi}$ components of **V** enter in the boundary conditions for the mean flow. In equation (9.1c) the turbulent dissipation function has been neglected; in equation (9.1d) repeated indices are summed and we have assumed that all the energy is carried by convection.

Let δp, $\delta\rho$, δT be the perturbations in the pressure , density and temperature introduced by rotation, then $\delta p/p_u$, $\delta\rho/\rho_u$, $\delta T/T_u$ (p_u, ρ_u, T_u denote unperturbed values) are much smaller than unity (see, f. ex, Durney 1976, equations (16) and (17)); $\delta S/S_u$, on the other hand, can be of the order of unity: the total entropy is given by [see the equation for δS above (3.3)]

$$S = S_u + c_p(\frac{1}{\gamma}\frac{\delta p}{p_u} - \frac{\delta\rho}{\rho_u}); \quad S_u = c_v \, \ell n\left[\frac{p_u}{p_{uc}}\left(\frac{\rho_u}{\rho_{uc}}\right)^{-\gamma}\right] \tag{9.2}$$

where p_{uc}, ρ_{uc} are the values of p_u, ρ_u at $r = R_c$ the lower boundary of the SCZ; the constant in S_u has been chosen so that $S_u = 0$ at $r = R_c$. Since the SCZ is essentially adiabatic, S_u remains "small" over most of the SCZ and clearly $\delta S/S_u < 1$ can be easily violated. A practical way to calculate S_u is to evaluate S_u' from the expression for the energy flux, Eq (9.1d), in the absence of rotation.

It is convenient to chose δp as one of the thermodynamic variables but it is apparent from the above discussion that there is no particular advantage to chose δS as the other thermodynamic variable. If we substract the equation for hydrostatic equilibrium ($-\nabla p_u + \rho_u \underset{\sim}{g} = 0$) from equation (9.1a), then the last two terms in this equation are replaced by $-\nabla\delta p + \delta\rho \underset{\sim}{g}$ where $\delta\rho$ can easily be expressed in terms of S and δp . In the other terms containing ρ and T in equations (9.1a) - (9.1d) and in the continuity equation, ρ , T can be replaced by ρ_u, T_u .

It remains to specify the velocity correlations. As shown at the end of section 5, $< v'^i v'^j >$ is of the following form $<v'^i v'^j> = f_{ij}(\ell, p_2^2/p_1^2, k_\phi^2/p_1^2)$.

Appealing relations are the following

$$\ell = c_1 | p_u/p_u' |; \quad c_1 = 1.5 \tag{9.3a}$$

$$p_2^2/p_1^2 = c_2 f(\alpha) + c_3\alpha; \quad k_\phi^2/p_1^2 = p_2^2/p_1^2; \quad c_2 = 4, \; c_3 = 1$$

$$f(\alpha) = (\alpha - 1)^2 \text{ for } \alpha < 1, \; f(\alpha) = 0 \text{ for } \alpha > 1 \tag{9.3b}$$

As is custommary, the mixing length is related to the pressure scale- height. Equations (9.3b) are some of the simplest possible expressions for p_2^2/p_1^2 and k_ϕ^2/p_1^2 containing the essential physics: a) for $\alpha << 1$, $p_2^2/p_1^2 = k_\theta^2/k_r^2 = k_\phi^2/k_r^2 = c_2$, and, therefore, $k_\theta = k_\phi$ which must be satisfy for the case of no rotation. Furthermore $\ell_r/\ell_\theta = 2$, i.e., in the surface layers the eddies are elongated in the radial direction. b) for $\alpha >> 1$, $p_2^2/p_1^2 = k_\phi^2/p_1^2 = p_{pe}^2/p_{pa}^2 = \ell_{pa}^2/\ell_{pe}^2 = c_3\alpha$ (ℓ_{pa}, ℓ_{pe} are the eddy's dimensions parallel and perpendicular to $\underset{\sim}{\Omega}_o$ respectively). For $\alpha >> 1$, the eddies are aligned along the axis of rotation and equation (9.3b) defines the eddy's elongation along this axis which should, indeed, be a function of α. c) at $\alpha = 1$, $p_2^2/p_1^2 = 1$ and the singularity at the equator is avoided (cf section 6i). It should be noticed that for very large values of $\alpha^{1/2}$, the mixing length itself must become θ -dependent being smaller at the equator than at the poles.

The equations to be solved numerically are now completely specified.

It is of interest to discuss other functional relations of p_2^2/p_1^2 on α. Assume first that in equation (9.3b), $c_3 = 0$, i.e., that the eddies are not elongated at all along the axis of rotation. Equation (5B.7a) for $1/\tau^2$ contains the factor $1 - \alpha \langle k_y^2 \rangle / \langle K^2 \rangle$. Since the SCZ has to convect the energy (rotation could stabilize convection at most in a very thin layer at the lower boundary), the factor above, must remain positive. If $c_3 = 0$, $\langle k_y^2 \rangle / \langle K^2 \rangle$ is of the order of unity, therefore, α ($= 4\Omega_o^2 T/g\nabla\Delta T_r$) cannot become large as Ω_o increases and in consequence $\nabla\Delta T_r$ has to increase as Ω_o^2. Since the radial component of the convective flux, F_r, must remain approximately constant as $\nabla\Delta T_r$ increases, τ must also increase as Ω_o^2 as the following relations show: $F_r \approx \tau \langle v_r'^2 \rangle \rho c_p \nabla\Delta T_r = (\tau^2 \langle v_r'^2 \rangle / \tau)\rho c_p \nabla\Delta T_r = \ell^2 \rho c_p \nabla\Delta T_r/\tau$. Convection is strongly stabilized. It is furthermore easy to show that $\langle v_\theta'^2 \rangle \approx \langle v_\phi'^2 \rangle \approx (2\Omega_o\tau)^2 \langle v_r'^2 \rangle$, i.e., large toroidal velocities are present.

Assume now, f. ex., that $p_2^2/p_1^2 \approx \alpha^{1/2}$, i.e., the eddy elongates as $\alpha^{1/2}$. It can then be readily shown that α has to be smaller than $\approx (1 + \sqrt{2})^2$ if $1/\tau^2$ is to remain positive (we assume $p_2 = k_\phi$ and take $\delta = -\theta$, $\beta = 0$ corresponding to the large α -behaviour for δ).

The above "elongation laws" are clearly physically unappealing. Consider now the case $p_2^2/p_1^2 = c_3\alpha$ for large α's [as in equation (9.3b)]. It can then be shown that for $\alpha >> 1$ (recall that $\sin\beta \sim \sin\theta \cos\delta/\alpha$ for $\alpha >> 1$; we take $\delta = -\theta$ and $k_\phi^2/p_1^2 = p_2^2/p_1^2$)

$$1 - \alpha \langle k_y^2 \rangle / \langle K^2 \rangle = [\sin^2\theta + c_3\cos^2\theta(\alpha - \sin^2\theta) + \alpha(c_3 - 1)]/[\sin^2\theta + \alpha c_3(\cos^2\theta + 1)] \tag{9.4}$$

This expression is always positive for $c_3 \geq 1$; $\alpha = 4\Omega_o^2 T/g\nabla\Delta T_r$ can become large as Ω_o increases, convection is less stabilized and the above large toroidal velocities are absent. In fact $\langle v_\theta'^2 \rangle \approx \langle v_\phi'^2 \rangle \approx \langle v_r'^2 \rangle$; $\langle v_\phi' v_r' \rangle \approx \langle v_\theta' v_\phi' \rangle \approx \langle v_r'^2 \rangle /(2\Omega_o\tau)$. Consider the case $c_3 < 1$: at the poles, the numerator in equation (9.4) is given by $\alpha(2c_3 - 1)$, whereas it is equal to $1 + \alpha(c_3 - 1)$ at the equator. For $1/\tau^2$ to be positive at the poles $c_3 > 1/2$. If $1/2 < c_3 < 1$, α can become large at the poles but at the equator α must remain smaller than $1/(1 - c_3)$. The above discussion shows that there is surprisingly little freedom in the choice of the "elongation law".

Equations (9.1) will be solved by an expansion in Legendre polynomials of the following form:

$$U_r = \sum_{L=1} \frac{\psi_L(r)}{r^2\rho} P_L(\cos\theta); \quad U_\theta = \sum_{L=1} \frac{\psi_L'(r)}{L(L+1)r\rho} \frac{\partial P_L}{\partial\theta} \tag{9.5a}$$

$$\Omega = \Omega_o \sum_{L=0} \omega_L(r) P_L(\cos\theta); \quad \delta p = \sum_{L=0} \delta p_L(r) P_L(\cos\theta); \quad S = \sum_{L=0} S_L(r) P_L(\cos\theta) \tag{9.5b}$$

The meridional velocities are expressed in terms of stream functions, and the continuity equation is then automatically satisfied. We will assume, furthermore, that there is symmetry across the equator, in which case, only even Legendre polynomials in the expansions (9.5) need to be considered. Equations (9.1) provide us with the equations for $\psi_L'''(r)$, $\omega_L''(r)$, $\delta p_L'(r)$, and $S_L''(r)$ in terms of lower order derivatives. It is clear that to obtain these equations some integrals (with respect to θ) of the Reynolds stresses multiplied by appropriate Legendre polynomials need to be evaluated.

We proceed now to a discussion of the boundary conditions.

These are:

1) $U_r = T_{r\theta} = T_{r\phi} = 0$ at $r = R_c,\ R_o$ i.e., free surface boundary conditions with the flow confined to the SCZ

2) Uniform energy flux at $r = R_c,\ R_o$ The uniform energy flux at $r = R_o$, is an approximation of Spruit's boundary condition (Spruit 1977a), namely $\delta F = 4\sigma T_u \delta T$, i.e., the perturbations in the flux must be radiated away.

3) $\delta p_o(r) = S_o(r) = \omega_o(r) = 0$ at $r = R_c$

In relation to boundary conditions 3) we observe that due to conservation of mass, energy and angular momentum the boundary conditions 1) and 2) above are not sufficient to specify the problem. For example, since $U_r = 0$ at $r = R_c, R_o$ and $T_{r\phi} = 0$ *at* $r = R_c$ it follows from equations (2.4) that at

$r = R_o$, $\int_0^\pi d\theta \sin^2\theta\, T_{r\phi} = \int_0^\pi d\theta \sin^2\theta (<v_r' v_\phi'> - \nu_t r \sin\theta\, \Omega_o\, \partial\Omega/\partial r) = 0$. Since $\sin^2\theta = 2[1 - P_2(\cos\theta)]/3$ this equation shows that if the free stress boundary condition is satisfied by ω_2' *at* $r = R_o$, then it will be also satisfied by ω_o'. No boundary condition should be imposed on ω_o' and this allows us to specify the total angular momentum of the SCZ, which we replace by the simpler boundary condition, $\omega_o(r) = 0$ *at* $r = R_c$.

The numerical solution of the full mean equations is a difficult problem and in this paper we restrict ourselves to the two following cases

9A. Numerical Solution of the energy equation.

The large-scale velocities, pressure and density are assumed to vanish. The equation to be solved is, therefore, $div\tilde{F} = 0$, with $\tilde{F}$ given by equation (9.1d), i.e., (recall that $\tau^2 <v_r'^2> = \ell^2$ and see Equations (5C.3) and (5C.5) for the expressions of $<v_r' v_\theta'>$ and $<v_\theta'^2>$)

$$F_r = \frac{\kappa(r)}{\tau}\left[-\frac{\partial S}{\partial r} + \frac{<k_r k_\theta>}{r<K^2>}\frac{\partial S}{\partial \theta}\right];\quad \kappa(r) = \rho T \ell^2 \tag{9A.1a}$$

$$F_\theta = \frac{\kappa(r)}{\tau<K^2>}\left[<k_r k_\theta>\frac{\partial S}{\partial r} - \frac{1}{r}\frac{\partial S}{\partial \theta}(<k_r^2 k_\theta^2/K^2> + (2\Omega_o \tau)^2 <k_y^2 k_\phi^2/K^2>)\right] \tag{9A.1b}$$

In equations (9A.1), τ and ratios like $<k_r k_\theta>/<K^2> \cdots$ are functions only of $\alpha = -4\Omega_o^2 c_p/g \frac{\partial S}{\partial r}$. It is clear that $div\tilde{F} = 0$, results in a system of second order coupled differential equations for the $S_L(r)$ defined in equation (9.5B). At the surface $F_r \sim -\frac{\kappa(r)}{\tau}\frac{\partial S}{\partial r}$ where τ *is* θ-independent and the boundary condition for uniform flux is simply $S_L' = 0$ *at* $r = R_o$. At the inner boundary, $r = R_c$, this boundary condition must be implemented as follows: S_0 ($= 0$), S_2, $S_4 \cdots$ are assumed known and α is calculated for several values of θ so that F_r, given by the whole equation (9A.1a) is the solar flux at $r = R_c$. Since α is known, $\partial S/\partial r$ is known as a function of θ and, therefore, S_0', S_2', S_4'.. are known at $r = R_c$; the S_2, $S_4 \cdots$ are determined by the outer boundary condition.

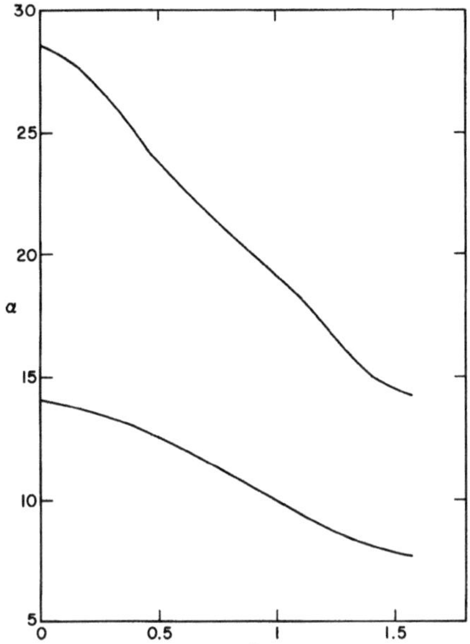

Fig. 3

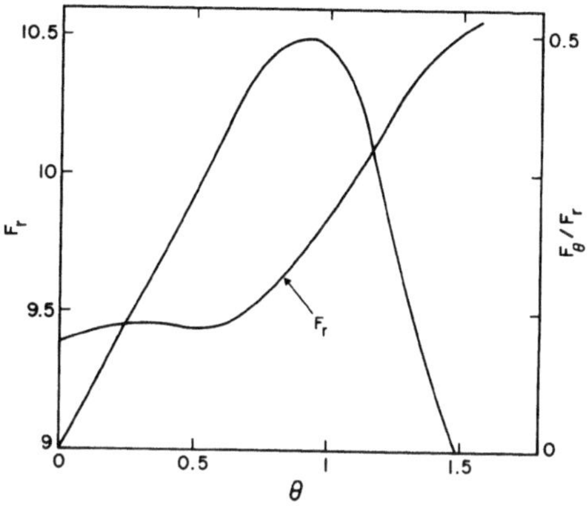

Fig. 4

Fig. 3 α ($= 4\Omega_o^2 T/g\nabla\Delta T_r$) vs θ for $r = R_c = 5.1\ 10^{10}cm$ (upper curve) and $r = 5.62\ 10^{10}cm$ for the elongation law $p_2^2 = k_\phi^2 = p_1^2\ (\alpha + 1)$

Fig. 4 $F_r/10^{10}$ (cm^2s^{-1}) and F_θ/F_r vs θ for $r = 5.63\ 10^{10}cm$ and $r = R_c$ respectively

Table 1

Values of S_2/r, S_4/r, S_0', S_2', S_4' at $r = R_c$ for different values of c_3 for elongation laws of the form $p_2^2/p_1^2 = c_3\alpha + 1$. All values are multiplied by 10^9 and cgs units have been used.

c_3	S_2/r	S_4/r	S_o'	S_2'	S_4'
0.8	1.1	0.65	−12.9	9.02	−4.8
1	2.3	0.70	−9.67	3.86	−1.2
1.2	3.4	0.85	−8.70	2.27	−0.4

In Fig. 3, α is plotted vs θ for $r = R_c$ and $r = 5.63\ 10^{10}cm$. All even Legendre polynomials up to P_{10} were included. The θ-dependence of α is due to the stronger stabilizing effect of rotation on convection at the equator than at the poles: it was shown above, that for large values of α, the factor $1-\alpha<k_y^2>/<K^2>$, contained in the equation for $1/\tau^2$, is given by equation (9.4). At the poles this factor is $(2c_3-1)/2c_3$ ($=0.5$, for $c_3=1$, used in the numerical calculations), whereas it is equal to $[1+\alpha(c_3-1)]/(1+\alpha c_3)$ ($=1/(1+\alpha)$ for $c_3=1$) at the equator, i.e., always smaller than at the poles. This increase in the stabilizing effect with increasing θ, results in an increase of the superadiabatic gradient with θ as shown in Figure 3. For $c_3>1$, the dependence of α on θ is weaker than that shown by figure 3, whereas it is stronger for $c_3<1$ as it is apparent from Table 1 (recall that $\alpha=-4\Omega_o^2 c_p/g\frac{\partial S}{\partial r}$ and $\nabla\Delta T_r=-\frac{T}{c_p}\frac{\partial S}{\partial r}$). Table 1 shows that S_2' decreases with increasing c_3 and that S_2/r increases.

In Fig. 4, F_θ/F_r and F_r are plotted vs θ for $r=R_c$ and $r=5.63\ 10^{10}cm$, respectively.

Uniform flux at $r=R_c$, has resulted in considerable latitudinal variations of the superadiabatic gradient with θ as well as latitudinal fluxes.

The contribution of the second term of equation (9A.1a) to F_r is not large and in consequence $\kappa(r)/\tau\approx-F_r/(\partial S/\partial r)$. From equations (5C.4f) and (5C.4g) we have, $<k_r k_\theta>/<K^2>\ \approx-\frac{2}{9}\frac{\partial P_2}{\partial\theta}$ for large values of α; therefore, $F_\theta/F_r\sim\frac{2}{9}\frac{\partial P_2}{\partial\theta}+\frac{\kappa_{\theta\theta}}{r}\frac{\partial S}{\partial\theta}/\frac{\partial S}{\partial r}$, where the definition of $\kappa_{\theta\theta}$ follows immediately from equation (9A.1b). Figure 4 shows that the second term of this equation is larger for $r\approx R_c$ than the first one. For $r>\approx 5.62\ 10^{10}cm$, the opposite is true and the pole-equator difference in flux that builds up at $r=5.62\ 10^{10}cm$ is smoothed out and vanishes at the surface.

So far the numerical calculations; these calculations have shown that uniformity in the flux does not necessarily implies that S_2' and S_2/R_c are small. This being the case it is of considerable interest to study the implications of the Taylor-Proudman theorem. In particular, do these values of S_2' and S_2/R_c, allow for rotation laws that differ from the $\Omega=f(r\sin\theta)$ law? If equation (3.1) are valid, i.e., for $r>R_c+\ell_{TP}$ (see section 8), then S_2, S_4, S_2', S_4' are given by [cf. Equations (3.8a), (3.8d)]:

$$S_2/r=-\beta B_1;\quad S_4/r=-\frac{9}{5}\beta B_2 \tag{9A.2a}$$

$$S_2'=-\beta(rB_1)';\quad S_4'=-\frac{9}{5}\beta(rB_2)' \tag{9A.2b}$$

$$\beta=2\Omega_o^2 c_p/21g=4.23\ 10^{-9}(cgs) \tag{9A.2c}$$

where the variation of $g(r)$ with r, is neglected $[g=g(R_c)]$; B_1 and B_2 are defined by equation (3.8c). At $r=R_c$, on the other hand, S_2, S_4, S_2', S_4' are determined by the boundary conditions imposed on equation (9.1) at $r=R_c, R_o$. (l_{TP} is the thickness of the boundary layer). In the region $R_c<r<R_c+l_{TP}$, $S_2/r\neq-\beta B_1$, $S_2'\neq-\beta(rB_1)'$.... As r approaches R_c+l_{TP} meridional circulations in the equation for U_ϕ change the rotation law so that for $r>R_c+l_{TP}$, equation (9A.2) are satisfied. Rotation in cylinders for $r>R_c+l_{TP}$ follows only if $S_2=S_4=0$. Clearly, this does not need to be the case. It was argued in section 9 that $\omega_o'<0$ for $R_c<r<R_c+l_{TP}$. Can ω_o' continue to be negative for $r>R_c+l_{TP}$? Since $B_1=12\omega_2+r\omega_2'+7r\omega_o'$, if $\omega_o'<0$, then

$-\beta B_1 > -\beta(12\omega_2 + r\omega_2') \approx 1.2\beta \approx 5.1\ 10^{-9} cgs$ where we have neglected ω_2' and taken for ω_2 the value, $\omega_2(R_c) = (R_c/R_o)^2\omega_2(R_o)$ corresponding to a rotation in cylinders with the observed differential rotation at the surface $[\omega_2(R_o) = -0.2]$. It follows from equation (9A.2a) that if $S_2/r > 5.1\ 10^{-9} cgs$ at $r = R_c$, then we expect that ω_o' will continue to be negative beyond $r = R_c + l_{TP}$. As an indication of the values of S_2/r at $r = R_c + l_{TP}$ we take those of Table 1. These are somewhat smaller than 5.1 $10^{-9} cgs$; however, for $c_3 = 2$, $S_2/r = 7.2\ 10^{-9} cgs$. Since the meridional motions have been neglected the details are clearly unimportant. what is important, though, is that the numerical calculations suggest that *latitudinal variations in the superadiabatic gradient could extend the region where the Taylor-Proudman balance fails to hold and in consequence, conciliate theories of differential rotation and dynamo theories.*

9B. Numerical Solution of the energy and angular velocity equations

We assume that the meridional motions, U_r and U_θ vanish. The energy equation remains $div\ \underset{\sim}{F} = 0$. The superadiabatic gradient is, therefore, known. Since the Reynolds stresses are functions only of $\nabla\Delta T$, the Reynolds stresses are also known and their contribution can be included in the equation for Ω. This case clearly illustrates the role of the Reynolds stresses in the generation of differential rotation. It completely by passes the difficulties associated with the Taylor-Proudman theorem.

The numerical results (cf Table 2) can be easily understood with the help of equations (8.6a) and (8.11): since ω_2 varies little with r these equations show that

$$\omega_2(r) = -\frac{1}{\Omega_o \nu_t} \int_{R_c}^{R_o} \rho r^2 dr \int_o^{\pi/2} cot\theta < v_\theta' v_\phi' > d\theta \Big/ \int_{R_c}^{R_o} \rho r^2 dr \tag{9B.1a}$$

$$\omega_o(r) = \frac{1}{\Omega_o \nu_t} \int_{R_c}^{r} \frac{dr}{r} \int_o^{\pi/2} < v_r' v_\phi' > d\theta \tag{9B.1b}$$

in good agreement with the numerical values (the contribution of the term in $P_2(\cos\theta)$ in equation (8.6a) can be neglected).

Table II

Value of ω_{2n} at the boundaries of the SCZ for the elongation law $p_2^2/p_1^2 = \alpha + 1$; $\nu_t = 10^{13} cm^2 s^{-1}$

r	ω_o	ω_2	ω_4	ω_6
R_c	0	−0.224	0.05	−0.013
R_o	−0.375	−0.23	0.05	−0.011

Equations (9B.1) are clearly incompatible with a rotation which is constant along cyclinders. It appears, from present day observations, that to alleviate the constraints imposed by the Taylor-Proudman balance, the solar $\omega_o(r)$ differs substantially more than the solar $\omega_2(r)$ from the above expressions. *The question is why. Why not the inverse?* (despite the uncertainties concerning the observations, an angular velocity increasing inwards as given by equation (9B.1b) is probably ruled out). The answer to the above question lies, perhaps, in the matching of S_2/r (at $r = R_c$) with $-\beta B_1$(at $r = R_c + \ell_{TP}$) as the following oversimplified example shows: assume that

$S_2(R_c)/R_c = 7.2\ 10^{-9} cgs$ (see § 9A). At $r = R_c + \ell_{TP}$ (the assumed value of r at which the Taylor-Proudman balance starts to be verified) $S_2(r)/r$ must be given by $-\beta B_1$ (see equations (9A.2)) with $\beta = 4.23\ 10^{-9} cgs$ and $B_1 \approx 12\ \omega_2 + 7r\omega_o'$ (we neglect ω_2'). Therefore, $B_1 \approx -7.2/4.23 = 1.7$; if $-7r\omega_o' < 1.7$ a negative value for ω_2 would be favored.

10. CONCLUSIONS

Clearly the main conclusions of this paper are that in order to achieve a satisfactory understanding of the solar differential rotation and its bearing on the dynamo theory a very careful treatment of the following two subjects is necessary:

i) the dynamics and thermodynamics of the region $R_c < r < R_c + \ell_{TP}$. At the lower boundary of the SCZ, the mean entropy, $S(R_c, \theta)$, and $\partial S(r, \theta)/\partial r$ are determined by boundary conditions imposed on the conservation equations at $r = R_c, R_o$. If for $r > R_c + \ell_{TP}$, the Taylor-Proudman balance holds then, in this region (the surface layers excepted), $S(r, \theta) - \langle S(r, \theta)\rangle_\theta \equiv Q(r, \theta)$ and $Q'(r, \theta)$ are determined by the angular velocity. Here the bracket denotes an average with respect to θ and the prime a derivative with respect to r. For a rotation constant in cylinders $Q(r, \theta)$ vanishes. In this region the viscous forces are important due to the rapid variation of the turbulent viscosity with r.

ii) The expression for the convective flux. If $Q(r, \theta)$ differs from zero, Ω cannot be constant along cylinders. It is not, however, the superadiabatic gradient that is physically meaningful, but the energy flux. It is clear that a realistic functional dependence of $\underset{\sim}{F}$ on $\underset{\sim}{\nabla}\Delta T$ must be used. The numerical solution of the energy equation has shown that surprisingly large values of $Q(r, \theta)$ can be compatible with a uniform flux. *It is essential to notice that for compressible fluids the Taylor-Proudman theorem does not require* Ω *to be constant along cylinders.* Assume that at the lower boundary the energy flux is uniform and consider rotation laws with Ω not constant along cylinders (NC-laws). *As a consequence of the Taylor-Proudman theorem some of the NC-laws can be in better agreement with the requirements of uniform flux than a rotation which is constant along cylinders.*

A careful treatment of the dynamics and thermodynamics of the region $R_c < r < R_c + \ell_{TP}$, unfortunately means using the turbulent viscosity as calculated from equation (4.9) which is quite involved in particular for the lower boundary since, here, the turbulent viscosity depends on the solution of the problem. The subtle variations of Ω in the lower SCZ, able to reconcile theories of differential rotation and dynamo theory are going to present a challenge even to future observations as those planned by NSO within the GONG project. (NSO 1984)

The understanding of the variation of Ω in the surface layers looks more promising. Here the turbulent viscosity must be of the Kippenhahn-type (Kippenhahn, 1963; see the Appendix) and the direct detection of subsurface meridional flows is perhaps not impossible.

ACKNOWLEDGEMENTS

NASA grant W-15, 076 is gratefully acknowledged.

APPENDIX

$$\tau < v_\theta' v_\theta' > = \tau < v_\phi' v_\phi' > = s\tau < v_r' v_r' > = s\nu_t \qquad \text{(A.1a)}$$

$$< v_i' v_j' > = 0 \text{ for } i \neq j \qquad \text{(A.1b)}$$

Equation (4.10) gives

$$\nu_{rr} = -2\nu_t \partial U_r/\partial r \qquad \text{(A.2a)}$$

$$\nu_{\theta\theta} = -2s\nu_t r^{-1}(\partial U_\theta/\partial\theta + U_r) \qquad \text{(A.2b)}$$

$$\nu_{\phi\phi} = -2s\nu_t r^{-1}(csc\theta\, \partial U_\phi/\partial\phi + U_r + U_\theta cot\theta) \qquad \text{(A.2c)}$$

$$\nu_{r\theta} = -\nu_t r^{-1}(r\,\partial U_\theta/\partial r + s(\partial U_r/\partial\theta - U_\theta)) \qquad \text{(A.2d)}$$

$$\nu_{r\phi} = -\nu_t r^{-1}(r\,\partial U_\phi/\partial r + s(csc\theta\, \partial U_r/\partial\phi - U_\phi)) \qquad \text{(A.2e)}$$

$$\nu_{\theta\phi} = -s\nu_t r^{-1}(\partial U_\phi/\partial\theta + csc\theta\, \partial U_\phi/\partial\phi - cot\theta U_\phi) \qquad \text{(A.2f)}$$

REFERENCES

Chan, K. L., 1987, this volume.

Chandrasekhar, S., 1961, *Hydrodynamic and Hydromagnetic Stability* (Oxford: Clarendon Press)

Cowling, T. G., 1951, *Ap. J.*, **114**, 272.

Durney, B. R., 1976, *Ap. J.*, **204**, 589.

Durney, B. R., 1981, *Ap. J.*, **244**, 678.

Durney, B. R., 1983, *Ap. J.*, **269**, 671.

Durney, B. R., 1985, *Ap. J.*, **297**, 787.

Durney, B. R., and Spruit, H. C., 1979, *Ap. J.*, **234**, 1067 (DS)

Flasar, F. M., and Gierasch, P. J. 1978, *Geophys. Ap. Fluid Dyn.*, **10**, 175.

Glatzmaier, G. S., 1987, this volume.

Hathaway, D. H., *Ap. J.*, **276**, 316.

Hill, F., 1987, this volume.

Hill, F., Gough, D., Toomre, J., and Haber, D., 1986, in Advances in Helio- and Asteroseismology, IAU Symp. 123, ed. J. Christensen-Dalsgaard (Dordrecht: Reidel) in press.

Kippenhahn, R., 1963, *Ap. J.*, **137**, 664.

Kuhn, J. R., 1987, this volume.

Landau, L. D., and Lifschitz, E. M., 1959, Fluid Mechanics (Reading: Addison Wesley).

NSO, 1984. National Solar Observatory. National Optical Astronomy Observatories: a proposal to study the solar interior by measuring global oscillations with a world-wide network of instruments, NSO-Preprint (unpublished).

Parker, E. N., 1955, *Ap. J.* **122**, 293.

Parker, E. N., 1975, *Ap. J.*, **198**, 205.

Pedloski, J., 1979, Geophysics Fluid Dynamics (New York: Springer)

Rudiger, G., and Tuominen, I., 1987, in this volume.

Schmitt, J. M., and Rossner, R., 1983, *Ap. J.*, **265** 901.

Schussler, M., 1984, in The Hydromagnetics of the Sun, Proc. 4th European Meeting on Solar Phys. ESA SP-220, p. 67.

Schussler, M., 1987, this volume.

Schwarzshild, M., 1958, Structure and Evolution of Stars (Princeton: Princeton University Press)

Spruit, H. C., 1977a, *Astr. Ap.*, **55**, 151.

Spruit, H. C., 1977b, *Solar Phys.* **55**, 3.

Stix, M., 1987, this volume.

Tayler, R., 1973, *M.N.R.A.S.*, **165**, 39.
Weiss, N. O., 1986, in Seismology of the Sun and distant Stars, ed. D. O. Gough (Dordrecht: Reidel), p. 253.
Yoshimura, 1975, *Ap. J.*, **201**, 740.

A REVIEW OF WHAT NUMERICAL SIMULATIONS TELL US ABOUT THE INTERNAL ROTATION OF THE SUN

Gary A. Glatzmaier
ESS-5 MS F665
Los Alamos National Laboratory
Los Alamos, NM 87545
USA

ABSTRACT. The simulated solar differential rotation from two independent numerical modeling efforts agree with each other and with present solar observations. The models solve the nonlinear, three-dimensional, time-dependent, anelastic equations of motion for thermal convection in a stratified, rotating, spherical shell. The simulated angular velocity in the convection zone is constant on cylinders coaxial with the rotation axis, maximum at the equator and decreasing with depth. The latitudinal variation of this angular velocity at the surface is in agreement with Doppler measurements of the solar surface rotation rate. The radial variation through the convection zone is consistent with the analysis of the rotational frequency splitting of solar oscillations.

1. INTRODUCTION

Two numerical models of global convection in the sun will be discussed. One was developed by Peter Gilman (Gilman & Miller 1986) based on an earlier Boussinesq model (Gilman 1977); the other was developed by the author (Glatzmaier 1984). Except for construction of the anelastic equations in spherical coordinates (Gilman & Glatzmaier 1981), these models have been developed completely independently. They represent an approach to the theoretical investigation of the internal dynamics of the sun that maximizes the role of physics and minimizes the role of parameterization. This approach, although expensive in terms of both computer resources and code development, is becoming more feasible with the continued improvement and availability of supercomputers.

After briefly describing the two models, I will discuss some properties of the numerical solutions and compare them to solar observations. Particular emphasis will be placed on the simulated differential rotation. I will finish with a discussion of model validation.

2. DESCRIPTION OF MODELS

These two models are very similar in many ways and very different in many others. Both models solve the nonlinear, three-dimensional, time-dependent, anelastic equations of motion for thermal convection of a density-stratified perfect gas in a rotating, spherical shell. These models also serve as self-consistent dynamo models when the magnetohydrodynamic equations are solved (Gilman & Miller 1981; Gilman 1983;

B. R. Durney and S. Sofia (eds.), The Internal Solar Angular Velocity, 263–274.

Glatzmaier 1984, 1985a,b).

The anelastic approximation (Gough 1969; Gilman and Glatzmaier 1981) is based on the assumption that the convective velocities are small compared to the local sound speed and that the resulting thermodynamic perturbations are small compared to the respective local thermodynamic variables. These are good approximations for the deep solar convection zone. The anelastic gas is compressible on a convective time scale, but not on an acoustic time scale. In other words, the sound speed is assumed to be infinite. The anelastic approximation has the advantage of being able to account for large radial stratifications in the thermodynamic variables, which is not accounted for in the Boussinesq approximation, while not requiring a very small time step which is needed in fully compressible calculations to resolve sound waves.

Both models have been designed as numerical analogs of global convection in the sun by constraining the mass, luminosity, average rotation rate, and radius to solar values. In practice, only the solar convection zone is simulated and only out to 90% of the solar radius (93% in Glatzmaier's model) due to severe numerical resolution requirements near the solar surface resulting from the small pressure scale-heights there. Stress-free, impermeable, constant heat flux boundary conditions are applied at the inner and outer spherical boundaries.

The basic reference states of the two models differ somewhat. Gilman's model convection zone spans five pressure scale-heights and is superadiabatic everywhere. Glatzmaier's model spans seven pressure scale-heights and is superadiabatic in the upper two thirds (in radius) of the shell and subadiabatic in the lower third. This latter case is designed to simulate convective penetration into a stable region below. However, due to the additional numerical resolution that would be required to simulate a stratification like that in a standard solar model, the modeled stable region is only slightly subadiabatic.

Present supercomputers have made three-dimensional numerical simulations possible today; however, the affordable numerical resolution in space and time is still very far from what would be required to resolve the many orders of magnitude in scale that exist in solar convection. Consequently, these global models resolve as many of the large scales as possible and parameterize, via eddy diffusion, the transport of momentum and heat by all the smaller unresolved scales. This parameterization is the weakest part of the models. However, the more modes that are resolved, i.e., the higher the numerical resolution, the less significant the unresolved modes should be in determining the structure and time-dependence of the resolved modes.

The parameterized viscous and thermal diffusivities in these models are usually specified time-independent functions of radius (either constant in radius or increasing with radius). Glatzmaier has also employed a time-dependent eddy diffusivity that depends on the local shear of the large-scale resolved velocity. In addition, the ratio of the viscous to thermal diffusivity is specified. Gilman typically sets this ratio to one; Glatzmaier usually makes it somewhat less than one. The effects of different eddy diffusivity parameterizations will be discussed in section 4.2.

The most significant difference between the two models is the numerical solution technique. In Gilman's model, Fourier expansions describe the longitudinal dependence of the functions while a finite difference scheme on a staggered grid is used in radius and latitude. The spatial resolution is 90 grid points in latitude, 20 grid points in radius, and 24 longitudinal wave numbers. The equations are integrated in time via an explicit leap-frog scheme with 1.36 hour time steps.

In Glatzmaier's model, the longitudinal and latitudinal dependence of the functions is described by spherical harmonic expansions while the radial dependence is described by Chebyshev polynomial expansions. A spectral transform method is employed which computes all spatial derivatives analytically in spectral space and all nonlinear terms in physical space each time step. The spatial resolution is 32 grid points in latitude, 33 grid points in radius, and 64 grid points in longitude. Semi-implicit time-integration treats the nonlinear terms with an explicit Adams-Bashforth scheme and the linear terms with an implicit Crank-Nicolson scheme. A three hour time step is used.

3. NUMERICAL RESULTS

Although the numerical techniques employed in these models are quite different, the numerical solutions are qualitatively the same. Several properties of the numerical solutions from both models will now be described.

3.1. Time and wave number dependence

The initial conditions, for both models, are small random entropy perturbations with zero velocity relative to the rotating frame of reference. Due to buoyancy forces resulting from the entropy perturbations in the superadiabatic environment, convective velocities quickly develop and organize into large-scale cellular convection. The amplitude of this motion grows exponentially in time, due to the supercritical linear terms in the equations of motion, until the nonlinear terms become large enough to terminate the growth. Kinetic energy increases at the expense of the entropy stratification which becomes less superadiabatic.

A time trace of kinetic energy is displayed in Figure 1 for one of Gilman's solutions. (Glatzmaier's model produces very similar kinetic energy traces.) The energy in the differential rotation (the axisymmetric part of the longitudinal component of velocity relative to the rotating frame of reference) is plotted separately. The nonaxisymmetric convective structure organizes within about two rotation periods (~1000 time steps); however, the differential rotation amplitude and structure, which is maintained primarily by the nonlinear interaction of the nonaxisymmetric convective modes, requires about 30 to 40 rotation periods to become fully established.

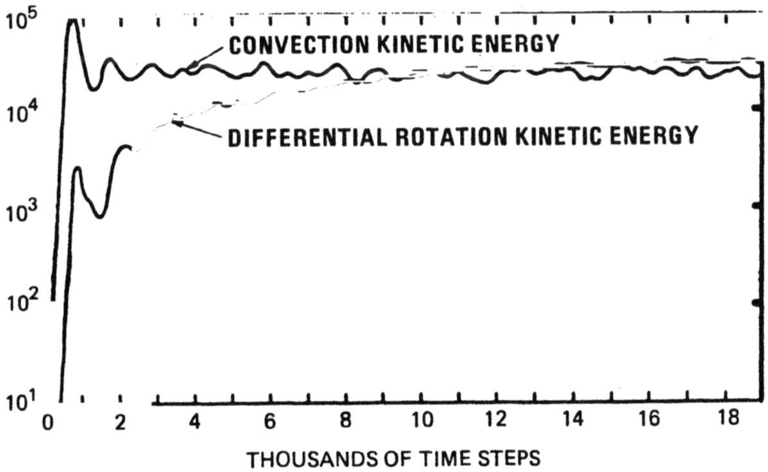

Figure 1. Kinetic energy of convection and differential rotation as a function of time (one time step = 1.36 hours; Gilman & Miller 1986). A dimensionless energy scale is used.

Energy spectra illustrate the large number of degrees of freedom associated with these three-dimensional simulations. The kinetic energy and thermal energy (entropy variance) integrated in radius and latitude are plotted as functions of longitudinal wave number *(m)* in Figure 2 for one of Glatzmaier's fully developed solutions at one instant in time. (A kinetic energy spectra, averaged in time, is illustrated in Figure 2 of Gilman & Miller 1986.) The peak in the axisymmetric *(m=0)* velocity mode is mainly due to the induced differential rotation; whereas, the peak in the axisymmetric entropy perturbation mode represents the tendency for convection to decrease the mean entropy stratification. In addition, broad peaks exist around *m=10* reflecting the most unstable, nonaxisymmetric, convective modes.

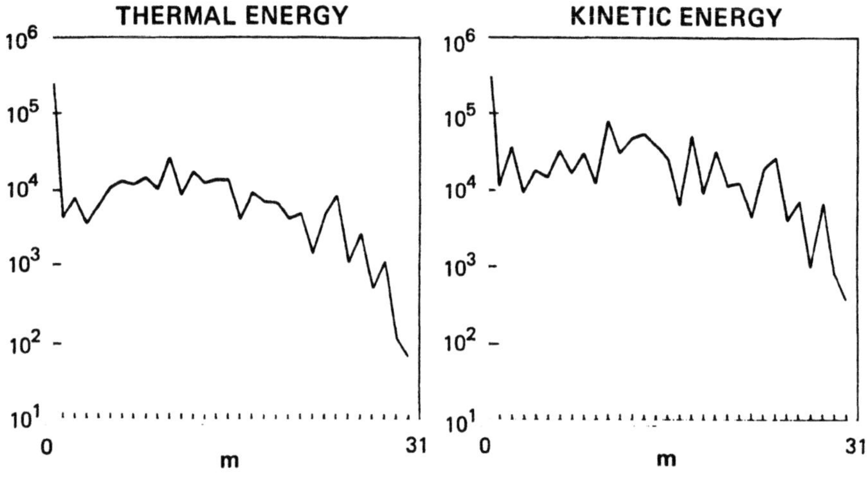

Figure 2. Thermal energy (variance of entropy perturbation) and kinetic energy integrated in radius and latitude and plotted as functions of longitudinal wave number (Glatzmaier 1985a). A dimensionless energy scale is used.

3.2. 3D convective structure

Some properties of the nonaxisymmetric components of velocity, which are responsible via nonlinear interactions for the maintenance of differential rotation, will now be examined. Simulated global convection in a rotating spherical shell tends to take the form of north-south (banana) rolls. Banana rolls develop because motions perpendicular to the axis of rotation are favored, especially at low latitude, due to the near balance that can be achieved among the pressure gradient, buoyancy, and Coriolis forces, all of which tend to be perpendicular to the axis of rotation at low latitude. The resulting small accelerations mean that this banana-roll structure is relatively stable compared to other possible convective structures. Figure 3a portrays these banana rolls with a plot of the radial component of velocity on a spherical surface just below the top boundary. Similar plots (extending 180° in longitude) from Gilman's model are illustrated in Figure 7 of Gilman & Miller (1986). This banana-roll structure has also been seen in a recent space-lab experiment of thermal convection in a rotating hemispherical shell with radial gravity (Hart, et al. 1986; Hart, Glatzmaier, & Toomre 1986; and Toomre, these proceedings).

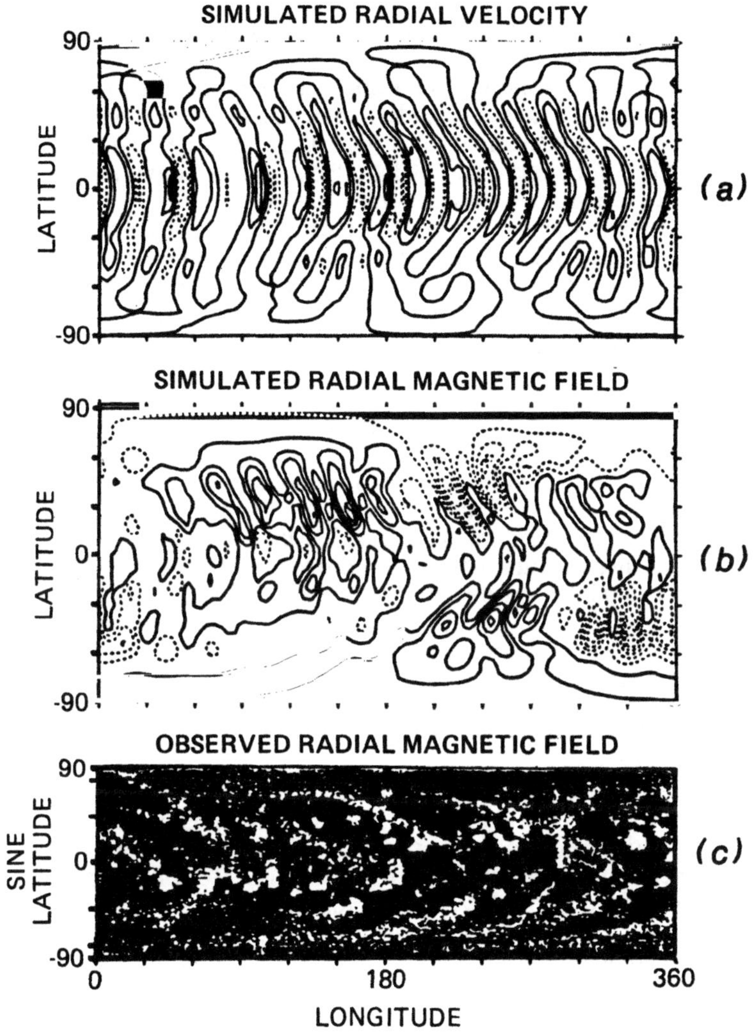

Figure 3. (a,b) Radial component of the simulated velocity and magnetic field at a given time step in a spherical surface just below the outer boundary (Glatzmaier 1984). Solid (broken) contours represent outward (inward) directed fields. (c) A Kitt Peak synoptic map of the large-scale magnetic flux at the solar surface (Carrington rotation 1705, Feb. 1981; courtesy of Dr. J. W. Harvey). Black (white) represents positive (negative) magnetic polarity.

However, there has been no direct evidence of the existence of banana rolls in the sun. If they do exist, as the numerical simulations predict, they would be very difficult to detect because the global pattern is constantly changing. Different modes propagate in longitude with different phase velocities; and these phase velocities depend on latitude. This time dependence of the global pattern is easily appreciated when viewing movies of both the space-lab experiment and the numerical simulations of the experiment (Toomre, these proceedings). It can also be seen in Figure 7 of Gilman & Miller (1986) which illustrates the change during one rotation period. If long time averages are made on Doppler measurements of the solar surface velocity in order to filter out granulation and super granulation, the global convective pattern may also be filtered out. In addition, the breakup of large-scale convection into smaller horizontal scales near the surface, where the density scale-height becomes very small (Chan, these proceedings), may be responsible for the non-observance of banana rolls at the solar surface.

This latter reason may also explain why a recent analysis of the movement of young sunspots (Ribes et al. 1985) suggests the existence of axisymmetric east-west rolls, not north-south banana rolls. It is possible that this is evidence of shallow meridional circulation just below the surface of the sun, above a deep banana-roll structure. It is certainly also possible that global-scale banana rolls do not exist in the sun and that the models are unsuccessful because they lack some important physical process or just do not have enough numerical resolution. On the other hand, the covariance of longitudinal and latitudinal motions of sunspots, measured on Mt. Wilson white light plates, is very similar to that predicted by the models (see Figure 17 of Gilman & Miller 1986).

There may be indirect evidence of banana rolls based on the measurement of large-scale solar magnetic fields. Figure 3b shows the simulated radial component of the magnetic field (on the same spherical surface and at the same time step as in Figure 3a) generated with Glatzmaier's model running as a dynamo (Glatzmaier 1984). Notice how both polarities of the simulated magnetic field (Figure 3b) tend to be concentrated in the downdrafts of the large-scale convection (Figure 3a) where the horizontal velocity converges. This numerical simulation of the large-scale magnetic field can be compared to a Kitt Peak synoptic map of the large-scale magnetic flux at the solar surface (Figure 3c). Notice the "chevron-shaped" structure that is prominent in both the simulation and the observation. It is tempting to consider this similarity as indirect evidence of banana rolls in the deep solar convection zone.

Banana rolls are also seen in plots of the horizontal components of simulated velocity (Figure 9 of Gilman & Miller 1986). Fluid moving (relative to the rotating frame) in the direction of rotation tends to drift toward the equator and vice versa. This is due to the spherical geometry and rotation (Glatzmaier 1985a). Coriolis forces acting on the circulating fluid tend to increase or decrease the cross section of a banana roll depending on the sense of the circulation. Since the cross sections of the rolls decrease at higher latitude due to the spherical geometry, the circulating fluid drifts away from the equatorial plane if the Coriolis forces are trying to decrease the cross section of the roll. Likewise, fluid drifts toward the equatorial plane if it has the opposite sense of circulation. Therefore, as illustrated in Figure 1 of Glatzmaier (1985a), there is a net latitudinal flux of angular momentum toward the equator in both hemispheres. In addition, due to mass conservation and the density stratification, this latitudinal flux is greatest in the outer part of the convection zone where velocities are large because density is small.

As a result, angular velocity also decreases with depth. This radial differential rotation shears the nonaxisymmetric convective rolls (Gilman & Miller 1986) producing cells tilted in radius and longitude in such a way that rising fluid tends to move in the direction of rotation and sinking fluid in the opposite direction. This is illustrated in Figure 8 of Gilman & Miller (1986) and in Figure 3 of Glatzmaier (1984). Consequently, there is a net upward radial flux of angular momentum. However, this feedback process can only go so far before it finds a nonlinear equilibrium because the larger the tilt, the less efficient the convection and therefore the smaller the convective velocities which help maintain the radial differential rotation which produces the tilt.

3.3. Meridional circulation and differential rotation

The axisymmetric part of the velocity field is divided between differential rotation, i.e., the axisymmetric longitudinal velocity relative to the rotating frame of reference, and the meridional circulation, i.e., the radial and latitudinal components of the axisymmetric velocity. The total kinetic energy in the meridional circulation is usually two orders of magnitude less than that in the differential rotation.

A typical profile of the simulated meridional circulation appears in Figure 4a. Streamlines of the mass flux illustrate how the axisymmetric flow in the equatorial region is outward in the outer part of the zone and inward in the inner part. This is mainly due to the Coriolis forces resulting from the differential rotation (Figure 5a). Note that there are also inward flows around ±50° latitude.

The outward flow at the equator causes the surface temperature to be slightly higher there; while the inward flows at ±50° latitude produce slightly lower surface temperatures (Glatzmaier 1985a). The corresponding axisymmetric part of the simulated surface temperature perturbation is illustrated in Figure 4b showing a $2K$ maximum at the equator, relative to the reference state surface temperature of $4\text{x}10^5K$, and minimums at ±50° latitude. A similar profile has been obtained for the latitudinal temperature variation in the photosphere using the Princeton solar distortion telescope (Kuhn et al. 1985; Kuhn, these proceedings). They see temperature minimums at ±53° latitude. The absolute temperature difference simulated is larger than that observed because the model surface is well below the photosphere. However, the ratio of the maximum temperature difference to the average temperature in that surface is somewhat less than that observed.

Finally the differential rotation, which is of major interest in these proceedings, will be discussed. In these simulations, differential rotation in latitude and radius is found to be maintained, against Coriolis and viscous forces, by the angular momentum flux which converges in the outer part of the convection zone and in the equatorial region. As described above, the angular momentum flux is a nonlinear product of nonaxisymmetric velocities which are in the form of banana rolls due to the effects of the spherical geometry, rotation, and density stratification. A typical profile of the simulated differential rotation is plotted in Figure 5a with contours of angular velocity relative to the rotating frame of reference. A similar profile, but of linear velocity, is illustrated in Figure 10 (step 18000, deep layer) of Gilman & Miller (1986). These profiles illustrate how angular velocity tends to be constant on cylinders coaxial with the rotation axis. Also plotted (Figure 5b) are the simulated latitudinal differential rotation velocities at the surface for both Gilman's and Glatzmaier's models and the solar profile obtained from Mt. Wilson Doppler measurements (Howard et al. 1983). Both models produce profiles that are in fair agreement with the observed surface differential rotation. In addition, the simulated radial differential rotation rate in the equatorial plane from Figure 5a is plotted in Figure 5c (solid curve) and compared to the rotation profile obtained from the analysis of the rotational frequency splitting of solar oscillations (dots in Figure 5c; Duvall et al. 1984; see more recent results in these proceedings). These profiles demonstrate that the simulated radial differential rotation is consistent with current observational estimates.

One point that should be emphasized is that, although the observational results show only a slight decrease (15%) in rotation rate with depth through the solar convection zone, that is all the numerical simulations have been predicting. Since the simulated profile

(Figure 5a) has angular velocity constant on cylinders, the decrease of the rotation rate at the surface from the equator to about 45° latitude is essentially the same as the decrease of the rotation rate in the equatorial plane from the surface to the base of the convection zone. The differential rotation portrayed in Figure 5a appears very substantial because it plotted relative to the rotating frame of reference. However, the absolute variation in radius simulated by the models and plotted in Figure 5c is actually only about 15% of the surface rotation rate.

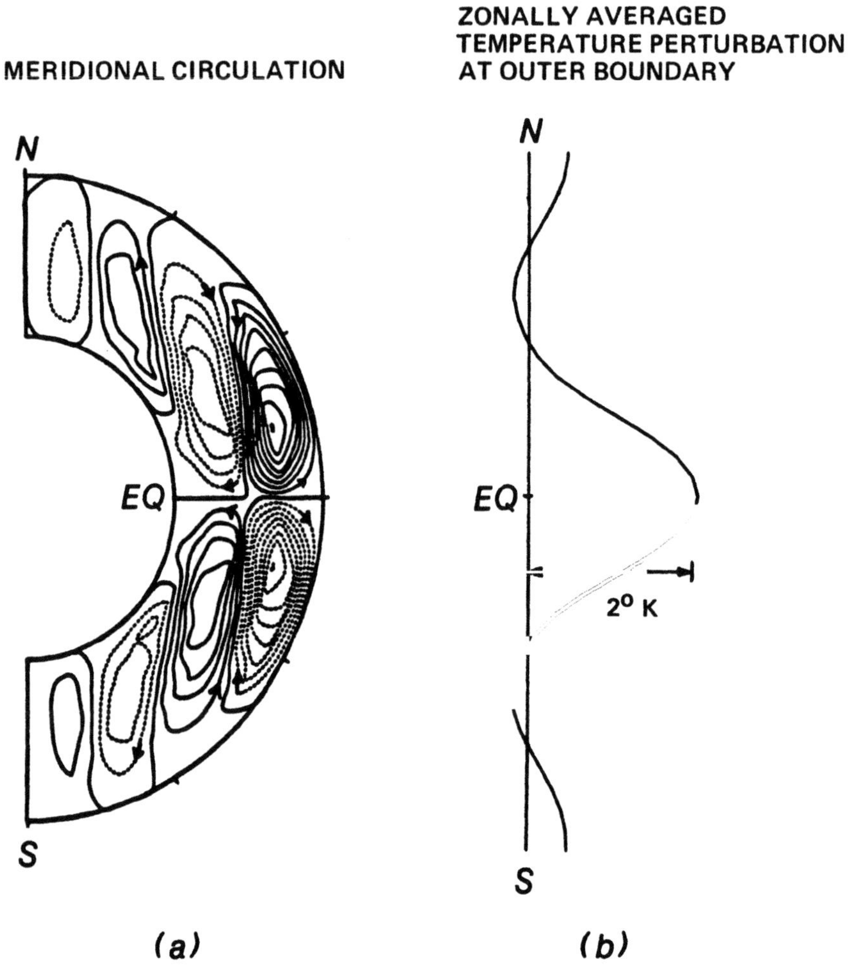

Figure 4. (a) Streamlines of the axisymmetric mass flux plotted in a meridian plane. (b) The latitudinal variation of the axisymmetric temperature perturbation at the outer boundary (93% of solar radius) relative to the reference state boundary temperature of 4×10^5K (Glatzmaier 1985a).

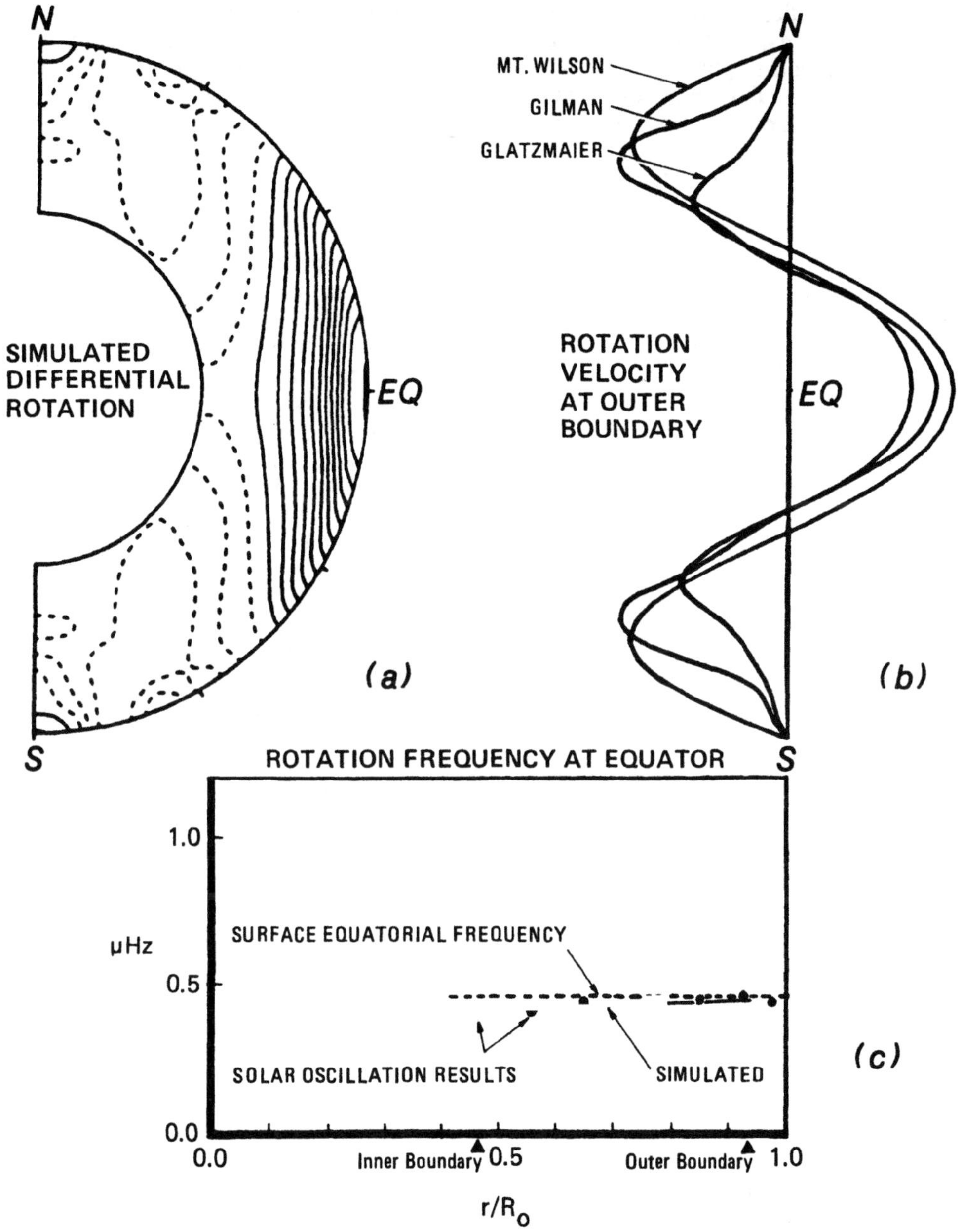

Figure 5. (a) Simulated angular velocity relative to the rotating frame of reference (Glatzmaier 1985a). (b) Simulated surface rotation velocity (Gilman & Miller 1986; Glatzmaier 1985a) and observed solar surface rotation velocity at Mt. Wilson (Howard et al. 1983). (c) Simulated angular velocity as a function of radius (solid curve, Glatzmaier 1985a) and estimates from the analysis of the frequency splitting of solar oscillations (dots, Duvall et al. 1984).

4. DISCUSSION OF MODEL VALIDITY

Having briefly described the three-dimensional models and compared their solutions to solar observations, an important question arises. How well do these numerical solutions represent the actual physical processes in the sun? Actually, this question should be broken into two. Do the codes accurately solve the model equations and do the model equations adequately describe the solar physics?

4.1. Do the codes accurately solve the model equations?

Although there will always be some degree of error in these numerical solutions due to numerical truncation and limited numerical resolution in space and time, it is hoped that enough accuracy is achieved that certainly the qualitative features and, to a good extent, the quantitative features do not depend on the particular details of the numerical solution technique. Fortunately, this has been the case for the two models described here. These models, having been developed completely independently and having employed very different numerical techniques, have produced essentially the same numerical solutions. These are solutions to a nonlinear system of equations that describe how the three components of velocity and three thermodynamic variables are updated on well over 50,000 grid points for tens of thousands of time steps. The probability is quite small that both of these very different numerical codes are generating erroneous solutions that agree so well with each other. This is indirect evidence that the codes are accurately solving the model equations.

Direct evidence now also exists. Very good qualitative agreement has been found between a recent space-lab experiment of thermal convection in a rotating hemispherical shell and the three-dimensional numerical simulations of it done with a modified version of Glatzmaier's code (Hart et al. 1986; Hart, Glatzmaier, & Toomre 1986; Toomre, these proceedings). Different scenarios, defined by the temperature boundary conditions, the strength of the central gravity, and the rotation rate, produced quite different convective structures including banana rolls, spirals, "soccer ball" patterns, and triangular waves. In all cases that could adequately be resolved numerically, the model produced numerical solutions in agreement with the experiment without any tuning or model adjustments whatsoever. However, one must realize that the convecting fluid in this experiment was a liquid, not a stratified gas like the sun. Also, the actual molecular diffusivities of the liquid were use in the model, not parameterized eddy diffusivities as are required for simulations of the sun. Yet this demonstrated agreement is direct evidence that the computer code is accurately solving the three-dimensional nonlinear system of equations that describes this thermal convection experiment.

4.2. Do the model equations adequately describe the solar physics?

The answer to this question is not so obvious. There are several concerns stemming from the limitations of three-dimensional numerical simulations.

There is concern about the inner and outer impermeable spherical boundaries that are imposed in both models. The inner boundary could be removed by solving the equations everywhere in a full sphere. This poses additional numerical complications, but not difficult ones. This would also allow the three-dimensional simulation of gravity waves in the radiative solar interior and the theoretical investigation of differential rotation in the center of the sun. The outer boundary, as mentioned above, is not placed at the photosphere, but rather 5% to 10% of the solar radius below the photosphere in order of avoid the large additional number of grid points that would be required to adequately resolve this region of small scale-heights. Consequently, it is possible that the models neglect some very important physical processes near the solar surface.

Another concern is the depth of the convection zone which is specified in these models. Both models have demonstrated that the deeper the convection zone, the better the resulting surface differential rotation profile matches the observed solar profile and the less tendency there is to form a polar vortex which has not been observed. On the other hand, the solar abundances of lithium and beryllium place a limit on the depth of the solar convection zone which is close to the depths of the superadiabatic regions chosen for the two models.

However, the major concern about these models is the parameterized eddy diffusivities which are treated as scalar quantities. As mentioned above, these diffusivities are either specified time-independent functions of radius or time-dependent functions of the local shear of the large-scale resolved velocity fields. Several more sophisticated parameterizations should be tested. For example, diffusion in the radial direction probably should be a function of the local Richardson number which accounts for both local stability (radial entropy gradient) and local wind shear. Also, accounting for Coriolis forces on the subgrid-scale eddies via a tensor formulation like that suggested by Durney (these proceedings) may introduce new effects.

The effects of different specified scalar diffusivities have been investigated to some extent with the three-dimensional models. Both models have demonstrated that the more the diffusivities vary in radius the less solar-like the surface differential rotation. Also, as the viscous and thermal diffusivities decrease, for a given spatial resolution, differential rotation kinetic energy increases faster than convection kinetic energy. If the ratio of viscous to thermal diffusivity decreases, the ratio of differential rotation kinetic energy to meridional kinetic energy increases. However, it should be pointed out that, although the models can produce an angular velocity increasing with depth (with a special combination of eddy diffusivities and convection zone depth), the latitudinal differential rotation at the surface in these cases looks nothing like the sun's. Whenever a solar-like surface differential rotation is obtained, the simulated angular velocity decreases with depth as illustrated in Figure 5.

One way to deal with the problem of ad hoc eddy diffusivity parameterizations is to increase the spatial and temporal resolution of the model in order to resolve more scales and decrease the amplitude, effect, and significance of the eddy diffusivities which parameterize the effects of the unresolved scales. That is, increase the role of physics in the numerical simulation while decreasing the role of parameterization. Of course the limitations of computers will always limit numerical resolution.

On the other hand, as seen in Figure 14 of Gilman & Miller (1986), most of the nonlinear work to maintain the simulated differential rotation is done by resolved modes with longitudinal wave numbers between 8 and 20. Therefore, whereas an increase in spatial resolution would provide better information about the small scales, it probably would not significantly change the large-scale structure of the simulated differential rotation.

Although a banana-roll structure has not been observed at the solar surface, the fact that the numerical simulations from both models are in fair agreement with several other solar observations provides some confidence that the physics pertaining to deep global convection is adequately described by the equations. As discussed above, the covariance of longitudinal and latitudinal sunspot motions measured on Mt. Wilson white light plates is very similar to the covariance predicted by the models. The surface structure of the large-scale magnetic flux simulated by the model when in a dynamo mode is quite similar to that seen in Kitt Peak magnetic synoptic maps. (However, problems still exist with magnetic cycle simulations (Gilman 1983; Glatzmaier 1985a,b).) The surface temperature variation with latitude is in fair agreement with observations made with the Princeton solar distortion telescope. The latitudinal differential rotation at the surface is also in fair agreement with Mt. Wilson Doppler measurements. The radial differential rotation through the convection zone is consistent with the analysis of the rotational frequency splitting of solar oscillations. These observations have been explained with relatively simple physical arguments once the numerical models demonstrated, within the limitations described above, that these arguments are consistent with the global, three-dimensional, nonlinear equations that describe mass, momentum, and energy conservation.

ACKNOWLEDGMENT

I would like to thank Dr. Peter Gilman for helpful discussions concerning this review.

REFERENCES

Duvall, T.L., Dziembowski, W.A., Goode, P.R., Gough, D.O., Harvey, J.W. and Leibacher, J.W. 1984, *Nature,* **310**, 22.
Gilman, P.A. 1977, *Geophys. Ap. Fluid Dyn.,* **8**, 93.
_____. 1983, *Ap. J. Suppl.,* **53**, 243.
Gilman, P.A. and Glatzmaier, G.A. 1981, *Ap. J. Suppl.,* **45**, 335.
Gilman, P.A. and Miller, J. 1981, *Ap. J. Suppl.,* **46**, 211.
_____. 1986, *Ap. J. Suppl.,* **61**, 585.
Glatzmaier, G.A. 1984, *J. Comp. Phys.,* **55**, 461.
_____. 1985a, *Ap. J.,* **291**, 300.
_____. 1985b, *Geophys. Ap. Fluid Dyn.,* **31**, 137.
Gough, D.O. 1969, *J. Atmos. Sci.,* **26**, 448.
Hart, J.E., Glatzmaier, G.A. and Toomre, J. 1986, *J. Fluid Mech.,* (in press).
Hart, J.E., Toomre, J., Deane, A.E., Hurlburt, N.E., Glatzmaier, G.A., Fichtl, G.H., Leslie, F., Fowlis, W.W. and Gilman, P.A. 1986, *Science,* **234**, 61.
Howard, R., Adkins, J.M., Boyden, J.E., Cragg, T.A., Gregory, T.S., LaBonte, B.J., Padilla, S.P. and Webster, L. 1983, *Solar Physics,* **83**, 321.
Kuhn, J.R., Libbrech, K.G. and Dicke, R.H. 1985, *Ap. J.,* **290**, 758.
Ribes, E., Mein, P. and Mangeney, A. 1985, *Nature,* **318**, 170.

Moment Closure for Thermal Convection: A Viable Approach?

Jackson R. Herring [α]

[α] National Center for Atmospheric Research *
P.O. Box 3000, Boulder, CO 80307

Abstract

We examine here the usefulness of two-point moment closures as a potential tool in computing turbulent convection, considering the eventual application to stellar convection. Confining our attention to the Boussinesq limit, we first present a formal discussion of how such procedures are constructed, and then discuss—briefly—how the complex two-point formalism may be used as the basis for much simpler single-point closures of the sort used in engineering applications. Noting that the underlying assumption of the closure is a form of randomness or near Gaussianity, we discuss examples of systems related to convection, and the extent to which this assumption is validated for these systems. We examine three potential problems for such application drawn from numerical simulations and experiments: (1) counter-gradient transport, (2) the apparent failure of numerical simulations to adhere to closure scaling at low Prandtl numbers, and (3) the appearance of large-scale structures in large Prandtl number convection and simple one-dimensional models.

Introduction

Some understanding of turbulent convection is undoubtedly essential for the development of rational models of stars and their internal motions. It is frequently typical for the model builders to rely heavily on some form of mixing length theory, which extends to the astrophysical regime an engineering parameterization of turbulence originally developed by Prandtl and others. We should recall, however, that this regime is extremely remote from that studied in the laboratory, or for that matter measured in any terrestrial observations. Recall, in this connection, that laboratory fluids typically have a Prandtl number ($\sigma \geq\sim .01$), whereas the radiative analogue of the stellar Prandtl number is typically $\leq 10 \times 10^{-9}$. This remoteness holds also for full numerical simulations of turbulent convection by super-computers: the reason being that for a fixed overall instability (Rayleigh number $\sim R_a$), the Reynolds number (R_λ) increases as $\sigma \to 0$ ($\sim \sigma^{-5/9}$). Thus the intimate connection between the parameterization procedures (Prandtl, etc.) and experiments that proved invaluable in the engineering context is not available in astrophysics.

In this circumstance, it is reasonable to invest at least a modest effort into assessing and possibly improving the phenomenological theory through a comparison of its predictions to results drawn from laboratory experiments, terrestrial-meteorological observations, and large scale, accurate three-dimensional simulations of turbulent convection. Such a program was outlined many years ago by—for example, Spiegel (1966)—and has been examined by several investigators. The simplest system of this type is Boussinesq, thin-layer convection, without rotation. Such a simple system contains the basic nonlinearity,

* The National Center for Atmospheric Research is sponsored by the National Science Foundation

B. R. Durney and S. Sofia (eds.), The Internal Solar Angular Velocity, 275–288.

and in fact contains the essentials of early mixing-length convection calculations if the distance from the lower plate boundary replaces the pressure scale height in the stellar model. Moreover, some numerical-experimental and theoretical information is now available (Siggia and Zappelius, 1983, Herring and Jackson (1984)), which shows the low $\sigma \to 0$ regime to be surprisingly turbulent, even just above the critical Rayleigh number, R_c. In this paper, we shall discuss the applicability of simple closure procedures to thermal convection, comparing—where possible—results to the numerical results for slip boundary conditions and large σ (in lieu of the possibility of any data at $\sigma = 10^{-9}$). The moment closures are rationally derived procedures and thus free of any empiricism of the sort that enters the mixing-length theory, and they could, in principle, be used to derive or improve mixing-length estimates. There are, however, two related open questions, neither trivial: (1) Is moment closure a reasonable representation of the dynamics? (2) If so, can the resulting set of equations be simplified so as to be a practical tool for the astrophysicist?

Structure and Basic Physics of Two-Point Closure

To begin, we must recall what moment closures attempt (we do this here in the simplest possible context of a Boussinesq fluid). For dynamical variables ($[\mathbf{u} = (u, v, w)], T, P$) = (velocity, temperature, and pressure) we form ensemble mean values (denoted by $\langle X\rangle(\mathbf{x}, t)$, for variable $X(\mathbf{x}, t)$, which for the statistically steady state depends on a reduced set of spatial variables. For thin-layer convection above a flat plate, these are the distance above the plate, z, and time, t. Because of the bilinear nature of Navier-Stokes,

$$\nabla \cdot \mathbf{u} = 0 \tag{1a}$$

$$(\partial_t - \sigma\nabla^2)\mathbf{u} = -\nabla p - \mathbf{u}\cdot\nabla\mathbf{u} + \mathbf{g}\sigma R_a\theta \tag{1b}$$

$$(\partial_t - \nabla^2)\theta = -\beta w - \nabla\cdot(\mathbf{u}\theta - \langle\mathbf{u}\theta\rangle) \tag{1c}$$

($T \equiv \theta + \langle T\rangle$, and $\beta(z) \equiv -\partial\langle T\rangle/\partial z$), an equation of motion for $\langle T\rangle(z, t)$, will depend on $\langle\mathbf{u}T\rangle$, whose equation of motion through (1b,c) depends on $\langle\mathbf{u}T\mathbf{u}T\rangle$, etc. The pattern is clear; equations of motion at a given degree depend on the next higher, and so on ad infinitum. Moment closures derive a functional relationship between level n and $n+1$, thereby closing of a tractable set of equations to be solved. Because spatial and time derivatives enter (1a)–(1c), deductive moment closures generally contain moments of amplitudes $(\mathbf{u}, \theta)$ at different space-time points; in general this makes their structure much more complex than Navier-Stokes. Of course the moments possess smooth dependence on the $(y \equiv \mathbf{x}, t)$ arguments, and this must be exploited if closure is to be made practical. At the simplest possible level, these closures relate Reynolds stresses to mean gradients; the most famous example is an expression for the eddy heat flux, under non-buoyant conditions (passive scalar):

$$\langle w\theta \rangle = C\partial \langle T \rangle / \partial z \tag{2}$$

where C is as given by (Taylor, 1922):

$$C = \int_0^\infty ds \langle \mathbf{u}(y)\mathbf{u}(y(s)) \rangle \tag{3}$$

Here, ds denotes a Lagrangian integration along a parcel trajectory ($= s$). Of course (3) assumes a constant imposed β. In addition it contains a Lagrangian moment $\langle \mathbf{u}(y)\mathbf{u}(y(s)) \rangle$, not an Eulerian as its basic ingredient, contrary to our original proposal. The involvement of "Lagrangian information" noted here is endemic to all closures proposed at large R_λ: either the basis of the closure must be shifted to a Lagrangian framework, or relationships must be found (usually empirical) to relate Lagrangian relaxation times to the corresponding Eulerian ones.

Before discussing detailed closure proposals, we should at least state the need for extending the quality of existing calculations beyond mixing-length estimates. Of course the latter contains unknown constants which should be determined, but more importantly, it may miss some qualitative physics as well. Consider for example the fact that turbulent convection in all its known manifestations (i.e., in the laboratory, numerical simulations, and in the meteorological context) exhibits non-local effects, contrary to local transport ideas of mixing length theory. This means that convective transport, in some region of space, may be counter gradient (a region of negative β). Fig. 1 gives an example.

Here we plot $\beta = \langle T \rangle(z)$ for slip-boundary (turbulent) convection at $R_a = 560R_c$, where R_c is the critical Rayleigh number (the simulation uses 32^3 Fourier modes). This profile is clearly at variance with the simple diffusive concept of the convective process, which must have $\beta \geq 0$. Another example where this is known to happen is the early two-dimensional simulations of Moore and Weiss (1973), which showed strikingly stable core regions (the region exterior to the boundary layer) at all Prandtl numbers calculated. It was earlier thought that the $\beta \leq 0$ region occurs only at small R_a; however it is observed in the planetary boundary layer at much larger R_a and Reynolds numbers (Deardorff, (1966)). Either (or both) of the following seem a plausible explanation of this phenomenon. First, turbulent mixing may be similar to molecular diffusion—in the large mean free path limit—and mix fluid over a region larger than the length scale of the local gradients, thereby averaging (1a–1c) over both boundary and interior regions. This would allow $\langle w\theta \rangle(z)$ to remain positive even if $\beta \equiv -\partial \langle T \rangle(z)$ is negative in the core region. The idea that mixing length theory needs generalizing in this direction was first suggested by Spiegel (1963). Put formally, we rewrite (1c) as

$$(\partial_t - \nabla^2 + \mathbf{u} \cdot \nabla)\theta = -\beta(z)\theta \tag{4}$$

Then, we may solve (4) by introducing a Green's function, $G(y, y')(y \equiv \mathbf{x}, t)$ which inverts its left-hand side:

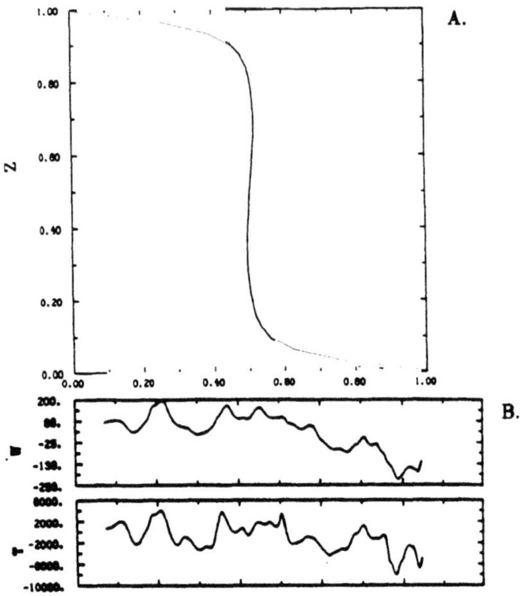

Fig. 1: A: Time-averaged $\langle T\rangle(z)$ for $R_a = 560R_c$, slip boundary convection. B: vertical velocity and temperature at a point near the mid-point of the flow. Here $\sigma = .7$.

$$\langle w\theta\rangle(z) = \int dy'\langle w(y)G(y,y')w(y')\rangle\beta(z') = \int dy'\langle w(y)w(y')\rangle\langle G(y,y')\rangle\beta(z') \quad (5)$$

where the second equality of (5) is an untested Gaussian-like approximation. We return to a discussion of the evaluation of $\langle G\rangle$ shortly.

The second possible explanation for such an "overshoot" phenomenon is more at variance with moment closures, but perhaps related to the above. The dynamics may dictate that the convection occurs in intense jets or thin sheets in a spatially (and perhaps temporally) intermittent fashion. In this case the flow is non-Gaussian, and approximations similar to that used in (5) are not well justified. A symptom of strong non-Gaussian flow would be signaled by values of Kurtosis of ($\mathbf{u}$ or $\nabla\times\mathbf{u}$) $\gg 3$. This has been demonstrated in two-dimensions both in decaying flows (Herring and McWilliams, (1984)), and in thermal convection (Moore and Weiss, (1968)).

To illustrate the structure of the closure and its level of complexity, we sketch how solutions for $\langle G\rangle$ may be obtained. The derivation is purely formal and readers wishing to follow only the physics of the arguments may wish to skip to the next section. The averaged Green's function $\langle G\rangle$ that enters (5) may be evaluated formally via a method of Keller (1965);

$$\langle(\partial_t - \nabla^2 + \mathbf{u}\cdot\nabla)\rangle^{-1}\langle G(y,y')\rangle = \delta(y,y') \quad (6)$$

We note that if the ensemble-averaged operator is expanded in powers of $\mathbf{u} \cdot \nabla$, a simple low Reynolds number eddy diffusion prescription results. To obtain approximations for (6) valid at high Reynolds numbers, we must resort to renormalized perturbation expansions, such as those introduced by Kraichnan (1964) and others (see e.g., Keller, (1965)). If this is done, what results is a generalized eddy conductivity prescription, whose form is:

$$(\partial_t - \nabla^2)\langle G(y,y')\rangle + \int dy''\langle \nabla' \cdot \mathbf{u}(y')\langle G(y'',y')\rangle \nabla'' \cdot \mathbf{u}(y'')\rangle dy'\langle G(y'',y')\rangle = 0 \quad (7)$$

We note that omitting the first $\langle G\rangle$ under the integral reproduces first-order perturbation theory (first order smoothing); the presence of this $\langle G\rangle$ serves to convert the original naive perturbation series, which is in ascending powers of Reynolds number, to one in which each term is nominally of the same order in R_λ . This equation is extremely complicated, but has a simple physical meaning: The second member on the left describes an eddy diffusion in which the transport is projected along a particle trajectory carried by the fluid motion. The order of magnitude of this term is estimated by the *operator*:

$$\langle \partial_i u_j \partial_i u_j \rangle \tau \quad (8)$$

where τ is the effective lifetime of an eddy, $L/(< w^2 >^{1/2})$. Combining (6), (7), and (8) then permits an estimate of non-local effects via closures.

We have touched upon counter gradient transport here not only to illustrate an effect that is not predicted by K-theory or mixing-length theory, but also because of the fact that there may be real inconsistencies in extending mixing-length predictions by using a combination of linear growth of perturbations on mean profiles with diffusive estimates of the self-limiting nature of the non-linear terms to compute convective (and momentum) transport in the general context of rotating M.H.D. flows. Our point here is that counter-gradient transport challenges the notion that instabilities on the mean fields supply turbulent energy, with self-interactions providing a destruction mechanism through random diffusion. It would appear that one physical explanation of the observed phenomenon is that the turbulent structures are, to a limited extent, self-organizing. This is the sinister possible explanation noted above.

Single Point Closure

The formalism sketched above is clearly too complicated to be used as a basis for computations. However, were we assured of sufficient accuracy, we might use it as a starting point to derive a simpler procedure, much as Navier-Stokes is "derived" from the too formidable Boltzmann equation. Such a program was suggested some time ago by Leslie (1973). A possible avenue is to split the two-point dependence of covariances, for example, $\langle u_i(\mathbf{x})u_j(\mathbf{x}')\rangle$ into slow $\Re \equiv \mathbf{x}+\mathbf{x}'$ and fast $\rho \equiv (\mathbf{x}-\mathbf{x}')/2$ variables, and assume that $\mid \partial_{\Re} u/\partial_\rho u \mid \ll 1$. Physically, this asserts that correlation lengths (integral scales) are much shorter than mean gradients of u and T. Following Leslie (1973) we call this approximation "semi-homogeneous"; it is equivalent to a form of the "WKB" approximation current in geophysics. This is a comfortable approximation with little basis in the real world. Carried to the limit in which the ρ dependence is $\sim \delta(\rho)$, it then yields a closed formalism in which (energy and temperature variance) spectral integrals occur as parameters of the closure. Such a recipe gives a procedure closely similar to engineering prescriptions (Hanjalic and Launder (1971), Lumley (1978)), and also the sub-grid-scale method of Deardorff (1972). More recently, Yoshizawa (1981) has made considerable progress in fleshing out the evaluation of spectral parameters in terms of the basic DIA formalism.

We discuss this procedure here briefly, following in part Deardorff's notation and equations. We motivate the equations by appealing to the underlying physics and only note in passing their (potential) derivation from closure. The basic idea is that the (stochastic) nonlinear interactions $(\mathbf{u}\cdot\nabla\mathbf{u}, \nabla p, \mathbf{u}\cdot\theta)$ impel fields $(\mathbf{u}, \theta)$ toward their statistical symmetry of isotropy-homogeneity. To describe the evolution of turbulence then requires (linear?) relaxation-rate equations for the moments (intensities), together with a prescription for how the relaxation rates relate to the turbulent intensities. Opposing these tendencies are viscous-conductive dissipation and boundary effects. For $\langle u_i u_j\rangle \equiv R_{ij}$, for example, we write:

$$\partial_t R_{ij}(\mathbf{x},t) = -(\langle u_i u_j u_k\rangle)_{,k} + \langle p(u_{i,j}+u_{j,i})\rangle - (\langle u_i p\rangle)_{,j} - (\langle u_j p\rangle)_{,i} + g(\delta_{i3}\langle u_j\theta\rangle + \delta_{j3}\langle u_i\theta\rangle) \tag{9}$$

$$-2\nu\langle(\partial_k u_i)(\partial_k u_j)\rangle.$$

For the heat flux $H_i(\mathbf{x}) \equiv \langle u_i\theta\rangle$,

$$\partial_t H_i = -(\langle u_i u_k\theta\rangle)_{,k} + R_{i3}\beta + \langle p(\partial_i\theta)\rangle - \partial_i\langle p\theta\rangle + \delta_{i3}g\Theta, \tag{10}$$

where the temperature variance $\Theta \equiv \langle\theta^2\rangle$ satisfies:

$$\partial_t\Theta = 2H_3\beta - (\langle u_k\theta^2\rangle)_{,k} - 2\kappa\langle(\partial_k\theta)(\partial_k\theta)\rangle. \tag{11}$$

So far, there is no approximation beyond Boussinesq, except that we have utilized the semi-homogeneous approximation to record only the dominant dissipative terms: $(\langle\theta\nabla^2\theta \sim \langle(\partial_k\theta)(\partial_k\theta\rangle$, etc.). We have also suppressed $\langle\mathbf{u}\rangle$, to keep things simple. The next step is to recall the role of the various right hand members of (9) – (11). Thus, the first RHS member of (9), diffuses turbulence intensity, R_{ij}, vanishing for homogeneous turbulence. In fact, this may be shown (Herring, (1973)) in the semi-homogeneous (WKB) approximation described above. Its somewhat simplified form is:

$$\langle u_i u_j u_k \rangle = -\Gamma\{R_{kn}R_{ij,n} + R_{jn}R_{ik,n} + R_{in}R_{jk,n}\}, \tag{12}$$

where

$$\Gamma \sim \int\int G_{ii}(\rho, \Re, s)d\rho ds. \tag{13}$$

Here, the Green's function integral is a relaxation time, which must either be taken from experiment or computed from closure. In the engineering context, it is usually taken to be E/ϵ, where E is the local energy density (R_{ii}), and ϵ is the energy dissipation rate. The next step is to relate ϵ to other known quantities. How to proceed here is more difficult, since ϵ shows strong intermittency at large R_λ. To get on with the story we follow the Hanjalic-Launder (1972) prescription:

$$\partial_t \epsilon = C_\epsilon \nabla \cdot \{(E^2/\epsilon)\nabla\epsilon\} - C_{\epsilon 2}(\epsilon^2/E) \tag{14}$$

In (14), we have substituted the time scale (13). We may record Hanjalic-Launder coefficients for completeness: $C_\epsilon = .07$, $C_{\epsilon 2} = 1.95$, although the values here are entirely empirical: Engineers adjust these values with no qualm, perhaps rightly so. The second term in (9) ($\langle p(u_{i,j} + u_{j,i})\rangle$) causes R_{ij} to tend toward isotropy. If anisotropy is small, we might imagine this term to be linearizable in the departure of R_{ij} from its isotropic form $Tr(\mathbf{R})\delta_{ij}/3$, and thus:

$$\langle p(u_{i,j} + u_{j,i})\rangle = -C_I(\epsilon/E)\{R_{ij} - \delta_{ij}Tr(\mathbf{R})/3\}. \tag{15}$$

The fifth term on the RHS of (9) $\sim \mathbf{H}$ belongs to the set of variables to be computed.

There remains the final term: viscous dissipation. This may not, of course, be neglected, since on the time and space average it must balance the heat flux. At large R_λ, traditional wisdom has it that scales contributing to $(\partial_x \mathbf{u})$ are very nearly isotropic. If this were so,

$$2\nu\langle(u_{i,k})(u_{j,k})\rangle \simeq \delta_{ij}\epsilon. \tag{16}$$

We have, of course, (14) for ϵ. Similar arguments may be made to plausibly evaluate the RHS's of (10) and (11). Thus,

$$\langle u_i u_j \theta \rangle = -C_\theta(E/\epsilon)(R_{in}\langle(u_j\theta)_{,n}\rangle + R_{jn}\langle(u_i\theta)_{,n}\rangle), \tag{17a}$$

and

$$\langle u_i\theta^2\rangle = -C_\Theta(E/\epsilon)R_{in}(\Theta^2)_{,n}. \tag{17b}$$

For the term that drives $H_i \to 0$ at thermal equilibrium,

$$\langle p(\partial_i\theta)\rangle = -C_s(\epsilon/E)\langle u_i\theta\rangle - \delta_{i3}\Theta/3. \tag{18}$$

The last (mean field) term in (18) derives from that part of the pressure linear in θ, and is evaluated as if its statistics were isotropic: thus the factor 1/3.

Although the averages (of $\mathbf{u}$ and T) entering these equations are ensemble means, it should be noted that the equations are very nearly the same as sub-grid-scale equations, in which the averages are over small volumes, Δ^3. In this case, $E^{1/2}/\Delta$ replaces E/ϵ, with some changes in the numerical coefficients. This method has the advantage of examining in dynamical detail scales larger than Δ, while reserving for approximation only scales $< \Delta$. It is clearly more computing, but has the advantage of more accuracy. Historically it preceded the engineering ensemble averaged prescription (Deardorff, (1973)).

We now make a few comments concerning the above prescription and its potential application to $\sigma \to 0$ convection. First, as presented, there is no allowance for the presence on boundaries, which would clearly upset applications of (15) and (16). Of course stellar convection zones have no "solid" boundaries. In practice, the "semi- homogeneous" formalism is applied to the region exterior to the viscous-conductive boundary layer, and empirical formulas (such as the "law-of-the-wall" or Priestley's law) are used to account for effects within the molecular boundary layers. At $\sigma \to 0$ the "conductive" boundary layer may extend over the whole domain, so it is clear the method should be generalized to treat this case.

Empirical prescriptions (14)–(17b) or their refinements (see Lumley, (1978) for an examination of this point) are not guaranteed realizable, as has been pointed out by Schumann (1977). Equation (15) is an example of where refinements are probably needed. The argument given here to suggest a linear relaxation presumes that return to isotropy is scale-independent. In reality more detailed spectral studies do not support this assertion. The two-point closure may be of importance here in properly generalizing such equations, since its realizability is more readily provided. Such equations are, however, correspondingly more complicated, so that some analysis is needed to arrive at more accurate equations at the level of complexity of (15).

There are also problems of a more fundamental sort, however, that should be examined in the low σ limit. Prior to a serious application we should assure that we are able to treat satisfactorily any significant "sub-sets" of the real problem, and this includes the passive scalar ($\beta = 0$) problem. At low σ, and in the homogeneous-isotropic context, moment closure is tractable, and yields the Batchelor-Howells-Townsend (1959) spectrum for $\Theta(k)$ ($\epsilon^{2/3}k^{-5/3}$ followed by $\epsilon^{2/3}\epsilon_\Theta\kappa^{-2}k^{-17/3}$ as $k \to \infty$). The closure dynamics implies that normalized measures of scalar transfer (the mixed skewness, $S_{\theta\theta u} \to \sigma^2$ as $\sigma \to 0$). This behavior seems now at variance with numerical simulations (Kerr (1987)), which indicates that $S_{\theta\theta u}$ is independent of σ as $\sigma \to 0$. The reason for the discrepancy is that the scalar collects at flow neutral-points, which inhibits the distortion of scalar blobs to small scales where they may be effectively dissipated. Such a dynamics was earlier suggested by Gibson (1968). The simulations so far are at modest R_λ.

Large Scale Collective Effects

An important but as yet uncharted aspect of convection is the dynamics of scales much larger than the buoyant input scale. These are undoubtedly strongly constrained by rotation and curvature effects, but there may also be collective (self-organizing) nonlinear dynamics operating which are important. Howard and Krishnamurti (1984), for example, have studied experiments in large aspect ratio two-and three-dimensional volumes. They reported that beyond a critical R_a there developed, in addition to the convective plumes, a circulation, whose dimensions filled the chamber, and which seemed to be driven by the convective motions. The orientation of the circulation was a symmetry-breaking accident of the initialization. Another perhaps similar collective phenomenon is the dynamics of the large scales activated by the convective turbulence that is observed in numerical studies of slip boundary (Boussinesq) convection (Herring and Jackson, (1984). Fig. 2a shows, for example, contour maps of the convective motion for $R_a = 15R_c$, $\sigma = 0.1$, and aspect ratio $\mathbf{r} = 4$.

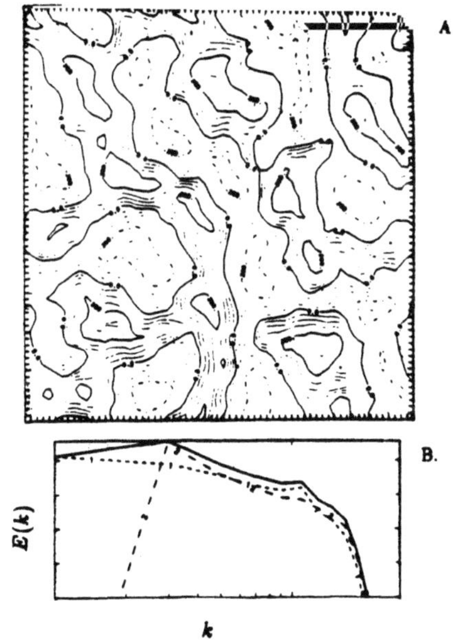

Fig. 2: A: Mid-plane contour plot of T (statistically steady) convection at $R_a = 15R_c, \sigma = .1$, for slip boundaries. B: variance spectrum of vertical motion field (labeled by "2"), and vertical vorticity field ("1"). Spectra are energy-normalized.

For $\mathbf{r} \simeq 1$, rolls are the stable configuration. Spectra for the vertical motion field and the vertical vorticity component of the energy are shown in Fig. 2B. Note that the vertical vorticity component peaks at the largest available scale. This implies that $\mathbf{r}$ should be substantially increased to estimate from numerics the $\mathbf{r} \to \infty$ statistics. The flow is turbulent, with both temporal and spatial chaos.

Of course, the above picture must change with increasing R_a and decreasing σ. But with $\sigma \to 0$, we know the vertical vorticity field becomes progressively destabilized (Siggia and Zippelius, (1983)), and we should then expect considerable activity at large scales, at least for slip boundaries. This large scale activity creates a real problem for direct numerical simulations. The calculations displayed in Figs. 11 and 12 are of modest resolution ($N_x = N_y = 64, N_z = 8$ spectral modes), but nevertheless take a significant time on a Cray 1 to proceed to a statistically steady state. If now R_a is increased so as to estimate asymptotics of heat transport, and the $\mathbf{r}$ is increased by a factor of 4, the calculation becomes prohibitive, even on a Cray-XMP. In these circumstances, we may expect the closure procedure to be of value, if correct, in estimating spectral values of the energy and temperature variance at large scales. The basic physics in the closure at these scales is the tendency toward equipartitioning among the available scales of motion (independent degrees of freedom). Closure estimates the flux of energy into the large scale region by assuming the convective scale structures (plumed) smash into each other randomly (the so-called beating terms). The geometrical constraints severely modify such transfer, and this produces technical problems, but there is no new physics. However, are these convective scale interactions sufficiently random to make the closure estimates useful?

Some insights into this question may possibly be obtained by examining a much simpler problem, for example the one space-dimensional Kuramoto equation:

$$(\partial_t + \partial_{xx} + \partial_{xxxx})u = -u_x u. \tag{19}$$

Here the negative viscosity term is the analogy of the buoyant instability, while the ∂_{xxxx} term is analogous to the viscous dissipation in Navier-Stokes. This equation has recently been the subject of intense study, both theoretically and numerically. A comparison of the numerical results with simple Markovian closures was originally made by Yamada and Kuramoto (1976). Numerical results of these authors and also of Hyman *et al.*, (1987), show surprisingly little energy at very large scales (i.e., larger than the "buoyant input" $k \sim 1$). The steady state energy spectrum has a large "bump" near $k \sim 1$, and drops abruptly to a level of about 1/10 of its maximum as $k \to 0$. The closure, on the other hand, has a $k \to 0$ energy level only slightly less than its maximum value near $k \sim 1$, as expected from equipartitioning ideas. As $k \to 0$, the closure is about a factor of 30 too small. The comparison is only slightly changed if the full DIA is used, or for that matter other, properly "Galilean invariant," theories. It appears that the "convective scales" form structures whose interactions are far from the Gaussian nature needed for the closure assumptions to be met. This model, of course, lacks the essential three-dimensionality needed for a realistic discussion of thermal convection, but it may indicate that an intensive study of the large scales *via* direct numerical simulations should be closely examined first.

Conclusions

Moment closure of Navier-Stokes dynamics as a tool for the computation of turbulent fluxes, decay rates, and spectra was first suggested over forty years ago. The original proposal of Millionshchikov (1941), equivalent to a random phase approximation, failed to incorporate eddy viscosity in a self-consistent manner, and consequently did not account for irreversible relaxation by non-linear interactions. This had as a consequence negative energy spectra for many important problems. Kraichnan (1959) introduced a self-consistent Green's function (along the lines of (7)), with the consequence that the closure became fully realizable. However, there remained with this theory (the DIA) more subtle difficulties (proper covariance to random Galilean transformations), which had as a consequence the wrong inertial range spectra ($k^{-3/2}$ instead of $k^{-5/3}$). Despite this unrealistic feature, this theory gave satisfactory results for moderate R_a convection (Herring, (1969), Dannevik, (1985)) and for small R_λ decaying turbulence. It also was able to compute inhomogeneous turbulence without introducing arbitrary parameters, although, to be sure, the construction of high R_a solutions poses considerable technical problems. However, the inertial range failure, noted above, probably makes it unsuitable as a tool to examine large R_a convection, even in the Boussinesq limit. This problem is cured by utilizing a Lagrangian description (the Lagrangian History DIA, Kraichnan, (1964)). (This method avoids the seemingly intractable difficulties of a purely Lagrangian formulation; instead, the LHDIA is posed in terms of a mixed Lagrangian-Eulerian basis.) Application of this technique to inhomogeneous problems has so far proved too difficult.

Instead of attempting direct solutions, it has been suggested (Leslie, (1973)) that the deductive closure may be used to justify single point closures of the sort invented and extensively explored in engineering applications (Hanjalic and Launder, (1974), Lumley, (1984)). Such methods have, of course, a great computational economy over two-point formalisms. However, this program is as yet incomplete; in fact, most engineering studies place reliance on experiments and direct numerical simulation studies. It should be noted that Yoshizawa (1984) has made considerable progress towards basing single point formalism on a more complete, deductive two-point approach. Here we have only noted briefly the physical arguments that lead to tractable, single point formulas. We have not indicated how such equations should be modified for the $\sigma \to 0$ limit. Even if these efforts to implement two-point closures to thermal convection succeed, there remain the problems associated with the propensity of Navier-Stokes to develop coherent structures and the effects these may have on transfer of heat and momentum. We have noted two respects in which this problem may upset the application of closure to large aspect ratio convection: (1) the possible problems with closure at small σ, and (2) the possible formation of large-scale structures, as was noted in the Kuramoto equation. Both difficulties are attributable to the development of structures which contradict the idea that the non-linear interactions are sufficiently disorganized so as to be treatable as near-Gaussian fields. In both cases, however, the contrary findings stem from high resolution numerical simulations at small R_λ—at least to date.

On the other hand, we should note that the sub-grid scale method briefly mentioned earlier may provide a computational tool intermediate between full simulations and second-order closure. This method has the advantage of bringing at least the large scales of the turbulent structures into direct computation, and thus the effects of intense structures are at least to a certain extent represented in the computations.

Finally, we should mention that the application of renormalization group methods to turbulence though the very recent work of Yakhot and Orszag (1987) seems, so far, astonishingly successful in predicting many of the features of turbulent flows, including scalar fields (Dannevik *et. al.*, (1987)). The method further seems more tractable than the global moment closures reviewed here; it has the flavor of the sub-grid-scale method, but without the empiricism needed in the latter.

From these considerations it seems reasonable to conclude that much work remains before moment closures may be taken seriously as a computational tool for understanding the quantitative aspects of thermal convection. Indeed, there is now emerging—largely through direct numerical simulations of modest R_λ and R_a—a suspicion that strongly non-Gaussian features of the flow may need to be brought within the purview of closure prescription before much quantitative progress can be made. A beginning in this difficult problem has ben made by Kraichnan (1985), but little practical results have yet emerged.

Acknowledgement

I am grateful to S. Zaleski for several enlightening discussions on the Kuramoto-Sivashinsky equation.

REFERENCES

Batchelor, G.K., Howells, I. D., and Townsend A. A., 1959: Small-scale variations of convected quantities like temperature in turbulent fluid. Part 2. The case of large conductivity. *J. Fluid Mech.*, **5**, 134–139.

Dannevik, W., 1985: *Two-point closure study of covariance budgets for turbulent Rayleigh-Bernard convection*, National Center for Atmospheric Research, NCAR/Ct-85.

Dannevik, W., Yakhot, V., and Orszag, S.A., 1987: Analytical theories of turbulence and the ϵ expansion. Preprint.

Deardorff, J. W., 1966: The counter gradient heat flux in the lower atmosphere and in the laboratory. *J. Atmos. Sci.*, **23**, 503–506.

Deardorff, J. W., 1973: The use of subgrid transport equations in a three-dimensional model of atmospheric turbulence. *Applied Mechanics and Fluids Engineering Conference*, Atlanta, Ga., June 20–22, 1973. 423–440.

Deardorff, J. W., 1982: Numerical investigation of neutral and unstable planetary boundaries. *J. Atmos. Sci.*, **29**, 91–115.

Gibson, C. H., 1968: Fine structure of scalar fields mixed by turbulence. I. Zero-gradient points and minimal-gradient surfaces. *Phys. Fluids*, **11**, 2316–2327.

Herring, J. R., 1973: Statistical turbulence theory and turbulence phenomenology. In *Proc. Langley Working Conf. on Free Turbulent Shear Flows.* NASASP321, Langley Research Center, Langley, VA (available from NTIS as N73–2815415GA).

Herring, J. R., and Kraichnan, R. H., 1979: A numerical comparison of velocity-based and strain-based Lagrangian-history turbulence approximations. *J. Fluid Mech.*, **91**, 581–597.

Herring, J. R., and Jackson, S., 1984: Thermal convection: Numerical experiments near the onset to turbulence and the statistical theory of turbulence. In *Turbulence and Chaotic Phenomena in Fluids*, Elsevier Science Publishers, T. Tatsumi, Ed., B. V. (North-Holland), IUTAM, pp. 111–116.

Herring, J. R., and Wyngaard, J. C., 1987: Convection with a first-order chemically reactive passive scalar. To appear in *Turbulent Shear Flows,* **5**, L. S. J. Bradbury, F. Durst, B. E. Launder, F. W. Schmidt, and J. H. Whitelaw, eds. (Springer-Verlag).

Howard, L. N., and Krishnamurti, R., 1984: Large-scale flow in turbulent convection turbulence and chaotic phenomena in fluids. In *Proceedings of the International Symposium on Turbulence and Chaotic Phenomena in Fluids, The International Union of Theoretical and Applied Mechanics.* (Elsevier Science Publishers B.V.) T. Tatsumi, Ed., pp. 535–542.

Hyman, J. M., Nicolaenko, B., and Zaleski, S., 1986: Order and complexity in the Kuramoto Sivashinsky model of weakly turbulent interfaces. Preprint.

Kerr, R. M., 1987: Kolmogorov and scalar-spectral regimes in numerical turbulence. NASA Technical Memorandum 867699.

Kraichnan, R. H., 1961: The structure of isotropic turbulence at very high Reynolds numbers. *J. Fluid Mech.*, **5**, 497–543.

Kraichnan, R. H., 1964: Lagrangian-history closure approximation for turbulence, *Phys. Fluids*, **8**, 575–598.

Kraichnan, R. H., 1985: Decimated amplitude equations in turbulence dynamics. *Theoretical Approaches to Turbulence*, *Appl. Math. Sci.*, **58**, (Springer-Verlag), New York, Berlin, Heidelberg, Tokyo. D. L. Dwoyer, M. Y. Hussaini, and R. G. Voigt, Eds., pp. 91–136.

Leslie, D. C., 1973: *Developments in the Theory of Turbulence*, Clarendon Press, Oxford.

Lumley, J. L., 1978: Computational Modeling of Turbulent Flows. *Advances in Applied Mechanics,* **Vol. 18**, Academic Press.

Millionshchikov, 1941: Theory of homogeneous isotropic turbulence. *Dokl. Akad. Nauk SSSR, 32,* **No. 9**, 611–614.

Moore, D. R., and N. O. Weiss, 1973: Two-dimensional Rayleigh-Bernard convection. *J. Fluid Mech.*, **58**, 289–312.

Schumann, U., 1977: Realizability of Reynolds stress turbulence models. *Phys. Fluids*, **20**, 721–725.

Spiegel, E. A., 1963: A generalization of the mixing-length theory of turbulent convection. *Astrophys. J.*, **138**, 216–225.

Spiegel, E. A., 1966: Energy Transport by Turbulent Convection. *Stellar Evolution*, R. F. Stein and A. G. W. Cameron, eds., Pleunum Press, pp. 143–174.

Taylor, G. I., 1921: Diffusion by continuous movements. *Proc. Roy. Lond. Math. Soc.*, **20**, 196.

Yamada, T., and Y. Kuramoto, 1976: A reduced model showing chemical turbulence. *Prog. Theor. Phys.*, **56**, 681–683.

Yakhot, V., and S. A. Orszag, 1987: Renormalization Group Analysis of Turbulence I. Basic Theory Preprint.

Yoshizawa, A., 1980: Statistical theory for Boussinesq turbulence. *J. Phys. Soc. Japan*, **48**, 647–652.

Zippelius, A., and E. D. Siggia, 1982: Disappearance of stable convection between free-slip boundaries. *Phys. Rev.*, **A26**, 1788–1790.

MAGNETIC ACTIVITY COMPLEXES, THERMAL RELAXATION OSCILLATIONS, AND THE DYNAMICS OF THE AZIMUTHAL MAGNETIC FIELD OF A STAR

E.N. Parker
Depts. of Physics and of Astronomy
Enrico Fermi Institute
University of Chicago
933 East 56th Street
Chicago, Illinois 60637

ABSTRACT. The large amount of magnetic flux cycling through an activity complex in a period of six months suggests that the azimuthal magnetic field in the convective zone of the sun is not less than 3×10^3 gauss and probably is considerable more. A field of 3 kilogauss or more has the effect of blocking the convective heat transport, producing a cool shadow above the azimuthal field and causing heat to accumulate below. The incremental weight of the cool shadow is sufficient to overpower the magnetic buoyancy, pressing the field against the bottom of the convective zone. The accumulation of heat underneath produces a thermal relaxation oscillation that erupts to the surface at intervals of the order of a week, supplying the intermittent emergence of magnetic flux observed in the activity complex.

1. INTRODUCTION

The appearance of bipolar magnetic regions at the surface of the sun is generally interpreted to imply a strong azimuthal field somewhere below, presumably in the convective envelope. The azimuthal fields have opposite signs in the two hemispheres and migrate toward the equator after appearing first at middle latitudes, providing the 11-year or 22-year magnetic cycle of the sun. There is a poloidal field of some 5 or 10 gauss observed at the surface of the sun, reversing at approximately 11-year intervals and apparently playing an essential role in the magnetic cycle. So far as total magnetic flux is concerned, the poloidal field, involving about 10^{23} Maxwells, is the minor partner compared to the azimuthal fields of 10^{24} Maxwells. Both are essential parts of the magnetic cycle, of course. The azimuthal field is generated by the interaction of the nonuniform rotation with the poloidal field, while the poloidal field is believed to be produced from the azimuthal field by the cyclonic motions within the convective zone (Parker, 1955b, 1957, 1970, 1971, 1979a, pp. 644-712, 746-764; Steenbeck and Krause, 1969a,b; Leighton, 1969; Köhler, 1970, 1973; Deinzer and Stix, 1971; Yoshimura,

B. R. Durney and S. Sofia (eds.), The Internal Solar Angular Velocity, 289–301.

1971, 1972a,b; 1973, 1975; Roberts, 1972; Roberts and Stix, 1972; Lerche and Parker, 1972; Stix, 1973, 1976a,b; Deinzer, Kusserow, and Stix, 1974; Gilman and Miller, 1981; Gilman, 1981, Glatzmaier, 1985) providing what is generally referred to as an $\alpha\omega$-dynamo. The precise form of the dynamo is not clear at the present time in view of the evidence (Duvall and Harvey, 1984; Duvall, et al, 1984; Brown, 1985, 1986; Libbrecht, 1986; Duvall, Harvey and Pomerantz, 1986) that the angular velocity Ω decreases downward through the convective zone. The simplest form of the theoretical $\alpha\omega$-dynamo requires that the angular velocity increase downward in order that the fields migrate toward the equator. It may be that the observed latitude gradient of the angular velocity is the important effect (Leighton, 1969; Lerche and Parker, 1972) or perhaps α has the sign expected for converging flows in the lower half of the convective zone (Yoshimura, 1975). Whatever the truth may be, the theory of the solar dynamo is presently in an unsettled state, and it remains for more precise determinations of $\partial\Omega/\partial r$ throughout the convective zone to clarify the picture.

Current ideas on the origin of the magnetic fields of other stars are based on the dynamo concept developed for the sun. The other stars are too distant to observe any details of their activity and their differential rotation. So the popular application of dynamo concepts to organizing the accumulating data on stellar X-ray emission, integrated Zeeman effect, angular velocity, and depth of the convective zone proceeds unhindered by the bewildering behavior of the sun. Indeed, detailed observation of the magnetic fields and the associated magnetic activity of the sun indicates there are many exotic effects and basic dilemmas hidden in the other stars.

It should not be assumed that the sun is entirely candid with us, in spite of its proximity. We can resolve the gross features of the magnetic activity but the essential small-scale structure of the fibril fields and their dynamical behavior lies below the limit of resolution of the present best ground based telescopes and must await observation either from a diffraction limited telescope of 1 m or more in orbit above the terrestrial atmosphere or from further improvement in active optics in ground based telescopes. What is more, even with the best observations, conditions below the visible surface are available only through theoretical inference. Even the gross static structure of the solar interior is not without questions, in view of the small detected neutrino flux. The fluid mechanics of the convective envelope of the sun is too complex to deduce from basic hydrodynamical principles. Much can be learned by applying the basic principles, of course, but the activity of the sun is a product of some of the more exotic aspects of the dynamical convection. In particular the interaction of the convection with the magnetic fields generated within it involves dynamical effects lying beyond present theory. Hence observation is the essential pilot for the theoretician, and it is just such a case that we present in this paper.

The recent observational studies of an activity complex by Gaizauskas, Harvey, Harvey, and Zwaan (1983) provide basic figures for the magnetic field that allow some inferences on the dynamical nature of the azimuthal field in the convective zone. Gaizauskas, et al studied a bipolar magnetic region - an activity complex - existing for well over a year before it faded into the background.

The bipolar active regions on the sun represent long lived sites of intermittent eruption of magnetic flux from the general azimuthal field in the convective zone below. Each bipolar active region is sustained by the emergence of fresh flux in amounts of the order of 10^{22} Maxwells at irregular intervals of about a week. Each eruption lasts a few days and, following each eruption, the flux in the bipolar region fades away, at a rate of the order of 10^{21} Maxwells per day. Thus, the bipolar magnetic region is not so much a fixed region of field as it is a site of emergence of field. The individual site is called an activity complex and may maintain its identity for a year or more. During that time there may be 30-60 eruptions of bipolar field. The point of emergence of field is restricted to small dimensions, of the order of 10^4 km x 10^4 km. There may be more than one point of emergence in a large activity complex. The activity complex may break in half to form two separate activity complexes drifting slowly apart at speeds of the order of 10^3 cm/sec.

Within an activity complex the erupting flux spreads out from the point of emergence, forming two rivers in opposite directions, each dominated by magnetic flux of a given sign, and flowing to the periphery of the activity complex (Brants and Steenbeck, 1985; Zwaan, Brants, and Cram, 1985; Zwaan, 1985). The magnetic field does not appear to be in an intense fibril state as it first appears at the surface (Zwaan, 1985; Brants, 1985a,b). The fibrils first appear with fields of the order of 500 gauss, which develop into the full intensity of 1-2 kilogauss within an hour or so. During the times of emergence the fibrils exhibit a strong tendency to cluster to form pores and sunspots, which appear at the outer ends of the two rivers of flux.

When the emergence of new flux ceases, the clustering tendency vanishes and the total flux passing through the bipolar activity complex begins to decline. Magnetic fibrils are observed to disperse outward across the surface of the sun (Leighton, 1964, 1969) and some are observed to retract back into the sun in the reverse of their emergence (Wallenhorst and Howard, 1982; Wallenhorst and Topka, 1982; Livi, Wang, and Martin, 1985; Martin, Livi, and Wang, 1985). The dispersal of the magnetic flux across the surface of the sun is presumably a result of the random walk of the individual magnetic fibrils in granules and supergranules (Leighton, 1969; Sheeley, DeVore, and Boris, 1985; DeVore, Sheeley, and Boris, 1985) providing a diffusion coefficient of the order of 10^{12}-10^{13} cm^3/sec. However, the total flux in the activity complex is observed to decline faster than can be accounted for by the transport of field across the quiet (weak field) regions surrounding the activity complex. As a simple example consider the decline of the magnetic flux Φ in a monopolar region of radius r with field B ($\Phi = \pi r^2 B$) as a consequence of outward radial motion of the fibrils across the surrounding quiet region where the mean field is b. The rate at which flux streams away across the quiet region is $\dot{\Phi} = 2\pi rvb$. Thus, an observed rate of decay of 10^{21} Maxwells per day in a region with a radius of 3×10^4 km surrounded by a quiet region with b=10 gauss, requires that v=0.5

km/sec. But no such systematic universal radial motions are observed, nor are they expected from the random 1 km/sec motion of the granules.

The magnetic flux sneaks away somehow, probably as a consequence of the unresolvability of the individual fibrils, so that changes in the diameter of the fibrils and in the thermal structure of the atmosphere within the fibrils provide a false interpretation of the unresolved integrated Zeeman effect. We have emphasized that the only alternative to dispersing the field outward from the activity complex is that it retracts back into the sun (Parker, 1984a). One should see magnetic fibrils of opposite sign streaming inward from the opposite ends of the activity complex, i.e. the opposite patches of photospheric field should meet somewhere in the middle and disappear. The effect is observed, but so far not in sufficient amounts to account for the fading of the total flux, probably for the same reasons that the outward dispersal of magnetic flux is not observed. This fundamental observational problem is of primary importance in understanding the dynamical evolution of the activity complex. As we shall soon see, the hypothesis that most of the flux is retracted, rather than dispersed, is the more conservative assumption. Consequently we suggest that it is the primary means by which the magnetic flux is able to disappear from the surface following its emergence.

Whatever may be the path of the field to disappearance, it is clear from the emergence, transient clustering, and eventual disappearance of the field that its behavior is controlled by the hydrodynamics of the convective zone. Now it was suggested many years ago (Parker, 1955a) that magnetic buoyancy brings the azimuthal field to the surface. The problem is then that the expected rate of rise of the field is so large ($\sim 10^3$ cm/sec) (of the order of the Alfven speed within the field) that the field would escape in a year or less, so that it could not be regenerated by any known dynamo process (Parker, 1979a, pp. 141-146, 1984b). In fact the observational evidence is that there is very little loss of azimuthal magnetic flux through the surface of the sun (Parker, 1984b) which has important consequences for the theory of the solar dynamo (Rai Choudhuri, 1985). It follows that even the spontaneous magnetic buoyancy is dominated and controlled by the hydrodynamics, in a manner that will appear shortly.

The problem is to identify the hydrodynamics from the observed behavior of the magnetic fields at the surface. Thus, for instance, it has been pointed out that the apparent mutual attraction of rising magnetic fibrils may be in part a consequence of the Bernoulli effect. The tendency for one fibril to follow in the path of another may be the effect of a wake in a convectively unstable medium (Parker, 1978, 1979a, pp. 151-161). Meyer, Schmidt, Weiss, and Wilson (1974; Meyer, Schmidt, and Weiss, 1979) have pointed out that the extreme clustering and compression of magnetic fibrils to form sunspots can be understood only as a consequence of a powerful converging flow at a depth of the order of magnitude of 10^4 km. The inertia (Reynolds stress) of the flow is sufficient to compress the field in opposition to the enormous magnetic pressure, i.e. $\frac{1}{2}\rho v^2 \gtrsim B^2/8\pi$. We have suggested that the converging flow is associated with a convective downdraft under the spot that also sweeps away the thermal energy that would otherwise accumulate beneath the spot, with a growing tendency to disperse the spot.

2. MAGNETIC FIELDS IN THE CONVECTIVE ENVELOPE

Consider, then, what can be learned from the extensive quantitative flux observations of an activity complex. Gaizauskas, Harvey, Harvey, and Zwaan (1983) provide a record of the total magnetic flux as a function of time for a period of almost a year. They show that repeated eruption of magnetic flux in amounts of $1\text{-}3\times10^{22}$ Maxwells at intervals of 1-3 weeks maintained the complex, with a mean flux level of about 0.6×10^{23} Maxwells for over half a year. The interesting aspect of this observation is that it allows an estimate of the strength of the azimuthal field in the convective zone below, and, as we shall soon see, this estimate is sufficiently large as to have important dynamical effects within the convective zone.

Consider the period of greatest flux ($5\text{-}8\times10^{22}$ Maxwells) during the six Carrington rotations 1664-1670, sustained by n=30 eruptions of magnetic flux, each eruption involving $\Phi = 10^{22}$ Maxwells or more. The total flux involved is $n\,\Phi = 3\times10^{23}$ Maxwells or more. This implies that there must have been at least 3×10^{23} Maxwells in some fairly narrow band of latitude beneath the activity complex. If the activity complex were able to draw on a belt 5° in width (about 5×10^{4} km in the lower convective zone), then the field strength in that belt must have been at least 6×10^{3} gauss over a depth of 10^{5} km, and probably substantially more.

There is a more conservative position, however, if we suppose that the magnetic flux is <u>retracted</u> following emergence. For in that case the flux can be recycled. No more than 8×10^{22} Maxwells was visible at any given time, so that with sufficiently parsimonious recycling it might be imagined that there is no more than twice this amount in total, i.e. 1.6×10^{23} Maxwells, requiring a field of 3×10^{3} gauss in the lower half of the convective zone. This number of 3×10^{3} gauss would seem to be, then, an absolute lower limit on the field strength in the lower half of the convective zone. We doubt that the hydrodynamics is so efficient as to supply half of the total flux in the 5° band at the surface at any given time. We expect that a number closer to 10^{4} gauss is probably more appropriate. However, a lower limit of 3×10^{3} gauss is sufficient to draw some important inferences.

The transport of heat through the convective zone is by the free turbulent mixing of the gas on scales ℓ (usually assumed to be comparable to the pressure scale height) at velocities u of a fraction of a km/sec (cf. Spruit, 1974). Formal expressions for the heat transport in terms of the convective velocity u are obtained using the mixing length representation of turbulent diffusion. Thus, if $\Delta dT/dt$ designates the magnitude of the excess of the actual temperature gradient over the adiabatic gradient, the heat flow I is written

$$I = \rho C_p \kappa \Delta dT/dz \quad \text{ergs/cm}^2\ \text{sec}.$$

with κ written as $\alpha u \iota$ where α is a number of the order of 10^{-1}. The convective velocity u is related to $\Delta dT/dz$ using the mixing length theory by assuming that $\frac{1}{2}\rho u^2$ is comparable in magnitude to the potential energy $g\iota\Delta\rho$ of a rising or falling cell of gas whose temperature differs by $\Delta T = \iota\Delta dT/dz$ from the ambient value. Then with $\Delta\rho = \rho\Delta T/T$ it follows that

$$u = \beta\iota \left(\frac{2g}{T} \Delta \frac{dT}{dz}\right)^{\frac{1}{2}}$$

where β is a constant of the order of unity.

The magnetic field interferes with this free turbulent mixing, as a consequence of the tension in the lines of force. Of course, one may imagine axisymmetric convective rolls of azimuthal field extending around the sun, in which the field is stationary in time and the tension does not interfere with the convective overturning. However, so well ordered a convective role is not very effective in heat transport, compared to the turbulent mixing of the unfettered fluid (with Reynolds numbers of the order of 10^{13} in the lower half of the convective zone). It follows, therefore, that a magnetic field whose energy density is as large or larger than the expected kinetic energy density $\frac{1}{2}\rho u^2$ of the free convection seriously hinders the convective transport of heat. It is readily shown from standard models of the convective zone that the equipartition field $B^2/8\pi = \frac{1}{2}\rho u^2$ rises to a maximum of about 3×10^3 gauss in the lower half of the convective zone and is general substantially less elsewhere. It follows, therefore, that even the lower limit of 3×10^3 gauss, inferred from observations of an activity complex, seriously blocks the convective transport of heat.

The essential point (Parker, 1987a) is that the broad bands of azimuthal field in the convective zone are thermal insulators. They act as barriers to the convective transport of heat. Hence heat must pile up beneath each band of field and there must be a cool shadow above. The theoretical dynamical consequences of the pile up and the shadow play a basic role in determining the properties of the active regions that appear at the surface above.

3. THERMAL SHADOWS

Fig. 1 is a plot of $\Delta dT/dz$ across the depth (2×10^5 km) of the convective zone of the sun (from Spruit, 1974). The total temperature deficit ΔT below a purely adiabatic atmosphere, starting at the bottom of the convective zone z=0 and integrating upward,

$$\Delta T(z) = \int_0^z dz \ \Delta(dT/dz)$$

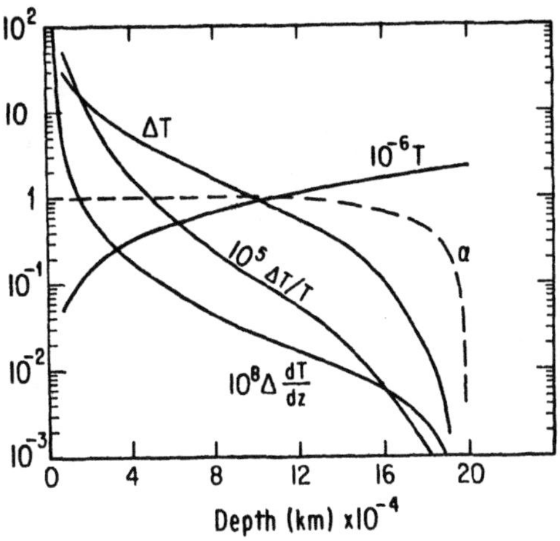

Figure 1. A plot of the difference $\Delta dT/dz$ between the actual temperature and the adiabatic lapse rate as a function of depth z below the surface of the sun. Also shown are the integral ΔT of $\Delta dT/dz$ upward from the base of the convective zone at a depth of 2×10^5 km, the temperature T, the ratio $\Delta T/T$, and the fraction α of the energy transported by convection.

is also shown. Note that ΔT, starting at zero at the base of the convective zone has a value of 1 °K at the middle $z=10^5$ km, rising to much larger values, of the order of 3×10^3 °K, toward the surface. The important point is that the cool shadow on the upper side of the magnetic field has a depth of the order of ΔT and not more.

It is obvious, of course, that a cool shadow on the upper side of the magnetic field causes a convective downdraft, and the heat transport in the downdraft tends to reduce the depth of the shadow. We have explored the theoretical properties of the convective heat transport around a thermal barrier (Parker, 1987a,b) and estimate the depth of the shadow to be $\delta T \sim 0.2$ °K, assuming that the field is confined below the mid plane of the convective zone. The net effect of the shadow is to push downward against the upper surface of the band of azimuthal field.

We expect the upper surface of the azimuthal field to be subject to the Rayleigh-Taylor instability from the density enhancement caused by both the presence of the magnetic field and the cool shadow (see discussion in Parker, 1987a,c). So the situation must be complicated. However, the net integrated effect is that the cool shadow presses downward on the field. If the vertical extent of the shadow is H, the net downward force per unit areaon the upper side of the field is

$$F_s \cong Hg\,\Delta\rho \text{ dynes/cm}$$

where $\Delta\rho$ is the density increase $\rho\Delta T/T$ caused by the shadow. This downward force opposes the magnetic buoyancy

$$F_B = hg\,\delta\rho \text{ dynes/cm}^2$$

where $\delta\rho = \rho(B^2/8\pi p)$ over the thickness ($h=10^5$ km) of the layer of field (assuming that the temperature within the field has the same value as the ambient fluid outside) is the density reduction produced by the magnetic field. The shadow F_s overpowers the magnetic buoyancy F_B provided that

$$B^2 < (H/h)\, 16\pi Nk\Delta T$$

for ionized hydrogen with N atoms/cm^3 (so that $p=2NkT$). Then with $N=0.7\times10^{23}/\text{cm}^3$ ($\rho \cong 0.1$ gm/cm^3) at $z=h=10^5$ km, it follows that $B < 7\times10^3$ gauss for $H = 1/2h$ and $\Delta T = 0.2$ °K. The cool shadow overpowers the magnetic buoyancy for fields up to 7×10^3 gauss.

Now the accumulation of heat underneath the field pushes upward and opposes the downward thrust of the shadow. However, it is evident from Fig. 1 that $\Delta dT/dz$, and hence the degree of pileup of heat, is significantly less than the accumulation of cool shadow above, so that it may be neglected.

We have an interesting dynamical balance, then, associated with each band of azimuthal field. Each band is capped by a cool shadow and convective downdraft that automatically blocks the buoyant rise of field (for $B \lesssim 7\times10^3$ gauss). The structure —band of field plus shadow — represents a dynamical system of considerable complexity. The system is continually disturbed by the surrounding convection and one can only guess at this preliminary stage how it responds to various perturbations. The cool shadow and the downdraft may be dislodged from their perch atop the band of field if the field is moved abruptly to the side. It should be noted that the characteristic thermal relaxation time of the shadow is only a few days or a week, as we shall see in the next section, so that the shadow can respond fairly quickly to any displacement of the field. Whether the system is stable to infinitesimal and/or finite perturbations is an interesting and important question, which impacts the general behavior of the field and the appearance of fields at the surface of the sun. About all that can be said at the moment is that the magnetic buoyancy is largely cancelled, so that we may understand why the azimuthal field of the sun does not float up to the surface and escape into space. It looks as though the cool shadow may keep the azimuthal field pressed against the bottom of the convective zone.

If, in fact, the azimuthal field is pressed down by the cool shadow that sits on top of it, the obvious question is why there are eruptions of flux coming up to the surface to produce and maintain the activity complexes. The answer seems to be that the accumulation of heat below the field, although smaller in magnitude than the shadow above, plays a decisive role.

4. THERMAL PILE UP

Consider the accumulation of heat beneath a broad band of azimuthal field. The accumulation takes place near the bottom of the convective zone. The heat flux I from below is about 10^{11} ergs/cm sec (compared to 0.6×10^{11} ergs/cm^2 sec at the surface). Thus, if the vertical dimension of the pile up is a, the thermal capacity per cm^2 is a ρC_p. The elevation of the temperature δT after a time t is (Parker, 1987a)

$$\delta T = It/\rho C_p a$$

The density declines by an amount $\rho\delta T/T$. The density within the field is reduced by the amount $\rho B^2/8\pi p$ by the magnetic pressure, providing the magnetic buoyancy. Note, then, that when the gas density in the accumulating heat beneath the field declines below the density within the field, the lower boundary is subject to the Rayleigh-Taylor instability. Detailed calculations of the pile up and of the ensuing Rayleigh-Taylor instability may be found in Parker (1987c). Tongues of hot field-free gas from below intrude upward through the field, producing thermal plumes that extend from the upper surface of the field to the surface of the sun. The calculations suggest that the tongues of gas have enough vigor to carry some of the field with them and to penetrate through the opposing downdraft to the surface (Parker, 1987d).

It is interesting, therefore, to consider the time t in which the heat piles up to a sufficient degree to force its way upward through the field. It is a surprisingly short time, because the temperature increment is so small, providing a thermal relaxation time of a few days that agrees with the observed intervals of the order of a week (6×10^5 sec) between emergence of hot gas and fresh field at the surface. To display a specific case, consider a field B of 3×10^3 gauss at the bottom of the convective zone where $\rho = 0.2$ gm/cm^3, $p = 6 \times 10^{13}$ dynes/cm^2, and $T = 2 \times 10^6$ °K. The ratio of the magnetic pressure to the gas pressure is 0.7×10^{-8}. The density within the field is reduced by the same fraction. Hence the density beneath the field must be reduced by this fraction to initiate the Rayleigh-Taylor instability, requiring an increase of the temperature by the same amount $\delta T/T = 0.7 \times 10^{-8}$. Hence

$$\frac{B^2}{8\pi p} = \frac{It}{\rho C_p aT}$$

and, with $\rho C_p = 2\gamma\, Nk/(\gamma-1)$, and $\gamma = 5/3$, we have

$$t = \frac{5}{16\pi}\,\frac{aB^2}{I}$$

Then if $a = 3 \times 10^4$ km, it follows that the characteristic accumulation time is only 3×10^4 sec. For a field of 10^4 gauss it is 3×10^5 sec.

Detailed calculation shows that the development of the Rayleigh-Taylor instability, which grows stronger with the passage of time and increasing δT, should occur in an interval of a few times the characteristic accumulation time, i.e. of the order of days or a week, in rough agreement with observation.

It appears then that the accumulated hot gas is able to push its way upward through the region of field, leaving the system to start the whole process over again. That is to say, the process has the appearance of a thermal relaxation oscillation with a period — probably irregular — of a few times the characteristic value t computed above.

5. DISCUSSION

It should be emphasized that the hydrodynamics associated with a broad band of strong azimuthal field in the convective zone of the sun must be fairly complicated. We expect the upper side of the field to be subject to a Rayleigh-Taylor instability because of the cool gas above and the reduced gas density within the field. We expect the lower side to be subject to an intermittent Rayleigh-Taylor instability as a consequence of the accumulating heat beneath (and within) the lower levels of the field. We suggest that the accumulation of heat below is responsible for driving a thermal relaxation oscillation with a period of the order of a week. We suggest that the intermittent emergence of magnetic field in an activity complex at the surface is a direct consequence of the thermal relaxation oscillation. Thus, the cool shadow above the field holds the field down against the bottom of the convective zone, while the accumulation of heat below causes intermittent escape of fluid and field to the surface. The whole structure operates as a single dynamical system which remains submerged in the deepest layers of the convective zone and meters out the field that is allowed to escape to the surface. We infer the existence and the general nature of the system from the "bubbles" that escape to the surface.

The system is presumably in a state of continual agitation as a consequence of its many unstable degrees of freedom. The calculations developed so far (Parker, 1987a,b,c,d) treat only the gross average behavior of the system. We expect that modification of the present simple picture will be in order as the basic theoretical dynamics of the system is worked out in more detail. We are particularly interested to understand how the dynamical properties of the system cause the retraction of the magnetic field from the surface once the eruption has ceased. We suggest that the counterflow, initiated by each eruption, and the general downdraft in the cool shadow play a central role in pulling the field back under the surface of the sun and restoring it (perhaps) to its place in the broad band of azimuthal field in the lower half of the convective zone. If it should ultimately prove that the downdrafts are not capable of retracting the major portion of the field, then our estimated lower limit of 3×10^3 gauss for the azimuthal field must be revised upward by at least a factor of two.

This brings us back to the question of the nature of the solar dynamo. The foregoing dynamical considerations do not alter the basic concept of the $\alpha\omega$-dynamo that is presumed to operate in some way in the lower convective zone of the sun. They indicate, however, that the simple mixing length kinematical models of the dynamo employed so far to explore the possibilities are much too simple. This surprises no one, of course, but it is now clear that the bands of azimuthal field are not passive in any way. Each band is a self-propelling dynamical entity whose shape and general motion is determined as much by the hydrodynamics of the ambient and locally induced convection as it is by the dynamo operation itself. The theoretical problem posed by this dynamical entity is a formidable one, involving the myriad internal degrees of freedom, many of which are unstable, as well as the external convective motions. It is obvious that we need all the help we can get from observation, as well as more detailed theoretical studies of the dynamics.

It is to be expected that helioseismology, in the next few years, may provide an accurate picture of $\partial\Omega/$or throughout the convective zone. Once that is available it should be possible to assess the theoretical dynamo problem, particularly the question of the separate contributions of $\partial\Omega/\partial r$ and $\partial\Omega/\partial\theta$. In this same direction it is essential to obtain an idea of the nature of the meridional circulation, giant cells, and convective rolls in the convective zone insofar as these motions manifest themselves at the surface of the sun. There has been important progress in theoretical and numerical studies of the large scale motions in recent years (summarized in other papers in these proceedings). Together with the variety of observational studies of the surface motions (also summarized in these proceedings) it is to be hoped that a comprehensive and reliable picture of the dynamical motions in the convective zone may eventually emerge. It is to be hoped that the dynamics of the bands of azimuthal field, and the hydrodynamics of the clustering of fields to form sunspots at the surface, and the hydrodynamics of the retraction (or other disappearance) of magnetic fields from the surface can be brought to a state of understanding. It is clear that detailed quantitative observational studies of the small-scale magnetic fields and the small-scale fluid motions associated with individual magnetic fibrils, as well as motions on all scales in and around magnetic active regions, play an essential role in developing any such understanding.

This work was supported in part by the National Aeronautics and Space Administration under NASA grant NGL 14-001-001.

REFERENCES

Brants, J.J. 1985a, Solar Phys. **95**, 15.
Brants, J.J. 1985b, Solar Phys. **95**, 197.
Brants, J.J. and Steenbeck, J.C.M. 1985, Solar Phys. **96**, 229.
Brown, T.M. 1985, Nature **317**, 591.
Brown, T.M. 1986, in 'Seismology of the Sun and Distant Stars,' ed. D.O. Gough, (D. Reidel, Dordrecht) pp. 199-214.
Deinzer, W., v. Kusserow, H.U., and Stix, M. 1974, Astron. Astrophys. **12**, 111.
Deinzer, W. and Stix, M. 1971, Astron. Astrophys. **12**, 111.
DeVore, C.R., Sheeley, N.R., and Boris, J.P. 1985, Solar Phys. **92**, 1.
Duvall, T.L., Dziembowski, W.A., Goode, P.R., Gough, D.O., Harvey, J.W. and Leibacher, J.W. 1984, Nature **310**, 22.
Duvall, T.L. and Harvey, J.W. 1984, Nature **310**, 19.
Duvall, T.L., Harvey, J.W., and Pomerantz, M.A. 1986, Nature **321, 500**.
Gaizauskas, V., Harvey, K.L., Harvey, J.W. and Zwaan, C. 1983, Astrophys. J. **265**, 1056.
Köhler, H. 1970, Solar Phys. **13**, 3.
Köhler, H. 1973, Astron. Astrophys. **25**, 467.
Lerche, I. and Parker, E.N. 1972, Astrophys. J. **176**, 213.
Libbrecht, K.G. 1986, Nature **319**, 753.
Gilman, P.A. 1983, Astrophys. J. Suppl. **53**, 243.
Gilman, P.A. and Miller, J. 1981, Astrophys. J. Suppl **46**, 211.
Glatzmaier, G.A. 1985, Astrophys. J. **291**, 300.
Leighton, R.B. 1964, Astrophys. J. **140**, 1547.
Leighton, R.B. 1969, Astrophys. J. **156**, 1.
Livi, S.H.B., Wang, J. and Martin, S.F. 1985, Austral. J. Phys. **38**, 855.
Martin, S.F., Livi, S.H.B., and Wang, J. 1985, Austral. J. Phys. **38**, 929.
Meyer, F., Schmidt, H.U., and Weiss, N.O. 1977, Mon. Not. Roy. Astron. Soc. **179**, 741.
Meyer, F., Schmidt, H.U., Weiss, N.O. and Wilson, P.R. 1974, Mon. Not. Roy. Astron. Soc. **169**, 35.
Parker, E.N. 1955a, Astrophys. J. **121**, 491.
Parker, E.N. 1955b, Astrophys. J. **122**, 293.
Parker, E.N. 1957, Proc. Natl. Acad. Sci. **43**, 8.
Parker, E.N. 1970, Annual Rev. Astron. Astrophys. **8**, 1.
Parker, E.N. 1971, Astrophys. J. **164**, 491.
Parker, E.N. 1978, Astrophys. J. **222**, 357.
Parker, E.N. 1979a, Cosmical Magnetic Fields, Oxford, Clarendon Press.
Parker, E.N. 1979b, Astrophys. J. **230**, 905.
Parker, E.N. 1979c, Astrophys. J. **232**, 291.
Parker, E.N. 1984a, Astrophys. J. **280**, 423.
Parker, E.N. 1984b, Astrophys. J. **281**, 839.
Parker, E.N. 1987a, Astrophys. J. (15 January).
Parker, E.N. 1987b, Astrophys. J. (submitted).
Parker, E.N. 1987c, Astrophys. J. (submitted).
Parker, E.N. 1987d, Astrophys. J. (submitted).
Rai Choudhuri, A. 1985, Astrophys. J. **281**, 846.
Roberts, P.H. 1972, Phil. Trans. Roy. Soc. London A. **272**, 663.

Roberts, P.H. and Stix, M. 1972, Astron. Astrophys. **18**, 453.
Sheeley, N.R., DeVore, C.R. and Boris, J.P. 1985, Solar Phys. **98**, 219.
Spruit, H.C. 1974, Solar Phys. **34**, 277.
Steenbeck, M. and Krause, F. 1969a, Astron. Nachr. **291**, 49.
Steenbeck, M. and Krause, F. 1969b, Astron. Nachr. **291**, 271.
Stix, M. 1973, Astron. Astrophys. **24**, 275.
Stix, M. 1976a, Astron. Astrophys. **47**, 243.
Stix, M. 1976b, "Basic Mechanisms of Solar Activity," IAU Symposium No. 71 (ed. J. Kleczek) D. Reidel, Dordrecht.
Wallenhorst, S.G. and Howard, R. 1982, Solar Phys. **76**, 203.
Wallenhorst, S.G. and Topka, K.P. 1982, Solar Phys. **81**, 33.
Yoshimura, H. 1971, Solar Phys. **18**, 417.
Yoshimura, H. 1972a, Solar Phys. **22**, 20.
Yoshimura, H. 1972b, Astrophys. J. **178**, 863.
Yoshimura, H. 1973, Solar Phys. **33**, 131.
Yoshimura, H. 1975, Astrophys. J. Suppl. **29**, 467.
Zwaan, C. 1985, Solar Phys. **100**, 397.
Zwaan, C., Brants, J.J., and Cram, L.E. 1985, Solar Phys. **95**, 3.

MAGNETIC FIELDS AND THE ROTATION OF THE SOLAR CONVECTION ZONE *)

M. Schüssler
Kiepenheuer-Institut für Sonnenphysik
Schöneckstr. 6
D-7800 Freiburg
Federal Republic of Germany

ABSTRACT. Arguments are presented in favour of the conjecture that magnetic fields in the convection zone are concentrated, highly fragmented and passive to velocity fields (convection, rotation) with the only exception of the uppermost layers, where buoyancy becomes dominant. The concept of an "anchor" or "coupling depth" is given some basis. It is concluded that large active regions have their origin in a region of dominating differential rotation at the lower boundary layer of the convection zone from where they are injected into the convection zone proper. It is shown that this view is consistent with observations of the evolution of active regions and favours the cluster model of sunspots. Information from magnetic fields about the rotation of deep layers of the convection zone can only be derived from observations of the very early phases of large active region evolution. These indicate that the rotation rate increases with depth in the upper layers of the convection zone but that the rotation of the source region of magnetic fields is slower than the surface value. It is argued that the observed rotation of magnetic tracers can be related to phase velocities of convective patterns as well as to real plasma velocities. Consequences for our picture of the solar dynamo are discussed briefly.

1. INTRODUCTION

The interaction of magnetic fields, convection and rotation in the solar convection zone has many aspects. Ignoring the magnetic fields, the hydrodynamic processes alone are far from being understood: We neither know the nature of convective flows in the deep convection zone nor do we have an accepted theory of the generation of differential rotation (Durney, 1985). Introduction of a magnetic field adds another significant source of complication to the problem. In such a situation and being unable to treat the problem by starting from first principles it seems adequate to isolate certain aspects whenever possible. The question I like to address here is, whether magnetic fields in the convection zone are "active" or "passive" with respect to velocity fields. This means, whether

*) Mitteilungen aus dem Kiepenheuer-Institut Nr. 266

B. R. Durney and S. Sofia (eds.), The Internal Solar Angular Velocity, 303–320.

Roughly interpreted in terms of a time series, the variation of Ω with the Zürich type shows a rotation rate which is decreasing with time (except for types B and C where rapid expansion dominates). The comparison between "new" A groups as the first signs of an active region and "old" ones as the late phase supports this view: Sunspot groups decelerate in time approaching the rotation rate of the background magnetic fields.

This is further confirmed by the faster rotation of small sunspots (being young in a statistical sense because of their short lifetime) compared to large sunspots (Howard et al., 1984) and by the observed deceleration of recurrent H- and J-spots (Balthasar et al., 1982; Lustig and Dvorak, 1984) and the deceleration of individual active regions in the first 10 days after appearance (Tuominen and Virtanen, these proceedings).

2.3 Interpretation

The observations show that active regions begin their existence somewhat decoupled from the dynamics of surface flows (see also the numerous descriptive publications on active region development, e.g. McIntosh, 1981; Brants and Steenbeck, 1985). However, after this initial phase they approach the rotation rate of the background fields and do not show significant reaction on subsurface dynamics. As another example we might refer to Gilman and Howard (1986) who found that individual sunspots within sunspot groups show just the rotation rate adequate for their location in latitude and no sign of a subsurface dynamics due to their membership to a group.

Our interpretation of the observations therefore is as follows: Young active regions during flux emergence are dominated by subsurface dynamics. During their evolution they become more and more coupled to and dominated by surface flows (rotation, convection) until they merge into the background field which is largely dominated by surface dynamics.

Consequently, we should look at the early phases of large active region development in order to derive information about the deeper parts of the convection zone. Tracers like pores and young sunspots (Ribes et al., 1985), young faculae and plages (Ternullo et al., 1981; Ternullo, 1986) seem most promising in this respect and have already revealed indications of a cycle dependent meridional circulation pattern and a latitudinally more rigid rotation of these tracers.

2.4 Cycle dependence

The dependence of tracer rotation on the solar cycle (e.g. Gilman and Howard, 1984; Balthasar et al., 1986) will not be discussed in detail here. We interpret this phenomenon as a reaction of the magnetic structures on variations of the velocity field of the surface plasma during the cycle as indicated by the synchronous variation of the Doppler rotation rate (Howard, 1976; Gilman and Howard, 1984). These motions are related to the dynamics of the dynamo process as has been shown successfully for the case of the "torsional oscillations" (Yoshimura, 1981; Schüssler, 1981; Tuominen and Virtanen, 1984).

3. MAGNETIC FIELD DYNAMICS IN THE CONVECTION ZONE

As we have seen in the preceding section, the observational results point to a complex interaction of magnetic fields, rotation and other flow patterns in the convection zone. In order to use observational information to derive conclusions about the interior of the convection zone we have to discuss the physics of magnetic fields in electrically well-conducting convective media. The key point of our problem is whether magnetic fields are "active" or "passive" with respect to flows (rotation, convection, circulation), i.e. whether the field is carried around and severely distorted or is able to exert sufficient forces to resist. To approach an answer we have to discuss a sequence of problems step by step in the following chapters. The results will enable us to put the concept of an "anchor depth" of magnetic structures on a firmer basis, to understand the different behaviour of young and old active regions and to support the cluster models of sunspots.

3.1 Concentrated vs. diffuse field

Decompose the magnetic field $\underline{B}$ in the convection zone into a suitably defined (e.g. in the sense of a two-scale-analysis, see Moffatt, 1978) spatial mean $\langle\underline{B}\rangle$ and a fluctuation $\underline{B}'$:

$$\underline{B} = \langle\underline{B}\rangle + \underline{B}' \qquad (3.1)$$

We will define then:

diffuse field : $|\underline{B}'|/|\langle\underline{B}\rangle| \lesssim 1$

concentrated field: $|\underline{B}|/|\langle\underline{B}\rangle| \gg 1$

The photospheric magnetic field is observed to be concentrated: magnetic structures with strong field are embedded in virtually field-free plasma. It is not obvious, however, that this should be so throughout the whole convection zone.

From the total flux emerging during the activity cycle, a value of $\langle\underline{B}\rangle < 100$ Gs can be derived for the mean toroidal field in the convection zone. For all depths this is much less than the equipartion field strength

$$B_e = v_c\,(4\pi\varrho)^{1/2} \qquad (3.2)$$

(v_c: convective velocity, ϱ: gas density). If the field were diffuse we would have

$$|\underline{B}| \simeq |\langle\underline{B}\rangle| \ll B_e$$

i.e. the magnetic energy density would be everywhere small compared to the kinetic energy density of the convective flows: The magnetic field would have to follow the motion of the electrically well-conducting gas without being able to exert significant forces. Consequently, diffuse fields in the convection zone are passive (to convective flows and differential rotation). If the field is concentrated, it depends on the properties of the magnetic structures (size, magnetic field, deformation, ...) whether they are passive or active with respect to convection and differential rotation (see below).

Although it is not proven rigorously, there are several arguments in favour of a concentrated field throughout the convection zone:

- The observed surface field is concentrated.

- Magnetic fields are not passive in the early phases of active region development: The (strictly obeyed!) polarity rules, the formation of spots by accumulation, fast rotation and separation of p- and f-parts cannot easily be reconciled with a passive diffuse field.

- Numerical simulations of magnetoconvection show the phenomenon of flux expulsion: The system separates in a field-free convecting and a strongly magnetic, nearly stagnant "phase" (Proctor and Weiss, 1982). This situation seems energetically favourable for convective transport (Parker, 1984).

Consequently, in what follows we will use as

Working hypothesis: The magnetic field in the convection zone is concentrated in the form of individual structures ("flux tubes") with $B \gg B_e$

It might well be that the large amount of magnetic flux to be stored in the lower half of the convection zone in order to feed complexes of activity leads to a dense packing of magnetic structures (B' only a few times $\langle B \rangle$) as argued by Parker (this volume). However, the problem is less severe because one could imagine large-scale flows collecting flux from an extended latitude range and lifting it up to build a complex of activity.

3.2 "Thick" vs. "thin" flux tubes

As we will see later, the dynamics of flux tubes in the convection zone depends crucially on the ratio of their diameter to other length scales of the system (scale height, eddy size of convective flows, scale length of angular velocity gradient, ...). A variety of instabilities and other dynamical processes exist (see Schüssler, 1984a) that fragment flux tubes into smaller and smaller filaments. Stabilizing effects like buoyancy, twisting and whirl flows around the tubes are efficient only in the surface layers (Schüssler, 1984b). Within the convection zone proper, fragmentation will proceed down to spatial scales where effects of resistive diffusion come into play: An instability will cease to be important if fragments merge by diffusion as fast as they are formed by the instability. Let us take as an example the interchange (fluting) instability. Its growth time, τ_f, is given (e.g. Schüssler, 1986):

$$\tau_f = \frac{(2aR_c)^{1/2}}{v_A} \tag{3.3}$$

(a: tube radius, R_c: radius of curvature of the interface between tube and surrounding medium, $v_A=B/(4\pi\rho)$: Alfvén velocity). The radius δ_f of a filament for which τ_f=diffusion time = $\delta_f^2\eta^{-1}$ (η: magnetic diffusivity) is then given by:

$$\delta_f = \left(\frac{2R_c\eta^2}{v_A^2}\right)^{1/3} \tag{3.4}$$

Taking equipartition fields ($v_A=v_C$) and R_c=L, the length scale of the dominant convective cells, we can estimate δ_f for different depths in the convection zone using Spruit's (1977) model. This is given in the following table:

Depth [cm]	η [$cm^2 s^{-1}$]	v_c [$cm s^{-1}$]	δ_f [cm]	δ_R [cm]	τ_R [s]	τ_{RAD} [s]
$1.0*10^8$	$1.5*10^6$	$1.3*10^5$	$3.0*10^3$	$3.4*10^4$	$7.7*10^2$	$3.9*10^1$
$1.0*10^9$	$1.0*10^5$	$3.2*10^4$	$2.7*10^3$	$5.6*10^4$	$3.1*10^4$	$3.9*10^5$
$1.0*10^{10}$	$7.8*10^3$	$9.1*10^3$	$2.5*10^3$	$9.3*10^4$	$1.1*10^6$	$2.9*10^3$
$1.8*10^{10}$	$7.5*10^3$	$4.9*10^3$	$4.4*10^3$	$1.7*10^5$	$3.7*10^6$	$3.4*10^3$

Table 2: Physical properties of magnetic filaments as a function of depth in the convection zone. δ_f: Diffusive scale for fluting, δ_R: resistive boundary layer thickness, τ_R: resistive diffusion time, τ_{Rad} radiative exchange time.

For this rough estimate we used the depth as typical size of the dominant convective cell. Table 2 also gives δ_R, the resistive boundary layer thickness:

$$\delta_R = \left(\frac{L\cdot\eta}{v_c}\right)^{1/2} \tag{3.5}$$

This is determined by the equality of resistive diffusion and advection by convection and gives the scale of structures formed by kinematic magnetic flux expulsion. Structures smaller than δ_R diffuse too rapidly to be held together by the convective cell and therefore cannot be sustained. For a spectrum of convective/turbulent eddies a spectrum of magnetic substructures with the respective boundary layer thicknesses would form, all of which would be smaller than δ_R given in Eq. (3.5). We see that both quantities, δ_R and δ_f, are fairly constant throughout the whole convection zone and that always $\delta_f \ll \delta_R$. <u>We therefore take δ_R as a reasonable upper</u>

limit of the sizes of magnetic structures after being shredded and disrupted by instabilities and dynamical fragmentation. The diffusion time

$$\tau_R = \frac{\delta_R^2}{\eta} \tag{3.6}$$

and the timescale of thermal radiative diffusion

$$\tau_{RAD} = \frac{\delta_R^2}{\eta_{RAD}} \tag{3.7}$$

with

$$\eta_{RAD} = \frac{16\sigma T^3}{3\varrho^2 c_v \kappa_R} \tag{3.8}$$

(σ: Stefan's constant, T: temperature, c_v: specific heat capacity, κ_R: Rosseland mean opacity) are also given in Table 2 for a structure of size δ_R. We find

$$\tau_{dyn} \equiv L \cdot v_c^{-1} = \tau_R \gg \tau_{RAD}$$

in the deep convection zone. Consequently, for magnetic structures which reside long enough ($\gg$ months) within the convection zone to be fragmented down to $\delta_R \simeq 1$ km, two points can be stated which become important in later sections:

- The magnetic diffusion time is small enough to allow mass exchange between magnetic structures and the surroundings during their dynamical evolution (e.g.stretching by differential rotation).

- The thermal diffusion time is small enough to ensure temperature equality between exterior and interior of the tube ($T_i = T_e$) during the dynamical evolution.

We can conjecture from the above considerations that instabilities and dynamical fragmentation processes lead to the formation of very small magnetic structures with sizes $\lesssim$ 1 km within the convection zone proper. Only in the (strongly stratified and superadiabatic) surface layers, stabilizing effects (buoyancy, twist, whirl flows) can sustain larger structures (see Ch. 3.5).

3.3 Dynamics of thin flux tubes

For magnetic flux concentrations whose length scale perpendicular to the field (e.g. diameter of a flux tube) is much smaller than any relevant longitudinal scale (pressure scale height Λ, radius of curvature R_c, scale length of large-scale flow patterns) the slender flux tube approximation (SFA) can be used to describe their dynamical evolution. The SFA has been derived in its general form by Spruit (1981). The forces which determine the dynamics of thin flux tubes, i.e. buoyancy, curvature force and drag force due to external flows, are discussed in Schüssler (1984a). The ratios of buoyancy and curvature forces, respectively, with the drag force are given by:

$$\frac{\text{Buoyancy force}}{\text{Drag force}} \approx \left(\frac{a}{\Lambda}\right)\left(\frac{B}{B_e}\right)^2\left(\frac{v_c}{v_\perp}\right)^2 \tag{3.9}$$

$$\frac{\text{Curvature force}}{\text{Drag force}} \approx \left(\frac{a}{R_c}\right)\left(\frac{B}{B_e}\right)^2\left(\frac{v_c}{v_\perp}\right)^2 \tag{3.10}$$

Λ: scale height

For equipartion fields $B \simeq B_e$ formed by convective flux expulsion and $v_\perp \simeq v_c$, i.e. component of external flow perpendicular to the tube of the order of the convective velocity, we see from Eqs. (3.9) and (3.10) that the drag force dominates the dynamics of a thin flux tube: $(a/\Lambda) \ll 1$ and $(a/R_c) \ll 1$ are in fact conditions for the applicability of the SFA. They are well met in the case of strongly fragmented fields: $a \simeq 1$ km, $\Lambda \simeq 10^4$ km; curvature forces come only into play for extreme distortions ($R_c \simeq a$) of the flux tube. Consequently, the magnetic field structured in this way will be passive to any large-scale flow with relative velocity of the order of the convective velocity, be it convection itself, differential rotation or, possibly, meridional circulation.

On the other hand, the increasing distortion of the flux tube path due to this passivity will increase the length of the tube. If the mass per unit length stays constant, this would lead to rapid increase of the magnetic field because the tube has to contract in order to maintain pressure equilibrium with the ambient gas. Take a length element l_0 which is stretched to l_1 due to distortion of the tube. If the mass within the element stays constant, the magnetic field will change from B_0 to B_1 given by

$$\frac{B_1}{B_0} = \frac{l_1}{l_0}\left(\frac{8\pi p_e - B_1^2}{8\pi p_e - B_0^2}\right)^{1/\gamma} \tag{3.11}$$

(p_e: external pressure, γ: ratio of specific heats) which for

$$\beta = \frac{8\pi p_e}{B^2} \gg 1 \qquad \text{for both } B_0,\ B_1$$

leads to:

$$\frac{B_1}{B_0} \approx \frac{l_1}{l_0} \tag{3.12}$$

Thus, the magnetic field would increase rapidly while the tube is distorted and the curvature force ($\propto B^2$) would grow even faster: very soon a dynamical equilibirum between drag force and curvature force would be established. This means, that magnetic structures would cease to be passive after a few dynamical timescales and one of the basic ingredients of dynamo models for the solar cycle, the winding of the toroidal field by differential rotation, would be rendered nearly inefficient. However, the assumption of constant mass per unit length does not hold:

- In the upper layers of the convection zone, the density is low and the deep layers represent a high mass reservoir: Matter can flow along the tube to fill the volume created by distortion of the tube; no significant contraction of the tube takes place, it stays passive.

- In the deep layers of the convective zone, obviously no further mass reservoir exists for tubes contained within the convection zone. However, fragmentation becomes crucial here: The distortion leads to fluting instability and the flow instability discussed by Tsinganos (1980). As discussed in the preceding section, this will lead to a filamentary field with a typical size of fragments $a \leqslant \delta_R \approx 1$ km. The diffusion time for such structures is comparable to the dynamical timescale of distortion and <u>mass can diffuse into the filaments on the same timescale as that of distortion</u>. Consequently, the volume created by distortion will be filled by diffusing mass while the magnetic field stays approximately constant: The filament continues to be passive to further distortion.

We conclude that due to instability and fragmentation the magnetic field structures that reside for longer times ($\gg$ month) in the convection zone are passive to large-scale flows (convection, differential rotation, circulation), i.e. are strongly coupled to and distorted/fragmented by these flows.

3.4 Interlude: "Passive" vs. "active" fields

The view of magnetic fields in the convection zone presented in the preceding sections is only partially in accordance with the observed properties of solar magnetic fields. The "quiet Sun" fields seem to confirm the idea of "passive" fields in view of their spatial relation to surface velocity fields (differential rotation, supergranulation network structure, granulation). Their large-scale, long-term behaviour can be described succesfully by the action of these velocity fields (possibly including meridional circulation) as shown by Sheeley et al. (1985). On the other hand, as discussed in Sec. 2., the early phases of active region evolution show a different behaviour: Here, magnetic fields are not passive to local surface velocities smaller structures accumulate to form larger structures, very large concentrations (sunspots) appear. Hale´s polarity rules which are nearly strictly obeyed show directly that the magnetic flux appearing at the surface has not been passively distorted and "brain-washed" by convection.

How can we explain this apparent contradiction? The clue seems to be temporal evolution: The "active" phase lasts only for the first few days of active region development (about as long as new flux emerges). Thereafter we observe fragmentation of the large structures and dispersion of the active region as a whole. The magnetic fields become gradually more and more passive and eventually merge into the background field network. But how can we understand that they are "active" in the beginning? In my opinion, this is one of the strongest hints that they do not originate in the convection zone proper. The polarity rules show, that differential rotation dominates over convection in the source region. More precisely, the velocity differences due to differential rotation on the scale of the dominant convective eddies must be larger than the respective convective velocities. Surface observations and results from rotational splitting of global oscillations (see contributions by Howard, Brown, Duvall/Harvey, Hill and others in these proceedings) indicate that this condition does not seem to be met within the convection zone. This is also supported by the 3D numerical simulations by Glatzmaier (these proceedings, see also Gilman and Miller, 1986).

A more favourable source region of active region fields is the layer of overshooting convection below the convection zone proper, where the convective velocities decrease with depth while vorticity expulsion leads to the formation of a shear layer of enhanced differential rotation (Schüssler, 1984a). In this way a region of dominating differential rotation can be formed, possibly the source region from where the toroidal fields responsible for active regions are ejected. In order not to be too seriously distorted and fragmented during their rise through the convection zone proper, this transition should be fast. In fact, Moreno-Insertis (1986) has shown that convectively unstable loops of toroidal field traverse the convection zone within a few days. The dynamics of this eruption process determines the surface evolution of active regions in their initial phases. Only then, information about the properties of the deeper parts can be derived from the observation of surface fields.

A certain degree of fragmentation is apparent even during this time as can be inferred from the detailed observations of flux eruption and accumulation of large sunspots out of smaller pieces (e.g. McIntosh, 1981). After the eruption phase has terminated, fragmentation proceeds and the magnetic structures become more and more coupled locally to the motion of the surrounding plasma. The only exception is the uppermost layer of the convection zone: Surface fields are not typical for the properties of magnetic fields in the convection zone, as will be shown in the next section.

.5 The peculiar state of surface fields

The (observable) surface layers are not typical for the state of the deep convection zone: They represent the transition region between the strongly superadiabatic uppermost convection zone and the stably stratified photosphere. Convective motions overshoot into the stable layer, temperature fluctuations are significant and magnetic fields can be concentrated by convective collapse far beyond equipartion flux densities. Partial evacuation of magnetic structures and the strong stratification of the region lead to a dominance of buoyancy over drag and curvature forces. Using Eq. (3.9) we can estimate

$$\frac{\text{Buoyancy force}}{\text{Drag force}} \approx \left(\frac{a}{\Lambda}\right)\left(\frac{B}{B_e}\right)^2 \begin{cases} \approx 10 \text{ for magnetic elements } (a=100 \text{ km}) \\ \approx 300 \text{ for sunspots } (a=10^4 \text{ km}) \end{cases}$$

with Λ=150 km and reasonable values for B. These values become even larger if we take into account the temperature reduction of the magnetic structures with respect to the non-magnetic photosphere.

It might be useful to illustrate the consequences of a dominating buoyancy force by way of a simple example (Fig. 2). Imagine a thin straight, rigid magnetic flux tube ($B/B_e \gg 1$) embedded in a non-magnetic surrounding medium of constant density ρ_e :

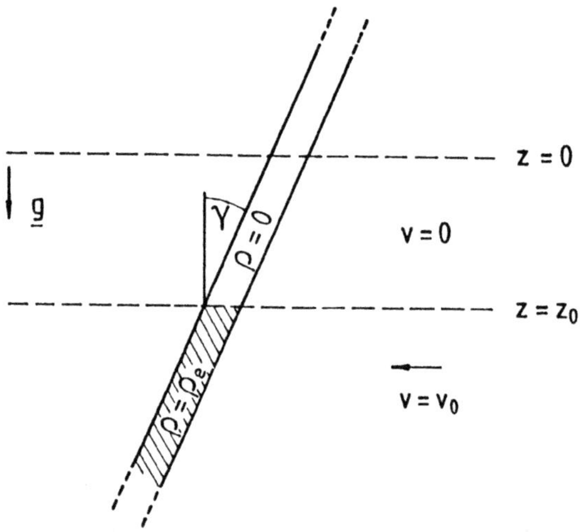

Fig. 2: Schematic sketch of a straight flux tube embedded in a constant density (incompressible) medium

The upper layer of the tube ($0 \leq z \leq z_0$) is assumed to be evacuated (ρ=0, "Wilson depression") while the lower part ($z > z_0$) is neutrally buoyant. In the chosen frame of reference the "surface" layer $0 \leq z \leq z_0$ of the external medium is static (v=0), while the lower part ($z > z_0$) moves with constant horizontal velocity v_0. The equilibrium state of the flux tube will now be described by two quantities:

- its inclination γ with respect to the direction of gravity,
- its velocity u (normal to the tube axis).

In the part $z > z_0$, the drag force cannot be balanced by any other force. Consequently, the tube has to float with the surrounding gas:

$$v_\perp = 0 \Rightarrow u = v_o \cos\gamma \tag{3.13}$$

In the upper part, the buoyancy force has to balance the drag force extexted on the floating tube:

$$g \cdot \sin\gamma = \left(\frac{C_D}{\pi}\right) \frac{v_o^2}{a} \cos^2\gamma \tag{3.14}$$

(C_D: drag coefficient)

For $\gamma << 1$ (justified a posteriori by the result) and $C_D/\pi \approx 1$. We have

$$\gamma = \frac{v_o^2}{a \cdot g} \tag{3.15}$$

If we take $v_o = 30\ ms^{-1}$, $a = 100$ km, $g = g_\odot$, the result is $\gamma = 3*10^{-5}$! Due to the strong buoyancy force, a very small inclination compensates the drag force. The tube will move with the velocity of the lower layer where the drag dominates (Foukal, 1977). In this sense the tube can be called "anchored" in the lower layer, although we prefer to use term "coupled".

Buoyancy is also partially responsible for another peculiarity of the surface layers: Magnetic structures with sizes much larger than the "diffusion limit" discussed in Ch. 3.2 can exist for times long compared with their dynamical time scales. The stabilization against fragmentation processes is probably due to buoyancy (Meyer et al., 1977) and whirl flows around magnetic structures (Schüssler, 1984b).

3.6 Coupling depths and the nature of sunspots

We have seen in the preceding sections that there are only two exceptions from the passive nature of magnetic fields in the convection zone: Young active regions during the emergence and the uppermost layers of the convection zone because of the dominating buoyancy force there. Only in these two cases we can potentially derive information about the rotation of deeper layers.

Let us first examine the deviation from local corotation (or, more generally, attachement to local flows) due to buoyancy. Moving to increasing depths, the superadiabaticity decreases and $B \rightarrow B_e$, the local equipartion field strength (the convective collapse mechanism ceases to operate). The coupling of a flux tube of radius a to the surrounding medium is thus determined by the ratio (Eq. 3.9):

$$\frac{\text{Buoyancy force}}{\text{Drag force}} = \left(\frac{a}{\Lambda}\right)$$

The flux tube will become dominated by drag forces approximately at that depth where its radius becomes equal to the local scale height. Fig. 3

shows $\Lambda(z)$ for the convection zone model of Spruit (1977).

For any given radius a, the depth z_a can be determined through $a = \Lambda(z_a)$. z_a could be called "anchor depth" because from there on the flux tube is attached to the motions of the surrounding medium, in particular to its rotation. However, because attachment is made to a medium which is by no means static, we prefer to use the term "coupling depth". We see that z_a depends strongly on a: While very small structures are already coupled to surface motions ($z_a \leqslant 1000$ km for $a \leqslant 400$ km), sunspots are coupled much deeper ($z_a \simeq 16000 \ldots 30000$ km for $a = 5000 \ldots 10000$ km), if they consist of a single flux tube (the "plug model"). Consequently, we

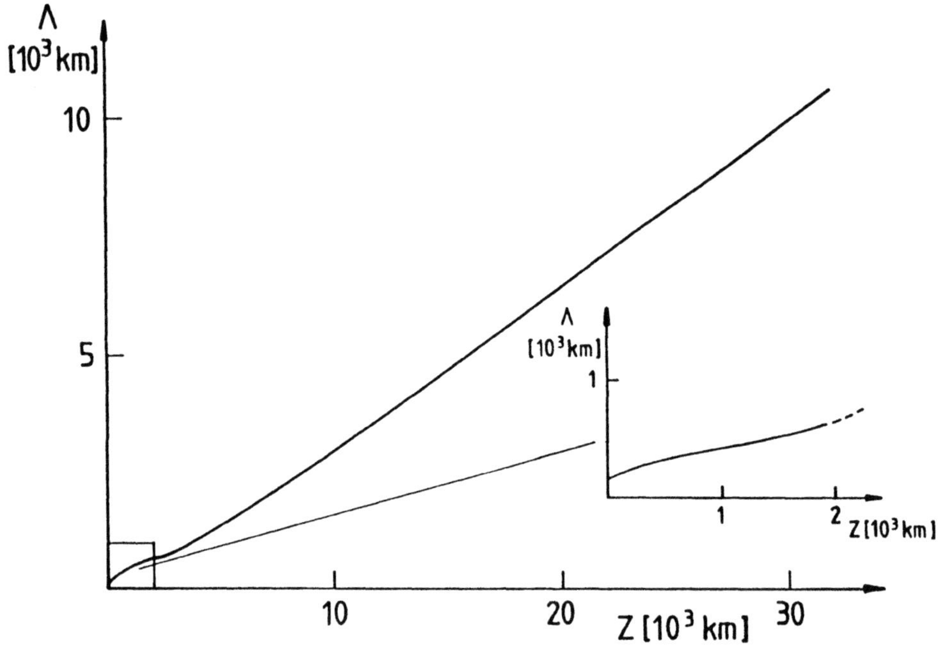

Fig. 3: Pressure scale height as a function of depth for Spruit's (1977) convection zone model. The uppermost layers are drawn separately (inset).

should expect a systematic dependence of the dynamic behaviour of flux tubes on their size. In particular, a depth-dependent differential rotation should manifest itself as a rotation rate depending on the size of a structure. This is not in agreement with observations: Different rotation rates of various "tracers" rather reflect different evolutionary stages of active regions. In particular, as discussed in Section 2, old H- and J-spots show the same rotation rate as the background magnetic fields in spite of a size which is a factor 100 larger than the diameter of a typical network flux tube. In the context of our line of argumentation this can be understood if we abandon the plug model and consider the cluster model (Parker, 1979; Spruit, 1981): Here the sunspot consists of many small fibrils ($a \approx 10^2 \ldots 10^3$ km) which are closely packed only in the uppermost $\sim 10^3$ km of the convection zone to form the visible sunspot umbra. Below this merging level, z_m, the spot splits into its fibril components. It is evident, that below z_m the coupling depth z_a is now determined by the size of the individual fibrils and not by the size of the spot. Consequently, z_a will be much smaller than the value calculated from the sunspot radius. The coupling depth will rather be the maximum of z_m and z_a (z_a calculated using a = fibril radius). Hence, for small fibrils ($a \approx 100$ km), <u>the coupling depth will be equal to the merging level of the cluster</u>. Because the fibrils will be coupled to local flows immediately below z_m, the following consequences ensue:

- the spot shows the rotation rate associated with the merging level z_m;

- z_m will decrease in time due to accumulating distortion and stretching of the fibrils below z_m: The rotation rate will approach the surface value;

- Sunspot decay and disruption involves all fibrils of a spot. This can be represented as a volume diffusion leading to the observed constant area decay rate of large sunspots (Meyer et al., 1974).

We see that the cluster model together with the concept of the coupling depth as introduced here is consistent with the various observational results as described to Section 2.

3.7 What do magnetic fields tell us about the rotation of the convection zone?

From the content of the previous chapters it is clear that we cannot draw definite conclusion at this stage. The physics of many processes is only conjectured and has to be worked out in much more detail before we can claim to stand on safe ground. However, the above chain of arguments seems consistent with the observations of surface magnetic fields and suggests the following picture of magnetic fields in the convection zone:

- Magnetic fields in the convection zone are passive and attached to local flows

- The only exceptions are the initial phases of active regions, i.e. new "injections" of ordered toroidal field into the convection zone proper, and the uppermost layers where buoyancy becomes dominant.

Hence, what can we learn about rotation?
The slightly larger rotation rate of the background magnetic fields as compared to the plasma Doppler rate of large sunspots indicates a rotation rate increasing with depth in the uppermost few 10^3 km of the convection zone (Gilman and Foukal, 1979). However, there is another possibility: Since magnetic structures are coupled strongly to the velocity fields of granulation and supergranulation, <u>they will show the rotation rate of these convection patterns</u>, a phase velocity which must not necessarily be identical to the material velocity of the plasma (see Duvall, 1980). Such

a drift of a convective pattern is observed by Glatzmaier (these proceedings) in his numerical simulations for the "cartridge belt" cells as well as in laboratory experiments (Busse and Carrigan, 1974; Toomre, these proceedings). Consequently, an increase of rotation with depth is not necessarily implied by the observed rotation rate of magnetic fields although a combination of both effects is possible: The results of F. Hill (these proceedings) point to an increase of rotation rate with depth in the upper layers of the convection zone.

The initial phases of active regions are the other potential source of information about rotation of deeper parts of the convection zone. This information has to be disentangled from the dynamics of the flux eruption itself which is not well understood presently (but see Moreno-Insertis, 1986). Again, the fast rotation of young active regions and young sunspots points to the existence of a region in the interior of the convection zone which rotates more rapidly than the surface. The location of this region is determined by a comparison between the timescales of rising through the convection zone and those of fragmentation and coupling. The high efficiency of the fragmentation processes leads to the conjecture that the relevant region is also situated in the upper parts of the convection zone. On the other hand, it has been reported (e.g. Bumba, 1981) that new active regions,especiallly in complexes of activity,appear predominantly eastward of older active regions (see also McIntosh, 1981). If these consecutive active regions originate from the same source region this would indicate that it rotates slower than the surface. However, we should bear in mind that all these rotation rates (and also the latitude dependence) might as well be related to phase velocities of convective patterns which can be different from the rotation rate of the same region.

We conclude that magnetic fields do not tell us very much about the internal rotation of the convection zone. They mainly seem to reflect the "jet" indicated by F. Hill´s observations of high degree p-modes, a region of faster rotation ($\Delta\Omega/\Omega \simeq 5\%$) a few 10^3 km deep. On the other hand, the source region from which active regions originate seems to rotate more slowly than the surface. Since we prefer to locate this region in the deep parts or below the convection zone proper, this would be in agreement with the results from p-mode splitting as well (Duvall and Harvey, these proceedings). However, as the magnetic fields seem closely coupled to the convective patterns, one must seriously consider the possibility that observed "magnetic" rotation rates reflect the phase velocities of convective patterns rather than mean plasma rotation rates. This might also be a clue to the confusing coexistence of rigid and differential rotation (e.g. Wilcox et al., 1970).

4. CONSEQUENCES FOR THE SOLAR DYNAMO

In the preceding sections we have presented arguments in favour of the conjecture that large active regions originate from a region of dominating differential rotation below the convection zone proper, the convective overshoot layer. The mechanism of vorticity expulsion produces and maintains differential rotation in this layer while the gradient of the convective velocity holds the magnetic at the same location (Schüssler, 1984a). The dynamo cycle is closed by either the "classical" α-effect of cyclonic convection (the αω-dynamo picture) or by the electromotive force due to the action of rotation on slow magnetostrophic waves propagating along the toroidal field. These waves are a natural consequence of the Rayleigh-Taylor instability of a magnetic layer in a rotating system and an effective α can be calculated which has a non-monotonic dependence on latitude (Schmitt, 1984; 1985). Dynamo models with this kind of α-effect account for the concentration of activity in low latitudes together with the "polar branch" of the solar cycle (Leroy and Noens, 1983) in a natural way (Schmitt, 1987).

We can conclude that our consideration of the dynamics of magnetic fields in a convection zone strongly favour the "boundary layer dynamo" as source of large active regions. The large degree of order in the solar cycle (in particular, the polarity rules; see Balthasar and Schüssler, 1983, 1984 for another aspect) cannot be understood in a region of dominating convection. On the other hand, the expulsion of magnetic fields and vorticity by themselves make the convective overshoot layer a very "attractive" site for the solar dynamo.

REFERENCES

Balthasar, H., Schüssler, M., Wöhl, H.: 1982, Solar Phys. 76, 21

Balthasar, H., Schüssler, M., 1983, Solar Phys. 87, 23

Balthasar, H., Schüssler, M.: 1984, Sol. Phys. 93, 177

Balthasar, H., Vazquez, M., Wöhl, H: 1986, Astron. Astrophys. 155, 87

Brants, J.J.,Steenbeck, J.C.M.: 1985, Solar Phys. 96, 229

Bumba, V: 1981, Bull. Astron. Inst. Czech. 32, 129

Busse, F.H., Carrigan, C.R.: 1974, J. Fluid Mech. 62, 579

Durney, B.R.: 1985, Astrophys. J. 297, 787

Duvall, T.L.: 1980, Solar Phys. 66, 213

Foukal, P.: 1977, Astrophys. J. 218, 539

Gilman, P.A., Foukal, P.: 1979, Astrophys. J. 229, 1179

Gilman, P.A., Miller, J.: 1986, Astrophys. J. Suppl. 61, 585

Gilman, P.A:, Howard, R.: 1986, Astrophys. J. 303, 480

Howard, R., Gilman, P.A., Gilman, P.I.: 1984, Astrophys. J. 283, 373

Howard, R.: 1984, Ann. Rev. Astron. Astrophys. 22, 131

Leroy, J.-L., Noens, J.-C.: 1983, Astron. Astrophys. 120, L1

Lustig, G., Dvorak, R.: 1984, Astron. Astrophys. 141, 105

McIntosh, P.S:: 1981, in "The Physics of Sunspots", L.E. Cram, J.H. Thomas (eds.), Sacramento Peak Observatory, p.7

Meyer, F., Schmidt, H.U., Weiss, N.O., Wilson, P.R.: 1974, Mon. Not. Roy. Astron. Soc. 169, 35

Meyer, F., Schmidt, H.U., Weiss, N.O.: 1977, Mon. Not. Roy., Astron. Soc. 179, 741

Moffatt, H.K.: 1978, "Magnetic field generation in electrically conducting fluids", Cambridge University Press

Moreno-Insertis, F.: 1986, Astron. Astrophys. 166, 291

Newton, H.W., Nunn, M.L.: 1951, Mon. Not. Roy. Astron. Soc. 111, 413

Parker, E.N.: 1979, Astrophys. J. 230, 905

Parker, E.N.: 1984, Astrophys. J. 283, 343

Proctor, M.R.E, Weiss, N.O.: 1982, Rep. Progr. Phys. 45, 1317

Ribes, E., Mein, P., Mangeney, A.: 1985, Nature 318, 170

Schmitt, D.: 1984, in "The Hydromagnetics of the Sun", Proc. 4th European Solar Meeting, Noordwijkerhout, ESA SP-220, p. 223
Schmitt, D.: 1985, thesis, Univ. Göttingen (in german)
Schmitt, D.: 1987, Astron. Astrophys., in press
Schröter, E.H.: 1985, Solar Phys. 100, 141
Schüssler, M.: 1981, Astron. Astrophys. 94, L17
Schüssler, M.: 1984a, in "The Hydromagnetics of the Sun", Proc. 4th European Meeting on Solar Physics, ESA SP-220, p. 67
Schüssler, M.: 1984b, Astron. Astrophys. 140, 453
Schüssler, M.: 1986, in W. Deinzer et al. (eds.), "Small scale magnetic flux concentrations in the solar photosphere", Vandenhoeck & Ruprecht, Göttingen, p. 103
Sheeley, N.R., DeVore, C.R., Boris, J.P.: 1985, Solar Phys. 98, 219
Snodgrass, H.B.: 1983, Astrophys. J. 270, 288
Spruit, H.C:: 1977, thesis, Univ. of Utrecht
Spruit, H.C.: 1981, Astron. Astrophys. 98, 155
Spruit, H.C.: 1981 in "The Physics of Sunspots", L.E. Cram, J.H. Thomas (eds.), Sacramento Peak Observatory, p. 98
Ternullo, M.: 1986, Solar Phys. 105, 197
Ternullo, M., Zappalá, R.A., Zucarello. F.: 1981, Solar Phys. 74, 111
Tsinganas, K.C.: 1980, Astrophys. J. 239, 746
Tuominen, I., Virtanen, H.: 1984, Astron. Nachr. 305, 225
Wilcox, J.M., Schatten, K.H., Tanenbaum, A.S., Howard, R.: 1970, Solar Phys. 14,255
Yoshimura, H.: 1981, Astrophys. J. 247, 1102

HYDROSTATIC ADJUSTMENT TIME OF THE SOLAR SUBCONVECTIVE LAYER

E.A. Spiegel
Astronomy Department
Columbia University
New York, N.Y. 10027.

ABSTRACT. One objection that has been raised against the suggestion that the solar cycle is driven by magnetohydromagnetic activity below the convective zone is that the Kelvin-Helmholtz time is a million years. In fact, as is argued here, the subconvective layer adjusts its hydrostatic structure in response to highly nonradial disturbances in a few years.

1. INTRODUCTION

The topic of this meeting is angular transport in the sun. Clearly the presence of strong magnetic flux tubes over large vertical extents will affect this process markedly. Such tubes are a manifest component of the sunspot cycle and it is therefore relevant to ask what is the vertical extent of mhd processes that we see as the solar cycle? My impression is that the current general opinion places the seat of solar activity near the bottom of the convection zone. Therefore, any discussion of differential rotation of the convection zone ought to take account of the large scale effects of prevelant strong flux tubes in the zone.

Having thus paid respects to the principal topic, I have raised the question of interest to me in this connection: Can strong fields form at depth in the convection zone? The impression one might have is that, once even a moderate field forms, magnetic buoyancy will bring it to the top before it can achieve sunspot strength. In these proceedings, and elsewhere, Parker battles magnetic buoyancy, his own creation, to find a means to allow strong fields to form in the lower depths of the convection zone. He has argued for a strong stable field down there. Perhaps it can be overstable with a period of eleven years.

The way that Weiss and I and van Ballegooien have tried to avoid the magnetic buoyancy problem is to place the strong-field formation process below the convection zone. Since the bottom of the convection zone is usually defined as the depth at which the thermal entropy gradient vanishes, all this needs to be understood in the light of the effect of strong fields on the stability. But once those fields are in place, overstability and nonlinear instabilities become part of the story, so I mean by "bottom of the zone" the bottom of the zone in a good theoreti-

B. R. Durney and S. Sofia (eds.), The Internal Solar Angular Velocity, 321–327.

cal model with no fields. Once this is understood, one may begin to look at the temporal dynamics of the subconvective magnetic layer -- a solar magnetocline -- to find the clock of the solar cycle and the source of strong field enhancement.

The subconvective picture does not require local dynamo action, though I suspect it is there. The first stage of the process is probably in a turbulent dynamo in the main convective zone. This will pump field into the subconvective layer to set up a magnetically driven instability, as Childress and I suggested in the proceedings of the previous conference, which Sabatino Sofia edited.

Naturally, there are some extensive, but incomplete, calculations that go with this, but I am not going to indulge myself by going into these matters. Rather I am interested in trying to decide whether a solar magnetocline is a useful notion. Weiss and I had guessed that the region of interest ought to be a local scale height thick. The chief objection that I have heard to this is that the Kelvin-Helmoltz time for such a layer is like 10^6 years. How do we then expect such a layer to vary on a time scale of eleven years? My aim here is to give a qualitative answer to this question.

II. THE IDEA BEHIND THE CALCULATION

The solar magnetocline is a layer below the convection zone with a predominantly toroidal magnetic field that is normally at the kilogauss level. In my present vision of it, waves of instability travel across it in latitude. These waves are amplitude-modulated and the envelope travels slowly, taking eleven years to go across. The passing amplitude maximum causes a bulge in layer thickness that intrudes into the convection. That intrusion is swept upward, to the surface. I ignore the longitudinal instabilities of this magnetic curtain as it is drawn up through the convection zone; they will be important at a more refined stage of this picture.

What is important here is that I am positing a thin vertical sheet at fixed latitude (in each hemisphere) of magnetic disturbance, running from the bottom of the subconvective layer up through most of the height of the convective zone. This layer produces perturbations in magnetic field, hence in pressure and convective activity, and in temperature.

If we ignore that the vertical sheet will soon give way to another, we can ask what the response time to a sudden intrusion of a single sheet will be. The general question is too complicated for the present treatment. I suggest the following pilot problem:

We have a plane-parallel, static fluid layer of semi-infinite horizontal extent. Suppose we have constant conductivity, as in Lamb's discussion of a polytropic atmosphere. This means that we have a linear temperature profile

$$T_o(z) = \beta z + C \qquad (1$$

ere z is measured downward. If the top of the layer, is at z = 0, C is he surface temperature. We shall always be at depths where C is negli-ible.

At the edge of the layer we have also $T = T_o(z)$ for $t<0$. At $t = 0$ we hange the edge temperature to

$$T(z) = (1+\varepsilon)\beta z \tag{2}$$

here $0<\varepsilon<<1$. This choice of disturbance is made for its simplicity, in he hope that it may give us a useful estimate to the time of response f a layer to an intruding magnetic curtain.

he problem is to estimate the time scale of hydrostatic adjustment o a change $T_o \rightarrow T$ on the edge of the layer. I give here a physical ar-ument like one that could be used to estimate spin-down time in an nalogous context. I have tried the arguments on geophysicists (in oods Hole in 1981), and I am still standing up. But then, I had the upport of Chris Bretherton.

II. THE SHAPE-UP TIME

here are two time scales that are important in estimating how long a ayer will take to adjust its vertical structure in response to forcing rom the side:

a) The thermal response time of the layer is

$$\tau_t \sim D^2/\chi \tag{3}$$

here the layer thickness is D and the (suitably averaged) thermal dif-usivity is χ,

b) The dynamic response time is

$$\tau_d \sim D/c \tag{4}$$

here c is the speed of sound. In a polytropic atmosphere, D is compara-le to the depth and to the local scale height, H.

As in the spin-down problem (see Greenspan's book on rotating fluids, r the warm-up problem of an incompressible fluid discussed by Veronis n Annual Reviews of Fluid Mechanics, 1970), we anticipate an answer hat is the geometric mean of these two times. In fact, whenever we see wo separate linear processes working at once, where one of them affects he working of the other, we get this geometrical law. This is a useful hing to recognize when the two times are disparate; here we have

$$\tau_d \ll \tau_t.$$

Of course, other times may enter (such as the magnetic response time), but I will not try to cover all such qualitative effects here.

The perturbation at the edge of the layer causes the edge itself to change its structure by the vertical propagation of a pressure wave. During the time τ_d required for this, the thermal effects propagate horizontally into the layer a distance δ such that

$$\delta \sim [\chi\tau_d]^{1/2}. \qquad (5$$

This is thermal boundary layer that is set up during the first hydrostatic adjustment period. It has time to put its own shape in order during this rapid transient phase. The thickness δ is analogous to the Ekman thickness of rotating fluids. We shall be taking

$$D \sim H = \text{local scale height},$$

so

$$\delta \sim [\chi H/c]^{1/2} \qquad (6$$

where g is the local gravitational acceleration.

The modification of the local gradient by a factor $(1+\varepsilon)$ produces a local change

$$\delta D \sim \varepsilon D \sim \varepsilon H \qquad (7$$

which is effected by a local expansion (or contraction) velocity

$$w \sim \delta D/\tau_d \sim \varepsilon Hc/H \sim \varepsilon c \qquad (8)$$

The continuity equation produces a corresponding suction (or pumping) horizontal flow u such that

$$u/\delta \sim w/D + w/H \sim w/H. \qquad (9)$$

There is a question of signs in the second member of (9). It is not delicate question if $D \sim H$ does not happen to turn into $D = H$; I am assuming it does not.

The horizontal velocity

$$u \sim \varepsilon c \delta / H \quad (10)$$

.ven by (8) and (9), though small, is the key to the process. The mag-
ɪtic curtain drawn up through the convection zone induces a suction (or
ɪmping) by its own local hydrostatic effects. The flow u starts a weak
ɪrge-scale motion that permits the new hydrostatic state to be reached.
ɪe velocity u is the analogue of the Ekman velocity.

A ring of fluid with horizontal symmetry axis (in the x-direction),
ɪ moved along at velocity u. Here we think of a ring in the main body
ɪ the fluid away from the boundary layer. In first approximation, the
ɪtion is isentropic. Since the specific entropy is

$$S = \log(p/\rho^{\gamma}), \quad (11)$$

ɪ get

$$\delta S = 0 = \delta p/\rho - \gamma \delta\rho/\rho. \quad (12)$$

ɪr constant molecular weight,

$$\delta p/p = \delta\rho/\rho + \delta T/T, \quad (13)$$

ɪ

$$\delta\rho/\rho = [1/(\gamma-1)]\delta T/T \quad (14)$$

ɪ get such a density change, we change the radius a of the ring by

$$\delta a \sim a\delta\rho/\rho \quad (15)$$

ɪ assume $a \sim D \sim H$, hence

$$\delta a \sim [H/(\gamma-1)]\delta T/T. \quad (16)$$

To take care of the hydrostatic adjustment that starts the process,
ɪamely,

$$\delta T \sim \varepsilon T, \quad (17)$$

we need

$$\delta a \sim \varepsilon n H \qquad (1$$

where $n=1/(\gamma-1)$. This part of the adjustment takes place mainly in mid-layer. The horizontal scale, L, there is given by factors that I hav left out. L is at most the circumference of the sun. It cannot be mor that the distance to the next magnetic curtain. I write

$$L = \alpha H \qquad (1$$

where α is a few.

The time for hydrostatic adjustment is

$$\tau \sim \delta a/w \qquad (2$$

where this w is the vertical velocity in the main body of layer and given by

$$w \sim \alpha u. \qquad ($$

From (10) and (18) we get

$$\tau \sim nH^2/(\alpha c \delta) \qquad (2$$

or

$$\tau \sim (n/\alpha)[H^3/(\chi c)]^{1/2} \qquad (23)$$

The shape-up time is the geometric mean of the two basic times:

$$\tau^2 = \tau_d \tau_t. \qquad ($$

IV. CONCLUSION

Calculations of solar hydrostatic adjustment to perturbations involvin the magnetic disturbances of the sun require the inclusion of sound pro pagation and of large scale non-radial effects. When these are includ ed, we see some interesting time scales with

$$\tau_d \ll \tau \ll \tau_t. \tag{25}$$

n the convection zone, where χ is χ_{eddy}, we have $\tau_t \sim 1$ month. In the ubconvective layer, that some think relevant to the cycle, we have $_t \sim 10^6$y. In both cases, $\tau_d \sim 0.5$ hr. Hence, the convection zone has response time of hours and the subconvective layer has a time of re- ponse given by $\tau \sim$ few years.

Can this number be eleven years? We cannot answer that but, it is ertainly possible. Would that be significant for fixing the time of the olar cycle? Not if there is some other mechanism at work to determine he time. Then the cycle can entrain the shape-up process and it will ollow the cycle. This will entail radius variations given by (18). If e use Livingston's data on δT, we get ε and can find $\delta R/R$.

If there is degeneracy in the basic mechanism of the solar cycle and broad band of frequencies is equally possible otherwise, then process- s such as the one discussed here may lift the degeneracy and influence he cycle time. That the shape-up process is the tail that wags the so- ar dog does not seem likely to me, but it is not excluded. At present, he shape-up time is the only time scale I know of that could be eleven ears.

I wrote these remarks at the Weizmann Institute, far from my solar hysics notes and beyond the deadline. Though I am happy for this ex- use to be informal, I am sorry to have omitted precise references. It s a pleasure to thank I. Procaccia for his hospitality in Israel, and . Durney and S. Sofia for their patience and kindness. The work that ays behind this discussion was supported by Grant PHY80-23721 of the .S.F.

MODELS FOR A DIFFERENTIALLY ROTATING SOLAR CONVECTION ZONE

Michael Stix
Kiepenheuer-Institut für Sonnenphysik
Schöneckstr. 6,
D-7800 Freiburg, FRG

ABSTRACT. This is a review of axisymmetric mean field models of the Sun's differential rotation. The driving terms in the mean equations are second order correlations of velocity and temperature fluctuations. Particular attention is given to "anisotropic viscosity" - a special form of driving by Reynolds stresses -, and to "latitude-dependent heat transport", which arises from the influence of rotation upon convection, generates meridional circulation and therewith differential rotation.

1. INTRODUCTION

In this contribution I shall present updated versions of relatively old models of the Sun's differential rotation. One class of these models, called "anisotropic viscosity" models, was first proposed by Kippenhahn (1963) and Köhler (1970), according to earlier ideas of Wasiutyński (1946) and Biermann (1951). The other class, known as "latitude-dependent heat transport" models, is based on an idea of Weiss (1965), and was presented in more detail by Durney and Roxburgh (1971) and Belvedere and Paternò (1977). The models described here are updated because they employ a "real" solar envelope, in contrast to the earlier models mentioned, which use spherical shells of constant density, or very thin shells, or shells with purely convective energy transport, or other simplifications. With "real" I mean that the model is the envelope part of a standard solar model, including the mixing length theory of convective energy transport according to Vitense (1953).

All models to be discussed here are mean (axisymmetric) models. The fluctuating (or non-axisymmetric) parts of the flow and temperature fields will be taken into account only in form of second-order correlations, which in the present case again can be understood as averages over longitude, ϕ. In particular, the terms driving the differential rotation are such correlations. In order to classify these "drivers" I shall follow some of the recent work of Rüdiger (1980, 1982ab, 1983 ab), in particular as far as the velocity correlations, or "Reynolds stresses" are concerned.

B. R. Durney and S. Sofia (eds.), The Internal Solar Angular Velocity, 329–342.

The work to be reported here was done in collaboration with G. Belverdere, L. Paternò, and R.M. Pidatella of the Osservatorio Astrofisico, Catania, and W. Schmidt of the Kiepenheuer-Institut.

2. CONSERVATION OF ANGULAR MOMENTUM

The law of conservation of angular momentum is derived from the azimuthal component of Euler's equation. In its averaged form, gravity and the pressure gradient do not appear:

$$\overline{\varrho \frac{\partial u_\phi}{\partial t}} + \overline{\varrho \left[(\underset{\sim}{u} \cdot \nabla) \underset{\sim}{u} \right]_\phi} = 0 \tag{1}$$

Let us also neglect density fluctuations. This means that acoustic waves are suppressed. We may then take ϱ out of the average (1); moreover, the equation of continuity simply is

$$div(\underset{\sim}{u} \varrho) = 0 \ . \tag{2}$$

We now divide u into its mean and fluctuating parts,

$$\underset{\sim}{u} = \overline{\underset{\sim}{u}} + \underset{\sim}{u}' \ , \tag{3}$$

where the mean,

$$\overline{\underset{\sim}{u}} = (v_r, v_\theta, r\Omega \sin\theta) \tag{4}$$

consists of an axisymmetric meridional circulation $\underset{\sim}{v}=(v_r, v_\theta, 0)$ and an azimuthal mean flow with angular velocity $\Omega(r, \theta, t)$. We multiply (1) by the lever arm, $a=r\sin\theta$, substitute (3) and (4), make use of (2), and obtain

$$\frac{\partial}{\partial t}(\varrho a^2 \Omega) + div\left(\varrho a^2 \Omega \underset{\sim}{v} + \varrho a \overline{u'_\phi \underset{\sim}{u}'}\right) = 0 \ , \tag{5}$$

which is Eq. (1.1) of Rüdiger (1980) in the inviscid case, $\nu=0$. This equation expresses the conservation of angular momentum: at any given place, the specific angular momentum, $\varrho a^2 \Omega$, changes either because it is transported by meridional circulation, $\underset{\sim}{v}$, or by means of Reynolds stresses. The latter appear in (5) as two components of the velocity correlation tensor,

$$Q_{r\phi} = \overline{u'_r u'_\phi} \ , \qquad Q_{\theta\phi} = \overline{u'_\theta u'_\phi} \ ; \tag{6}$$

they are called the turbulent radial and latitudinal zonal momentum fluxes by Rüdiger.

For the present purpose I shall neglect the time dependence of Ω. That is, I neglect both the braking of the Sun's rotation by the solar wind, and the torsional oscillation (Howard and Labonte, 1980). The latter is of small amplitude, and the former is a slow process; it is therefore reasonable to assume that these two effects have little relevance to the problem of maintaining the differential rotation in the convection zone. The neglect of solar wind braking is also consistent with the stress-free boundary condition

$$Q_{r\phi} = 0 \tag{7}$$

at $r=r_i$, $r_\odot$, the radii of the inner and outer spherical surfaces of the convective shell. Thus no angular momentum enters or leaves the convection zone; in addition, the condition $v_r=0$ at $r=r_i$, $r_\odot$ ensures that the meridional flow is confined to the shell.

3. REYNOLDS STRESSES

The main problem is to express the Reynolds stresses in terms of the mean flow. Substitution of such an expression into Eq. (5), and an analoguous procedure with the equations governing the meridional components, v_r and v_θ, would then lead to a closed system for the mean flow. I shall not follow through this entire program here. Only a few steps which have relevance to the simple models discussed below will be described.

3.1 Diffusive and Non-Diffusive Fluxes

One plausible effect of turbulence is mixture and exchange. Differences in angular velocity tend to be smoothed out, and the flux which does this must therefore go down the gradient of angular velocity:

$$(Q_{r\phi}, Q_{\theta\phi}) = -\nu_T r \sin\theta \, \nabla\Omega \quad . \tag{8}$$

This is the "diffusive" part of the Reynolds stresses (Rüdiger, 1980, 1983a) because it leads to a diffusion term in the angular momentum balance (5), just as if we had started from the full Navier-Stokes equation, including viscosity, instead from Euler´s equation. The molecular viscosity, ν, is simply replaced by the turbulent viscosity, ν_T.

If the Reynolds stresses can be expressed in the form (8), and if no meridional circulation is present, the asymptotic solution of (5) is uniform rotation, Ω=constant. The time required to reach this state is of order d^2/ν_T, where $d=r_\odot-r_i$ is the thickness of the convection zone. For the Sun, this diffusion time scale is 1 to 10 years, depending on the adopted value of ν_T. In the presence of meridional circulation Eq. (5) admits differential rotation. If the sense is positive, i.e. if the circulation rises in the polar and descends in the equatorial regions, the resulting rotation is faster near the equator, as observed on the Sun (Kippenhahn, 1964).

In a turbulent fluid which is exposed to gravity and rotation anisotropies prevail. These give rise to additional effects of the Reynolds stresses which, in general, are "non-diffusive"; they <u>maintain</u>, rather than smooth out, a non-uniform rotation. In particular, as

Rüdiger (1977, 1980) points out, terms proportional to Ω itself (instead to the gradient of Ω) must be included. Hence, a more general form is

$$
\begin{aligned}
Q_{r\phi} &= -\nu_T r \sin\theta \frac{\partial\Omega}{\partial r} + \Lambda_V \sin\theta\,\Omega , \\
Q_{\theta\phi} &= -s\nu_T \sin\theta \frac{\partial\Omega}{\partial\theta} + \Lambda_H \cos\theta\,\Omega .
\end{aligned}
\tag{9}
$$

The factor s is included in order to indicate that, in general, the turbulent exchange yields different diffusivities in the two directions.

As we shall see below, the Λ-terms drive differential rotation. The Λ's are generally functions of r and θ; in fact $\Lambda_H(r, \theta)$ must vanish at the poles (in second order), according to the definition (6) of $Q_{\theta\phi}$. Appropriate expansions are, therefore (Rüdiger, 1982a, 1983b)

$$
\begin{aligned}
\Lambda_V &= \nu_T \left[V^0(r) + V^1(r)\sin^2\theta + \dots \right] , \\
\Lambda_H &= \nu_T \left[H^1(r)\sin^2\theta + \dots \right] ,
\end{aligned}
\tag{10}
$$

where a factor ν_T has been added for dimensional reasons, and symmetry with respect to the equatorial plane is assumed.

The problem of determining the Reynolds stresses is now shifted to the problem to calculate the coefficients ν_T, V^0, V^1, H^1 ... . Knowledge or assumptions concerning the turbulence, in particuar how it is affected by rotation, is necessary for this purpose. An easier and often chosen path is to specify the coefficients in such a way that they reproduce observational results. That is, the properties of the Reynolds stresses are determined a posteriori. I shall follow this approach also in the present contribution, although a few results of direct calculations will be discussed below in Sect. 2.4.

3.2 Anisotropic Viscosity

If no other than theradial direction is distinguished in the turbulent velocity field, there should be no horizontal angular momentum flux (except the one following the gradient of Ω), and hence $\Lambda_H=0$. The simplest possible Λ_V is a constant, $\Lambda_V=\nu_T V^0$. Substitution into the steady form of (5) where, for the moment, we neglect meridional circulation, then yields

$$
\frac{1}{\varrho r^2}\frac{\partial}{\partial r}\left(r^4 \varrho \frac{\partial\Omega}{\partial r}\right) + \frac{s}{\sin^3\theta}\frac{\partial}{\partial\theta}\left(\sin^3\theta \frac{\partial\Omega}{\partial\theta}\right) = \frac{V^0}{\varrho r^2}\frac{\partial}{\partial r}\left(r^3 \varrho \Omega\right) . \tag{11}
$$

We now consider differential rotation as a weak effect, i.e. on the right of (11) we neglect $\partial\Omega/\partial r$ in comparison to Ω/r. The inhomogeneous nature of the problem thus becomes clear: the derivatives of Ω, on the left of (11), are driven by the inhomogeneous term on the right, containing Ω itself. Incidentally, as Rüdiger has observed, the driving term is identical to a term occuring in the angular momentum balance of Kippenhahn (1963). There the Reynolds stresses are expressed exclusively in a diffusive form similar to (8), but with an anisotropic tensorial turbulent viscosity. We must identify V^0 with $2(s-1)$ and use $v_T\rho=\eta$. Thus, it turns out that "anisotropic viscosity" comprises both diffusive and non-diffusive angular momentum fluxes. Since the latter drive, while the former dissipate differential rotation we may reasonably follow Rüdiger and consider the two as physically distinct in the first place. Rüdiger calls the non-diffusive fluxes the "Λ-effect" and points out that Biermann (1951) first proposed such an effect as the source of the Sun's differential rotation.

Let us solve (11). We write the angular velocity in the form

$$\Omega = \Omega^*\left[1 + \omega_0(r) + \omega_2(r)P_2(\theta)\right] \tag{12}$$

and - in consistence with our assumption of weak differential rotation - replace Ω on the right of (11) by the constant Ω^* (if $\omega_0(r_\odot)=0$, then Ω^* is the average of Ω over the solar surface). We use a prime to denote d/dr and obtain

$$\begin{aligned} r^2\omega_0'' + (4 + r\rho'/\rho)r\omega_0' - 2s\omega_2 &= (3 + r\rho'/\rho)V^0, \\ r^2\omega_2'' + (4 + r\rho'/\rho)r\omega_2' - 10s\omega_2 &= 0. \end{aligned} \tag{13}$$

Clearly $\omega_2=0$, $\omega_0'=V^0/r$ is a solution to (13); when the boundary condition is taken into account it can be seen that it is the only solution. Therefore, the observed latitudinal dependence of Ω cannot be obtained from constant vertical flux alone, at least not in the case $\underset{\sim}{v}=0$ considered so far. If, on the other hand, meridional circulation is admitted - in fact Kippenhahn (1963) shows that it must be admitted if the anisotropyparameter s is a constant and not 1 -, it can and will drive differential rotation with latitude dependent Ω. The circulation itself is driven by Reynolds stresses which appear in the mean meridional momentum balance in a similar manner as in (5), and which are related to the mean flow with the help of the anisotropic viscosity tensor already mentioned. The "AV-models" discussed in Sect. 3.5 below are of this type.

3.3. A Model with Vertical and Horizontal Flux

Although in Kippenhahn's model meridional circulation is essential in driving the differential rotatio n, this is not necessarily true in more general cases. As an example, let us again consider $\Lambda_V=v_TV^0=\text{const}$; but, in addition, let us retain the first term, $v_TH^1\sin^2\theta$, in the expansioin (10) of Λ_H. We assume again that meridional circulation is absent and arrive at equations for ω_0 and ω_2 with left hand sides identical to (13) but with additional driving terms, viz.

$$r^2\omega_0'' + \ldots. \qquad = (3 + r\rho'/\rho)V^0 + 2H^1/3\,,$$

$$r^2\omega_2'' + \ldots. \qquad = 10H^1/3\,. \tag{14}$$

An equivalent system of equations was considered by Schmidt and Stix (1983) in an attempt to generalize the concept of anisotropic viscosity. The two parameters s and σ must be related to the constants on the right of (14) by $V^0=2(\sigma-1)$ and $H^1=7(\sigma-s)/2$.

The solution to (14) is $\omega_2=-H^1/3s$, $\omega_0'=V^0/r$, and thus includes the θ-dependence. This result is plausible: positive V^0 means upward flux, and outwards increasing angular velocity; positive H^1 means equatorward flux, and therefore an accelerated equator ($\omega_2<0$). As we shall see in the following section, turbulent eddies which are elongated in latitude yield $H^1>0$. Such elongated structures ("bananas") apear also in non-axisymmetric numerical simulations; there also they play an essential part in the maintenance of equatorial acceleration (Glatzmaier, 1985; Gilman and Miller, 1986; s.a. Glatzmaier, these proceedings).

The interesting aspect of the present example is that it leads essentially to the same angular velocity distribution whether or not meridional circulation is included. In fact, numerical models showed circulations having equatorward or poleward direction, depending on the total depth of the convection zone, but always had nearly the same $\Omega(r,\theta)$. This also illustrates that meridional circulation is not essential, but rather a side-effect in this example.

3.4. Calculation of Reynolds Stresses

In order to calculate the correlations $\overline{u_i'u_j'}$ an equation governing the fluctuating part, $\underline{u}'$, of the velocity field must be solved. The difference between the Navier-Stokes equation and its average is, in cartesian co-ordinates

$$\frac{\partial u_i'}{\partial t} - \nu\nabla^2 u_i' + \frac{1}{\rho}\frac{\partial p'}{\partial x_i} + 2\varepsilon_{ijk}\Omega_j u_k' = f_i' + u_j'\frac{\partial u_i'}{\partial x_j} - \overline{u_j'\frac{\partial u_i'}{\partial x_j}}\,, \tag{15}$$

where f_i' is an external fluctuating force, and mean flow components other than the rotation, Ω_i, have been neglected (cf. Rüdiger 1983b, Eq. 2.4).

Two simplifications of the solution of (15) are common. One is the neglect of the non-linear terms on the right, called the quasilinear approximation, first order smoothing, or second order correlation approximation; the latter name is because, if said terms were retained, third and higher order correlations would appear in the process of calculating the Reynolds stresses. The other simplification is the assumption that the turbulent velocity field $\underline{u}'$ consists of eddies specified by single wave numbers, k_r, k_θ and k_ϕ (e.g. Gough, 1978; Durney and Spruit, 1979), or equivalently that the spectrum defined by the Fourier transform of (15) has the form of a δ-function (e.g. Rüdiger, 1980, 1983b).

The special case of slow rotation offers further simplification. The velocity coefficient of Ω_j in (15) can then be taken as an "unperturbed" field, u'_{ko}, with known or assumed properties. Slow rotation here means $\Omega\tau<1$, where τ is the correlation time of the turbulence, or the turnover time of the eddies. Except in the deepest part of the convection zone, where τ is of the order of one month according to the mixing length theory, the assumption of slow rotation is reasonable. In this case the result, in the form written down by Rüdiger, is

$$\Lambda_V = 2\tau[(k_\phi^2+k_\theta^2)\overline{u_\phi'^2} - (k_r^2+k_\theta^2)\overline{u_r'^2}]/k^2 \quad ,$$
$$\Lambda_H = 2\tau[(k_r^2+k_\phi^2)\overline{u_\phi'^2} - (k_r^2+k_\theta^2)\overline{u_\theta'^2}]/k^2 \quad . \tag{16}$$

In the case of slow rotation the radial direction, distinguished by gravity, determines the main anisotropy of the turbulence. Approximate symmetry in the two horizontal directions is, therefore, a consistent assumption. With $\overline{u_\theta'^2} \approx \overline{u_\phi'^2}$, expression (16) for Λ_H is positive if $k_\theta^2 < k_\phi^2$, in agreement with what was said about elongated eddies in the previous section. Such k_i values favour equatorial acceleration. However, we also see immediately that Λ_V tends to be negative, especially when the vertical velocity dominates (large $\overline{u_r'^2}$) or when the eddies are flat (large k_r^2); in these cases we have $\Lambda_V \approx -\tau\overline{u_r'^2}$, independently of latitude, which means an anisotropy parameter s<1. For the AV-models described presently this will be a difficulty, because driving a meridional circulation, and thereby an accelerated equator, will require s>1.

The case of rapid rotation, $\Omega\tau>1$, is more difficult. The angular velocity then enters in a non-linear way (Rüdiger, 1983b; Durney, 1983, 1985), and the eddies tend to be tilted towards the axis of rotation (Durney, 1983). Another problem is that, in a general theory, the turnover time τ is not a free parameter but should instead be determined from a dispersion relationship which contains the wave numbers, k_i, the rate of rotation, Ω, and properties of the mean steady state itself (Durney, 1985). For a attempt to obtain a consistent simultaneous solution of mean flow and Reynolds stresses see Durney (these proceedings). Here I stay with the simplified models.

3.5 The AV-Model of the Sun's Convection Zone

This model rests on the concept of anisotropic viscosity discussed by Kippenhahn (1963), and was exploited by Köhler (1970) in an incompressible, and by Schmidt (1982) in a compressible numerical calculation. It deals with mean quantities only; I therefore omit the overbars in this section. The equations describing the model are (Pidatella et al., 1986):

$$\nabla\cdot(\underset{\sim}{F}_R+\underset{\sim}{F}_C) + \rho T\underset{\sim}{u}\cdot\nabla S = 0 \quad ,$$
$$\nabla\cdot(\rho\underset{\sim}{u}) = 0 \, , \tag{17}$$
$$\rho(\underset{\sim}{u}\cdot\nabla)\underset{\sim}{u} = -\nabla P + \rho\underset{\sim}{g} + \underset{\sim}{R} \quad ,$$

where the radiative and convective energy fluxes are

$$\underset{\sim}{F}_R = -(16\sigma T^3/3\kappa\varrho)\nabla T \quad ,$$
$$\underset{\sim}{F}_C = -(\varrho c_p \ell v_c/2)(\nabla T - \nabla T^*) \quad . \tag{18}$$

The viscous force, $\underset{\sim}{R}$, contains the driving terms, which are proportional to s-1; ℓ is the mixing length, v_c the convective velocity, and ∇T^* the actual mean temperature gradient within the eddies (different from the adiabatic because of radiative heat exchange). Entropy and pressure are given by the equations of state

$$dS = c_p(dT/T - \nabla_{ad}\, dP/P) \quad ,$$
$$P = \varrho \mathcal{R} T/\mu(\varrho, T) \quad , \tag{19}$$

where the arguments of μ indicate that the mean molecular weight is determined by the ionization equilibrium as expressed by the Saha equations. Convection is treated in the mixing length theory, with ℓ proportional to the pressure scale height, H (Vitense, 1953).

To zeroth order $\underset{\sim}{u}$ vanishes, and therewith R. Then we have a standard solar envelope model. The parameter $\alpha = l/H$ determines the radius r_i of the inner boundary of the convection zone. Typical values are $\alpha = 1.25$, with $r_i = 0.75 r_\odot$, or $\alpha = 1.5$ with $r_i = 0.695 r_\odot$.

When the mean flow $\underset{\sim}{u}$ is included as a first order perturbation, we may linearize the equations and obtain an inhomogeneous system with driving terms $\sim$(s-1), as outlined in Sect. 3.2 above, and written down in detail by Pidatella et al. (1986). We calibrate the driving inhomogeneity so that the observed surface differential rotation of the Sun is obtained, i.e. that $\omega_2(r_\odot) = -0.189$ (Howard and Harvey, 1970). And these are the results:

(i) The calibration with $\omega_2(r_\odot)$ typically yields s=1.03. This differs only slightly from 1, which means that a small perturbation is sufficient and that, therefore, the linearization is justified. However, s is larger than 1. As already mentioned in Sect. 3.4, this means that the whole theory (including the deduction of s from the turbulence) is not consistent. To save the model it may be necessary to include the parameters V^1 and H^1 with their correct latitude dependence, or to include the effects of fast rotation on the eddies in the deeper part of the convection zone.

(ii) Typical temperature differences between pole and equator at the surface ar 1 K; typical surface velocities of the circulation are 1 m/s. This velocity is equatorward if there is only 1 cell filling the entire convection zone, and opposite if there are 2 cells above each other. The latter case, shown in Fig. 1, occurs for some of the deeper models. It may actually resemble the real Sun where according to Doppler shift measurements a poleward circulation is indicated (LaBonte and Howard, 1982; R.K Ulrich, 1986, private communication), but where sunspots, moving with the gas at some depth (see e.g. Schüßler, these proceedings), seem to travel equatorwards (at least at low latitude; see e.g. Howard, these proceedings).

(iii) In the deep part of the convection zone differences between the polar and equatorial (radial) energy fluxes of order 15% develope. They are smoothed out by the latitudinal convective energy flux, as first suggested by Spruit (1977), down to surface values of typically $5 \cdot 10^{-4}$. The temperature difference between the polar and equatorial regions is of (relative) order 10^{-5} to 10^{-4} everywhere (Stix, 1981).

(iv) The surfaces of constant angular velocity are more or less radial, cf. Fig. 1. The deviation from the cylindrical shape is consistent with the sizable flux difference mentioned, although this difference is still smaller than predicted by Durney (1976).

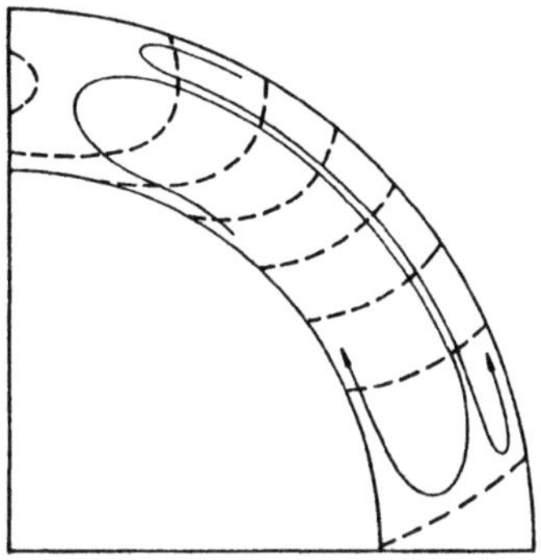

Figure 1. Angular velocity surfaces (dashed), and streamlines of meridional circulation for the AV-model with α=1.5

3.6. Observational Constraints

I now return to Rüdiger's expansion (10) of the vertical and horizontal non-diffusive fluxes. The terms V^1 and H^1 will be retained this time, but meridional circulation will be neglected again. Following Rüdiger, I assume a depth dependence of form $x^{-\lambda}$, where $x=r/r_\odot$, i.e.

$$V^1 = \tilde{V}^1 x^{-\lambda} , \qquad H^1 = \tilde{H}^1 x^{-\lambda} \tag{20}$$

(note that $\tilde{V}^1$, $\tilde{H}^1$ here are <u>surface</u> values). The free parameter λ should be chosen in such a way that the strong increase of the rotational interaction with convective eddies is simulated.

The solution of (5) can now be calculated essentially in the same way as in the earlier cases of Sects. 3.2 and 3.3. The expansion (12) of the angular velocity is however replaced by the equivalent expansion

$$\Omega = \Omega_P \left(1 + \Omega^1(r) \sin^2\theta + \dots \right) , \tag{21}$$

where $\Omega^1(r_\odot)$ is now the observed quantity which serves to calibrate the constants $\tilde{V}^1$ and $\tilde{H}^1$. In the case of a thick shell with strong density stratification Rüdiger (1982a; note the different notation) obtains

$$\Omega^1(r_\odot) = \tilde{V}^1 x_i^{-\lambda}/\lambda + (1 + x_i^{-\lambda})\tilde{H}^1/2s \quad . \tag{22}$$

We see that both the horizontal and the vertical momentum fluxes contribute to the equatorial acceleration. In order to calibrate both constants in a unique way, we need a second observational constraint. The horizontal correlation, $Q_{\theta\Phi}$, fulfills this purpose. Combining (9), (10), (20) and (21) we find

$$Q_{\theta\phi} = -\nu_T \Omega_p \sin^2\theta \cos\theta \, x_i^{-\lambda}(2s\tilde{V}^1/\lambda + \tilde{H}^1) \, . \tag{23}$$

Ward (1965), Belvedere et al. (1976), Schröter and Wöhl (1976) and Gilman and Howard (1984) give $Q_{\theta\Phi}$ as a function of latitude. Using $\Omega_p = 2\cdot 10^{-6}\,s^{-1}$ for the polar angular velocity, and $v_T = 10^9\,m^2/s$ for the turbulent diffusivity, we obtain from these observations the dimensionless "Ward-function".

$$W^1 \equiv x_i^{-\lambda}(2s\tilde{V}^1/\lambda + \tilde{H}^1) \approx 3 \quad . \tag{24}$$

Equations (22) and (24) can now be solved for $\tilde{V}^1$ and $\tilde{H}^1$. Let us take $s=1$, $x_i=0.7$, $\lambda=10$, and use $\Omega^1=0.62$ (derived from the result of Howard and Harvey, 1970). This yields

$$\begin{aligned} \tilde{V}^1 &= -\lambda\left[\Omega^1 + (1+x_i^{\lambda})W^1/2s\right] \approx -21.6 \quad , \\ \tilde{H}^1 &= 2s\Omega^1 + W^1 \approx 4.24 \quad . \end{aligned} \tag{25}$$

The fluxes are downward and equatorward. It would be interesting to see how these results were modified if meridional circulation was included, and if a solar envelope such as the one used in our AV-model was employed instead of Rüdiger's analytical calculation.

Of course, one could introduce more parameters, for example extending the expansion (10), or making similar expansions for the correlations which appear in the energy transport equation. Correspondingly more observational constraints are required. Rüdiger and Tuominen (these proceedings) present an example.

4. TURBULENT HEAT TRANSPORT

4.1. General Remarks

The convective energy flux (18) used in the mixing length theory is a special case of turbulent heat transport. A recent discussion of this problem is given by Rüdiger (1982b). In the framework of the second order correlation approximation the turbulent heat flux is

$$F_{ci} = \rho c_p \overline{u_i' T'} \tag{26}$$

As with the Reynolds stresses, the main problem is to relate the correlations $\overline{u_i' T'}$ to mean quantities. In a compressible medium a gradient of the specific entropy (or an excess of the temperature gradient over its adiabatic value) is essential for energy transport. A proper generalization of (18) therefore would be

$$F_{ci} = -\rho c_p \kappa_{ij} \left[\frac{\partial T}{\partial x_j} - \left(\frac{\partial T}{\partial x_j}\right)_S \right] \quad . \tag{27}$$

Among the large number of terms contributing to the conductivity tensor κ_{ij} in the presence of stratification and rotation (Rüdiger, 1982b) the simplest is proportional to δ_{ik}, but depends on latitude in general:

$$\kappa_{ij} = \delta_{ij} \left[\beta + \gamma P_2(\theta) \right] \quad . \tag{28}$$

This concept is related to an idea of Weiss (1965) used by Durney and Roxburgh (1970) in their model of the Sun's differential rotation: A scalar heat transport coefficient, depending on latitude, gives rise to meridional circulation which (in addition to its role in restoring the energy balance) transports angular momentum towards the equator and thus sustains the differential rotation. The concept was further developed by Belvedere and Paternò (1977), Belvedere et al. (1980), Schmidt (1982) and Moss and Vilhu (1983). It is also the basis of the latest calculations by Pidatella et al (1986), which I shall report now.

4.2. The HT-Model of the Sun's Convection Zone

The equations describing this model are formally the same as for the AV-model, Eqs. (17) to (19). Only the driving is different: The viscous force, $\underset{\sim}{R}$, contains no anisotropies; instead, a transport coefficient of form (28) is incorporated into the expression (18) for the convective energy flux. The zero order model, with $\underset{\sim}{u}=0$, is entirely identical. We simply set $\beta=\ell v_c/2$ and $\gamma=0$ in (28).

The perturbation enters in form of γ or, more precisely, as a latitude-dependent ratio α of mixing length to pressure scale height:

$$\alpha = \alpha_0 + \alpha_2(r) P_2(\theta) \quad . \tag{29}$$

Since α is the free parameter of mixing length theory, we think it is natural to use this parameter also in order to impose a perturbation (the coefficient γ in (28) is then $\gamma=\alpha_2 v_c H/2$). We assume that $\alpha_2(r)$ has its maximum at great depth in the convection zone. The increase with depth of the rotational effect is thus simulated.

As in the AV-model, all quantities are now expanded in Legendre polynomials including P_2, and the perturbed system is linearized. The driving inhomogeneity, $\hat{\alpha}_2$ (now denoting the maximum value of $\alpha_2(r)$), is then calibrated so that again $\omega_2(r_\odot)=-.189$, i.e. that the observed differential rotation is obtained. A difference to the AV-model is that the results now strongly depend on the type of boundary condition applied at the lower boundary, $r=r_i$, and also on the values chosen for the turbulent viscosity, ν_T, and for α_o, which determines the location of the lower boundary.

(i) For $\alpha_o=1.25$, $\nu_T=2\cdot10^8$ m^2/s, and a spherically symmetric pressure distribution at $r=r_i$ we find $\alpha_2=-0.2$, a pole-equator temperature difference $\Delta T=-0.5$ K and meridional surface motions $v_\theta=-15$ m/s. Flux differences between polar and equatorial regions are up to -22% at great depth, and $\sim -2\cdot10^{-4}$ at the surface. The negative α_2 means that the mixing length is larger in the equatorial part, and that the convective energy transport is more effective there. Again the angular velocity surfaces substantially deviate from cylinders, see Fig. 2.

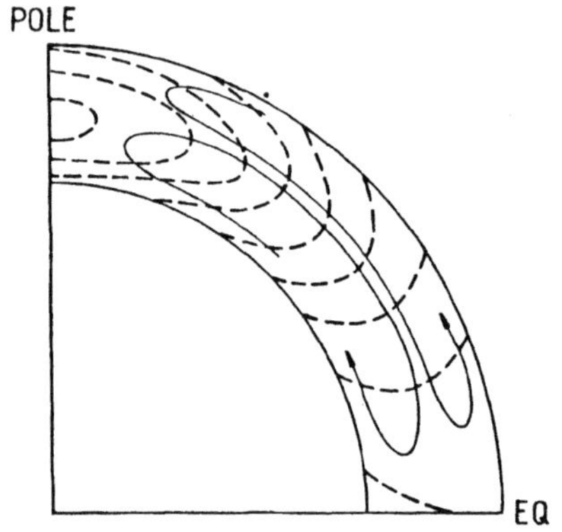

Figure 2. Angular velocity surfaces (dashed), and streamlines of meridional circulation for the HT-model with $\alpha=1.5$.

(ii) With the same α_o and ν_T, but a spherically symmetric energy flux at $r=r_i$, we obtain $\alpha_2=0.6$, $\Delta T=6$ K, and $v_\theta=36$ m/s. These perturbations are uncomfortably large. Nevertheless, the boundary condition used in this model must be considered as the realistic case, as was repeatedly pointed out by H.C. Spruit (private communication): The energy flow must be continuous across the lower boundary. A latitude dependence would therefore reach into the core where the transport is merely radiative, i.e. down the temperature gradient. Pole-equator temperature differences must then be exspected, and these would <u>not</u> be screened by the convection zone (Stix, 1981). A numerical experiment of Pidatella et al. (1986) confirms this argument: The model with the spherically symmetric pressure at $r=r_i$ converts itself into the model with spherically symmetric energy flux if the boundary is displaced only a short distance down into the radiative core.

It seems that the perturbations are larger (as compared to case (i)) for the following reason: The driving heat flux difference is suppressed at $r=r_i$, but must finally maintain the same differential rotation. It therefore works at lesser depth, with concomitant larger ΔT and v_θ at the surface. The sense of the circulation is consistent with this argument: v_θ is positive even if there are 2 cells; i.e. the upper cell is the driving cell in this case.

(iii) When the turbulent viscosity ν_T is increased, the surface values of ΔT and v_θ also increase. This makes the HT-model less attractive than the AV-model. Although in mixing length theory the heat diffusivity is not a constant, a value of order 10^9 m^2/s may be more representative than the $2 \cdot 10^8$ m^2/s which we have mostly used for ν_T. In other words, we have a Prandtl number of 0.2 typically, while Prandtl numbers closer to 1 are usually adopted for turbulence. Eschrich and Rüdiger (1983) estimate that the solar convection zone has a Prandtl number smaller than 0.3, but this conclusion rests again on the second order correlation approximation.

5. SUMMARY

"Anisotropic viscosity" is the simplest form of driving differential rotation by means of Reynolds stresses. However, only the radial gradient of Ω is driven directly; for θ-dependent angular velocity meridional circulation (driven by the same anisotropy) is necessary. A difficulty is that the required anisotropy parameter is s>1, while the theory of stratified rotating turbulent media rather suggests s<1.

The Reynolds stresses $\overline{u'_r u'_\phi}$ and $\overline{u'_\theta u'_\phi}$ directly drive a depth- and latitude-dependent angular velocity if their own θ-dependence is taken into account. In this case meridional circulation may be a byproduct only.

The models driven by Reynolds stresses seem attractive because even a very small perturbation is sufficient to sustain the observed differential rotation. The pole-equator temperature difference, and the meridional velocity at the surface, can therefore be small.

The latitude dependence of the convective energy transport coefficient is the simplest effect of turbulent heat transport. It drives a depth- and latitude-dependent angular velocity via meridional circulation, which itself is kept going by the balance of energy flux.

The models driven by latitude-dependent heat transport tend to produce too large pole-equator temperature differences and observable meridional velocities, in particular when the turbulent viscosity is of order 10^9 m^2/s or larger. The perturbation required for the driving is sizable.

All models discussed in this contribution have considerable - typically 20% - differences between the radial energy fluxes in the polar and equatorial regions. The difference is smoothed out by the θ-component of the convective flux, so that it is unobservable at the surface. Temperature differences are of order 10^{-5} - 10^{-3} (relative) at all depths. Consistent with the latitude dependence of the energy flux is a large deviation of the constant Ω surfaces from cylinders. These surfaces are rather more radial than parallel to the axis of rotation.

Acknowledgements. I thank the staff of the Sacramento Peak Observatory, and in particular Dr. B.R. Durney, for the stimulating discussions and the hospitality which I enjoyed during the week prior to this meeting. I am grateful to the Deutsche Forschungsgemeinschaft for travel support.

REFERENCES

Belvedere, G., Paternò, L.: 1977, Solar Phys. 54, 289

Belvedere, G., Paternò, L., Stix, M.: 1980, Geophys. Astrophys. Fluid Dyn. 14, 209

Belvedere, G., Godoli, G., Motta, S., Paternò, L., Zappalà, R.A.: 1976, Solar Phys. 46, 23

Biermann, L.: 1951, Z. Astrophys. 28, 304

Durney, B.R.: 1976, Astrophys. J. 204, 589

Durney, B.R.: 1983, Astrophys. J. 269, 671

Durney, B.R.: 1985, Astrophys. J. 297, 787

Durney, B.R., Roxburgh, I.W.: 1971, Solar Phys. 16, 3

Durney, B.R., Spruit, H.C.: 1979, Astrophys. J. 234, 1067

Eschrich, K.-O., Rüdiger, G.:1983, Astron. Nachr. 304, 171

Gilman, P.A., Howard, R.: 1984, Solar Phys. 93, 171

Gilman, P.A., Miller, J.: 1986, Astrophys. J. Suppl. 61, 585

Glatzmaier, G.A.: 1985, Astrophys. J. 291, 300

Gough, D.O.: 1978, Workshop Solar Rotation, G. Belvedere and L. Paternò (Eds.), Catania, p. 337

Howard, R., Harvey, J.: 1970, Solar Phys. 12, 23

Howard, R., LaBonte, B.J.: 1980, Astrophys. J. 239, L33

Kippenhahn, R.: 1963, Astrophys. J. 137, 664

Kippenhahn, R.: 1964, IV Cent. Nasc. Galileo Galilei, Atti Conv., Tomo 4, G. Barbèra (Ed.), Firenze, p. 291

Köhler, H.: 1970, Solar Phys. 13, 3

LaBonte, B.J., Howard, R.: 1982, Solar Phys. 80, 361

Moss, D., Vilhu, O.: 1983, Astron. Astrophys. 119, 47

Pidatella, R.M., Stix, M., Belvedere, G., Paternò, L.: 1986, Astron. Astrophys. 156, 22

Rüdiger, G.: 1977, Astron. Nachr. 298, 245

Rüdiger, G.: 1980, Geophys. Astrophys. Fluid Dyn. 16, 239

Rüdiger, G.: 1982a, Geophys. Astrophys. Fluid Dyn. 21, 1

Rüdiger, G.: 1982b, Astron. Nachr. 303, 293

Rüdiger, G.: 1983a, in Stellar and Planetary Magnetism, A.M. Soward (Ed.), Gordon and Breach, p. 133

Rüdiger, G.: 1983b, Geophys. Astrophys. Fluid Dyn. 25, 213

Schmidt, W.: 1982, Geophys. Astrophys. Fluid Dyn. 21, 27

Schmidt, W., Stix, M.: 1983, Astron. Astrophys. 118, 1

Schröter, E.H., Wöhl, H.: 1976, Solar Phys. 49, 19

Spruit, H.C.: 1977, Astron. Astrophys. 55, 151

Stix, M.: 1981, Astron. Astrophys. 93, 339

Vitense, E.: 1953, Z. Astrophys. 32, 135

Ward, F.: 1965, Astrophys. J. 141, 534

Wasiutyński, J.: 1946, Astrophys. Norvegica, Vol. 4

Weiss, N.O.: 1965, Observatory 85, 37

BEYOND THE MIXING-LENGTH THEORY : A TURBULENT DIFFUSIVITY APPROACH

W. Unno
Research Institute for Science and Technology, Kinki University
Kowakae, Higashi-Osaka-shi, Osaka 577, Japan

ABSTRACT. The large eddy simulation of the turbulent convection can be reduced to different types of treatment depending on the window size which filters out smaller scale eddies. In the largest eddy limit, it is reduced in a sense to Xiong's nonlocal mixing-length theory which is suitable for the construction of stellar convection zones. For large wave numbers the spectral theory is valid and gives correct estimates for the eddy diffusivities. The single mode simulation subject to the effective Reynolds number of about 10 is suggested from the nonlinear convection theory to simulate correct dynamics of large scale flow. However, since eddies of different sizes are segregated each other to different depths in the convection zone by means of high stratification of eddy diffusivities, the single mode simulation is also appropriate to study hydrodynamical properties in stellar convection zones.

1. INTRODUCTION

To avoid 10^{10} mesh points to simulate stellar convection, we need to adopt the Large Eddy Simulation (LES) in which a window size or the mixing length is introduced to express the diffusivity effect of smaller eddies. Although the diffusivity by smaller eddies is intuitively well understood, its consistency with the LES must be proven. This can be done using the Kolmogoroff scaling for smaller eddies. There is, however, the nonlocal effect of boundary which makes the effective Reynolds number of the turbulent convection smaller for the largest eddy than for smaller eddies. The nonlinear convection provides the clue for this nonlocality effect. Therefore, different approach are suitable for different purposes. Xiong's nonlocal mixing-length theory and the single mode approximation will be discussed briefly.

2. THE KOLMOGOROFF SCALING

The spectral theory of turbulent convection (Ledoux et al. 1961, Nakano et al. 1979) shows that the Kolmogoroff spectrum attains at high wave numbers. Therefore, the following Kolmogoroff scaling should be appropriate to describe small fluctuations separately from large scale motions,

$$x = \epsilon x' \quad \text{and} \quad t = \epsilon^{2/3} t'. \tag{1}$$

B. R. Durney and S. Sofia (eds.), The Internal Solar Angular Velocity, 343–346.

The velocity field u is expressed by the sum of large and small scale fields as follows,

$$u = v + \epsilon^{1/3} v'. \tag{2}$$

Similarly, we assume

$$\rho = \bar{\rho} + \epsilon^{\alpha}\rho', \quad T = \bar{T} + \epsilon^{\alpha}T', \quad (\alpha \geq 2/3), \tag{3}$$

anticipating sharp decrease of the density and temperature spectra towards large wave numbers. The Reynolds number is expressed as $\epsilon_{\min}^{4/3}$, and the viscous force can be neglected in the inertial subrange ($\epsilon >> \epsilon_{\min}$). Variables with prime and their derivatives are taken to be of the same order of magnitude as the corresponding large scale quantities.

The effect of small scale motions on large scale motions enters as $\epsilon^{\alpha+1/3}\partial<\rho' v_j'>/\partial x_j$, $\epsilon^{2/3}\partial<\bar{\rho} v_i' v_j'>/\partial x_j$ and $\epsilon^{\alpha+1/3}\partial<\bar{\rho} W' v_j'>/\partial x_j$ (W : specific enthalpy) in equations of conservation of mass, momentum and energy, respectively. To evaluate these terms and to express them in terms of large scale variables, the conservation equations for small scale motions should be solved. To the lowest order in ϵ, the latter equations appear to be (Unno et al. 1984)

$$\frac{\partial v_j'}{\partial x_j'} = 0, \quad v_j \frac{\partial v_i'}{\partial x_j'} = 0, \quad \text{and} \quad v_j \frac{\partial S'}{\partial x_j'} = 0. \tag{4}$$

The specific momentum v' and entropy S' of small scale fluctuations are, therefore, conserved along the large scale motion, and this conclusion apparently rends support to the assumption of the mixing-length theory. The evaluation of the mixing length or more directly the turbulent Reynolds stress and dissipation then requires the treatment of equations of higher orders in ϵ.

On the other hand, the eddy viscosity and eddy thermal conductivity are most simply formulated in the mixing-length theory (Wasiutynski 1946; Unno 1969),

$$<\epsilon^{2/3}\bar{\rho} v_i' v_j'> = \frac{1}{3}\epsilon^{2/3}\bar{\rho} v^2 \delta_{ij} - \bar{\rho}\nu_e \left[\frac{\partial v_i}{\partial x_j} + \frac{\partial v_j}{\partial x_i} - \frac{2}{3}\delta_{ij}\frac{\partial v_k}{\partial x_k}\right], \tag{5}$$

and

$$<\epsilon^{\alpha+1/3}\bar{\rho} W' v_j'> = - \bar{\rho}\kappa_e \bar{T}\partial\bar{S}/\partial x_j, \tag{6}$$

where the eddy viscosity ν_e and the eddy conductivity κ_e are represented by the Austausch, $\epsilon^{4/3}(vl)$, l being the representative cell size. More precise evaluation of κ_e and ν_e is due to the spectral theory. Nakano et al. (1979) obtained

$$\kappa_e = 2.5\nu_e, \quad \nu_e = R_e^{-1}<w^2>^{1/2}(\pi/k_z), \tag{7}$$

where $<w^2>$ denotes $<\epsilon^{2/3}v'^2>$ at the vertical wavenumber k_z and the effective Reynolds number R_e is about 20 depending little on wave number k_z .

A 3-D time-dependent large eddy simulation (LES) of the solar granulation has been forwarded by Nordlund (1978). In the present symposium, Glatzmaier (1986) has presented comprehensive study on the solar internal rotation by his LES calculations. There are, however, some simplifications of LES with which we can study various special problems efficiently. We will discuss these simplifications in the following sections.

3. THE XIONG THEORY

There are several theories that can be used to construct stellar convection zone models in more sound hydrodynamical basis than the usual local mixing-length theory (e.g., Spiegel 1963; Unno 1969; Ulrich 1970). Among them, Xiong's (1979) nonlocal mixing-length theory seems to be optimum in regard to the accuracy and the simplicity. As a matter of fact, Xiong's theory may well be understood essentially as a special case of the LES treatment for $\epsilon=1$, although working equations are written in terms of velocity and temperature correlations. Nonlinear characteristics in Xiong's theory are the divergence of the gradient diffusion flux of the auto- and cross-correlations of velocity and temperature fluctuations. Without these fluxes, the theory reduces to the usual mixing-length theory. In other words, these gradient diffusion fluxes represent nonlocal effects, while the dissipation through cascade (effects of small scale eddies) represents the local effect.

Xiong's (1975) theory was applied to the solar model construction (Unno et al. 1985). The agreement with the observed model in the top of the convection zone is excellent, and the thickness of the convection zone agrees with that given by the helioseismology. The biggest advantage of Xiong's theory over the local mixing-length theory is the much less sensitivity of the model to the choice of the mixing length. It appears to be the local treatment that made the model sensitive to the mixing length. In Unno et al. (1985), the mixing length l_e was taken to be the larger one of the pressure scale height and the distance from the convection boundary.

By virtue of the nonlocal nature, Xiong's (1979) theory can be applied also to the penetrative convection and mixing. The semi-convection of slightly evolved early type stars can be calculated unambiguously (Xiong 1981). The solar Li, Be problem can be studied as well (Kondo et al. 1987).

4. THE SINGLE MODE THEORY

Van der Borght (1971) was the first who formulated the cellular convection theory of compressible fluid. A similar problem for the Boussinesq fluid, however, can be solved analytically, if the stress-free fixed-temperature boundary conditions are adopted (Weiss 1981a, b; Rudraiah 1981; Proctor and Weiss 1982; Rudraiah et al. 1985a). Then, Rudraiah et al. (1985b) extended the use of the nonlinear convection calculation to the turbulent convection by use of the turbulent eddy diffusivities with which the effective Reynolds number is about 20. It turns out that the statistically steady turbulent convection is achieved with the effective Rayleigh number of four times the critical and with the Nusselt number of about 2. This means that the main convective motion becomes stationary if the eddy viscosity and conductivity are increased by a factor of 2 from the values estimated by the spectral theory. Then, the effective Reynolds number for the largest eddy becomes about 10. A simple interpretation for the reduction of the effective Reynolds number may be that the boundary conditions restrict the number of the largest eddies so that one parent eddy feeds twice more smaller eddies compared with the homogeneous turbulence.

On the other hand, Antia, Chitre and Pandey (1981) and Narasimha and Antia (1982) showed that eddies of different horizontal sizes are practically segregated to different depth in the convection zone by means of stratification of density and eddy diffusivities. Therefore, the single eddy simulation with the appropriate eddy diffusivities at each level can give an adequate representation of stellar convection zones. The method to construct stellar convection zones will then be as follows. Starting with the local mixing-length theory, we obtain the first approximation of the model and the eddy diffusivities. Next, choosing various hoizontal eddy sizes, we calculate the quasi-linear stability problem with twice larger eddy diffusivities. The velocity and temperature amplitudes of each eddy are fixed so that the resulting energy flux at the depth of amplitude maximum coincides with the given energy flux (Narasimha and Antia 1982). Then, the revised eddy diffusivities are obtained. The improved model should be obtained by adjusting the quasi-linear stability of each eddy to be neutral, if the energy cascade from larger eddies can be neglected by the eddy segregation.

Rudraiah et al. (1985b) applied the eddy segregation model also to the solar magnetohydrodynamic convection. Intensification of magnetic tubes in successive higher layers and the sunspot darkness caused by the large Chandrasekhar number at shallow layers are qualitatively discussed. Another qualitative prediction can be made for the overstability of a large eddy at the boundary of the convection zone (Kondo and Unno 1983). The eddy viscosity and conductivity drop abruptly at the edge of the convection zone, which is a situation favorable for a large eddy penetrating deeper in the stable layer. This type of overstability, however, requires deep penetration which is possible only if the stable layer is only mildly stable. Therefore, the possibility of solar 160 min oscillation by means of the overstable large eddy penetrating below the convection zone is not likely but needs further careful study.

In conclusion, the single mode theory or the largest eddy simulation is a useful approximation to treat the dynamics of convection. For chaotic variation of turbulent convection, we must go back to the LES with multiple eddy sizes coupling nonlinearly with one another.

REFERENCES

Antia, H. M., Chitre, S. M., and Pandey, S. K. 1981, *Solar Phys.*, **70**, 67.

Glatzmaier, G. A. 1986, in these proceedings.

Kondo, M.-A., and Unno, W. 1983, *Geophys. Astrophys. Fluid Dyn.*, **27**, 229.

Kondo, M.-A., Unno, W., and Xiong D.-R. 1987, in preparation.

Ledoux, P., Schwarzschild, M., and Spiegel, E. A. 1961, *Astrophys. J.*, **133**, 184.

Nakano, T., Fukushima, T., Unno, W., and Kondo, M. 1979, *Publ. Astron. Soc. Japan*, **31**, 713.

Narasimha, D., and Antia, H. M. 1982, *Astrophys. J.*, **262**, 358.

Nordlund, A. 1978, in *Astronom. Papers dedicated to B. Stromgren*, eds. A. Reiz and T. Andersen, Copenhagen Univ. Obs., Copenhagen.

Proctor, M. R. E., and Weiss, N.O. 1982, *Rep. Prog. Phys.* , **45**, 1317.

Rudraiah, N. 1981, *Publ. Astron. Soc. Japan*, **33**, 721.

Rudraiah, N., Kumudini, V., and Unno, W. 1985a, *Publ. Astron. Soc. Japan*, **37**, 183; 1985b, *ibid.*. **37**, 207.

Spiegel, E. A. 1963, *Astrophys. J.*, **138**, 216.

Ulrich, R. K. 1970, *Astrophys. Space Sci.*, **7**, 183.

Unno, W. 1969, *Publ. Astron. Soc. Japan*, **21**, 240.

Unno, W., Kondo, M.-A., and Chen, B. 1984, *Proc. Mini-Symp. on Astrophys. and Geophys. Convection*, eds. W. Unno and T. Asai, Dept. of Astronomy, Univ. of Tokyo, p.1.

Unno, W., Kondo, M.-A., and Xiong, D.-R. 1985, *Publ. Astron. Soc. Japan*, **37**, 235.

van der Borght, R. 1971, *Publ. Astron. Soc. Japan*, **23**, 539.

Wasiutynski, J. 1946, *Astrophys. Norw.*, **4**, 1.

Weiss, N. O. 1981a, *J. Fluid Mech.*, **108**, 247; 1981b, *ibid.*, **108**, 273.

Xiong, D.-R. 1979, *Acta Astron. Sinica*, **20**, 238. [= *Chinese Astron.*, **4**, 234, 1980]

Xiong, D.-R. 1981, *Sci. Sinica*, **24**, 1406.

MIXING-LENGTH, SHEARS, AND DIFFERENTIAL ROTATION

K. L. Chan
Applied Research Corporation

S. Sofia
Yale University

H. G. Mayr
NASA/ Goddard Space Flight Center

ABSTRACT. This paper discusses three related subjects whose conclusions can be summarized as follows. (i) Numerical experiments indicate that the mixing length is scaled by the pressure scale height, not by the density scale height. (ii) Numerical experiments indicate that the effective viscosity of convective turbulence on large scale shears can be roughly approximated by 1/3 the product of the root-mean-square (rms) vertical velocity with the pressure scale height. (iii) The differential rotation of the solar convection zone may be considered as an unstable axisymmetric mode which produces slower rotation at the surface.

1. INTRODUCTION

The mixing-length, the shears, and the differential rotation represent different aspects of convection in the solar envelope. They are concepts used to describe convective phenomena in different (ascending) length scales. At the present stage of theoretical development, no single analytical or computational effort can include all these aspects simultaneously in a unified model. In this paper, we satisfy our curiosity by looking at each of these subjects in a piecemeal fashion. References to previous works will be brief; the readers are referred to the other papers in this proceeding for a comprehensive overview of the problems and developments in related areas.

B. R. Durney and S. Sofia (eds.), The Internal Solar Angular Velocity, 347–360.

2. SCALING OF THE MIXING-LENGTH

According to conventional wisdom in stellar theory, the mixing-length represents a basic scale of convection which describes the local sizes of the convective "bubbles". Its most important application is at the top region of the convection zone through which the adiabat, the structure, and the depth of the whole convection zone are determined. In deeper regions, it is sometimes used to estimate the strength of the "background" convective turbulence for computation of motions of larger scales.

A long-standing question in the mixing-length theory of convection is whether the mixing-length is scaled by the pressure scale height or the density scale height. Through numerical experiments, we now know that the mixing length is scaled by the pressure scale height.

Our numerical experiments are performed by solving the fully compressible Navier Stokes equations for an ideal gas in three dimensions (Chan and Sofia, 1986). The domain of computation is a rectangular box with an aspect ratio of 1.5 (width / depth). It has periodic side boundaries; the top and bottom are both impenetrable and stress free. A quantitative illustration of the scaling effects of the pressure scale height is best given by looking at the longitudinal correlation of the vertical velocity:

$$\mathcal{A} = \langle V_1 V_2 \rangle / (\mathrm{rms}\langle V_1 \rangle \, \mathrm{rms}\langle V_2 \rangle) \quad (1)$$

Here, V_1 and V_2 are the vertical velocities at two different horizontal levels labeled by 1 and 2. The angle brackets denote the combined time and horizontal average. The variables discussed in this and the next section use units which make the original pressure, density, temperature at the top, and the depth all equal to 1. The correlation length over which $\mathcal{A}$ drops substantially (say, from 1 to 0.5) characterizes the local size of the energy containing eddies of the turbulence. In terms of the mixing length interpretation, this length is also the vertical distance over which a convective "bubble" can preserve its identity.

A variety of cases have been computed for comparison. The cases have different ratios of specific heats γ(= 5./3. or 20./17.) which make the ratio of the scale heights different (density scale height / pressure scale height ≃ γ). The input fluxes from the bottom boundaries (= 0.125 or 0.333) are different so that the effects of different strengths of convection can be studied. The numerical grids (20X20X37 or 20X20X46) are different so that the invariance of the results can be tested. Fig. 1a plots $\mathcal{A}$ versus the logarithmic pressure for the different cases. The similarity of the distributions at different depths (the plusses) shows that this correlation function is essentially a function of only the difference in the logarithmic pressures between levels 1 and 2. Overlaying the distributions of all cases (different symbols) together shows that they agree very well. The half width at half maximum of $\mathcal{A}$ is about 1. In Fig. 1b, the same cases are plotted against the logarithmic density. The cases with different values of γ show different widths. Therefore, the correlation length is not an invariant if measured in terms of the density scale height. It is scaled by the pressure scale height, not by the density scale height. Furthermore, the scaling seems to be independent of the magnitude of the convective flux.

3. MIXING-LENGTH EFFECTS ON SHEARS

Shifting our attention to larger scales, we ask the following question: "How would the convective turbulence affect a large scale mean shear flow?" This question is relevant because many studies of differential rotation in a convection zone assume that the action of the convective turbulence on a large scale mode of motion can be treated as an eddy viscosity. The value of this eddy viscosity is usually taken to be the product of the mixing length with the rms velocity of the convective turbulence, computed by the mixing length theory.

We study the effective viscosity of the convective turbulence on a large scale shear by observing the decay of shears in a numerical convection model ($\gamma = 5/3$). First, we evolve the model to thermal and statistical equilibrium. At one fixed instance, we suddenly impose a mean shear in the velocity field; the shear velocity is in the horizontal y direction (x is another horizontal direction and z is the vertical direction). Two kinds of shears have been studied. The first is a vertical shear with linear distribution of velocity. The second is a horizontal shear with one sinusoidal period across the x direction. We have so far computed two cases for each kind of shear, with perturbations imposed at two widely separated instances. The peak to peak amplitudes of all the shear velocities are 0.2.

If the effect of the convective turbulence on the shear is lumped into an effective viscosity, the decay of the mean shear velocity can be described by the equation:

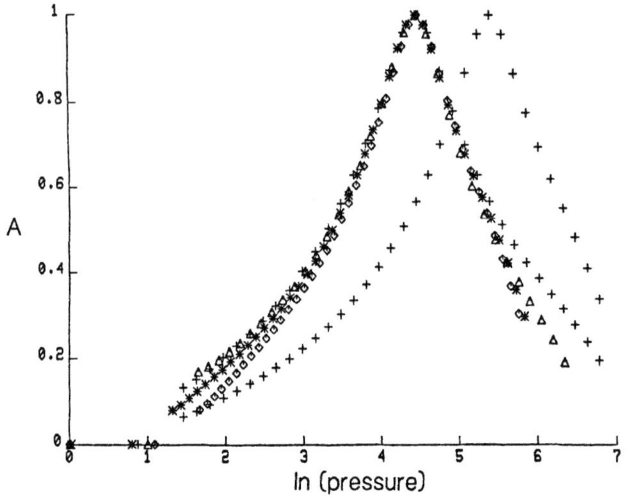

Fig. 1a Correlation vs. ln(pressure)

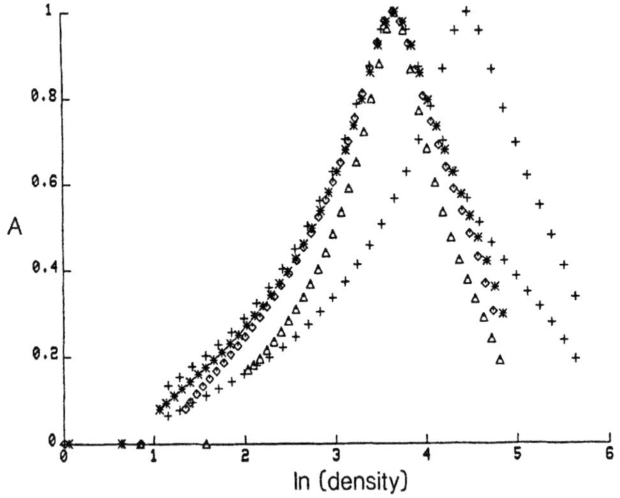

Fig. 1b Correlation vs. ln(density)

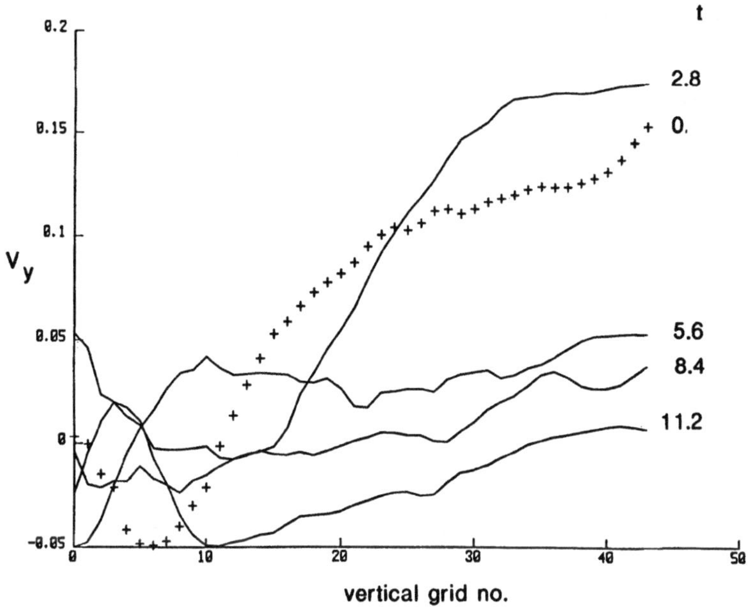

Fig. 2 Time evolution of the shear velocity

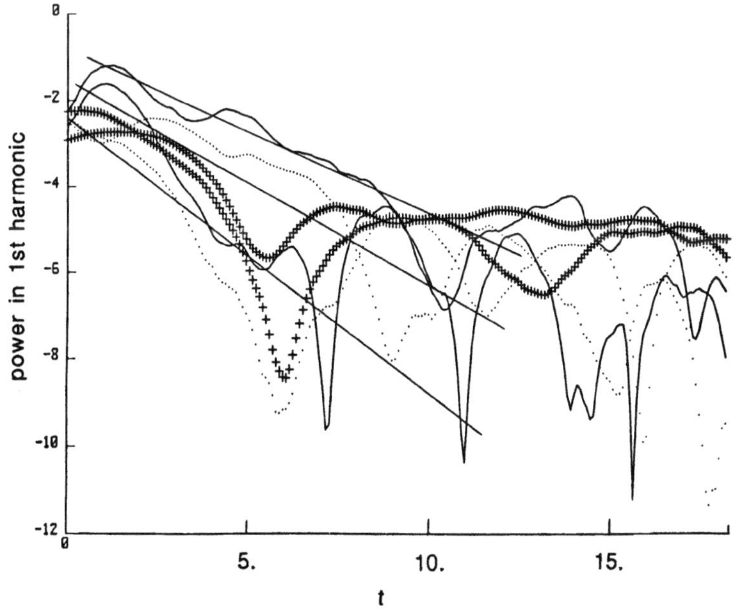

Fig. 3 Time evolution of the powers in the 1st harmonic modes

$$\partial_t \overline{V_y} = \nu_{eff}\, \partial_x^2 \overline{V_y} \tag{2}$$

If we can determine the e-folding time τ for the decay of $\overline{V_y}$ in a length scale λ, we can estimate the effective viscosity ν_{eff} by

$$\nu_{eff} = \frac{\lambda^2}{(2\pi)^2\,\tau} \tag{3}$$

An example of the decay of a vertical shear is illustrated in Fig. 2.

This figure shows the vertical profiles of the mean shear velocity $\overline{V_y}$ at several instances (labeled at the right end of the curves). Here, the mean (overbar) is taken on horizontal planes. The dotted curve shows the profile at the instance when the shear was first imposed. The solid curves show profiles at successively later instances. Between t=0 and t=2.8, there is a surge of $\overline{V_y}$ at the top region. Apparently, momentum has been transferred from lower levels to the top region in this period. As the density at the top is lower than that at the lower levels, the mean velocity shoots up. This initial transient may be caused by the adjustment of the pressure field to the perturbation. After this rapid event, the shear decays. The time scale for decay by a factor e is between 5.6 and 8.4. The evolution of the rms velocity associated with the shear also shows a similar decay in about 7 to 8 units of time. If we take the total depth of the linear shear to be half a wave length, the e-folding time of decay to be 7.0, and substitute these values into Eq. (3), we obtain a value of 0.014 for the effective viscosity.

Next, let us look at the cases with horizontal sinusoidal shears. Because of the horizontal periodic boundary conditions, Fourier transforms can be performed across the shears to study the evolution of the power spectra of $\overline{V_y}$. Fig. 3 shows the evolution of the first harmonic modes (in which the shears were originally introduced) at three different depths - the top region, the middle region, and the bottom region, for the two cases. The solid curves are for the top region, the dotted curves are for the middle region, and the plusses are for the bottom region. Note that the power is plotted in logarithmic scale. These curves are very noisy. It is hard to differentiate the rates of decay in different regions. May be the rate at the bottom is a little higher. On the other hand, one can have some confidence in drawing several lines to delimit the general rate of decay. The top straight line can probably represent a lower limit to the rate of decay; it gives a value of 0.011 for the effective viscosity. The bottom straight line is probably close to the upper limit; it gives a value of 0.018. The middle line seems to be a better estimate between the two extremes, and it gives a value of 0.014. This value is about the same as the one for the vertical shear. The value of the effective viscosity is much larger than (approximately 50 times) the sub-grid-scale turbulence viscosity used for the stability of the numerical computation. This is an assurance that the shears are mainly dissipated by the convective turbulence.

How does the value of this numerically evaluated viscosity compare with that from the mixing length theory? From the numerical model, the product of the rms vertical velocity (V_z) with the pressure scale height (H_p) has a range between 0.03 and 0.05 (top and bottom respectively). It is about 3 times larger than the effective viscosity for the decay of shears. Therefore, we can write the following estimate:

$$\nu_{eff} \simeq (1/3)\ \mathrm{rms}(V_z)\, H_p \tag{4}$$

This estimate is very rough, but from the range between the upper and lower limits, it seems that it should be accurate to within a factor of 2. One other point to note is that the density varies by a factor of 6 between the bottom and top regions. The effective kinematic viscosity does not vary by that much. If not totally independent, it is at least not a sensitive function of the density.

For turbulent convection in a Boussinesq fluid, Nakano et. al. (1979) have estimated that the effective kinematic viscosity is about 1/6 times the product of the rms velocity with their "velocity mixing-length". Considering the large differences between their approach and ours, the two estimates for ν_{eff} are not too far off.

4. AXISYMMETRIC MODES OF CONVECTION IN A ROTATING SHELL

Finally, we discuss a linear axisymmetric model for the differential rotation of the solar convection zone. Although highly simplified, this heuristic model seems able to make some useful qualitative predictions. It is a generalization of the Jovian circulation model proposed by Mayr, Harris, and Chan (1984). Much more information about the fundamentals of this model can be found in their paper.

Recent numerical models of the solar differential rotation suffer from producing a reversed butterfly diagram in the magnetic activity cycle (Glatzmaier 1986). This is mainly caused by the prediction that the mean rotational rate at the solar surface is larger than that at the interior. The simple model described below can avoid this problem by creating a sub-rotating solar convection zone [mean rate of rotation decreases outward; see Schatten, Mayr, Levine (1981) for a qualitative discussion] with the correct differential rotation.

Let Ω be the rotational rate of the basic reference state, d be the depth of the convection zone, r be the radius, ν be the kinematic eddy viscosity, and κ be the eddy diffusivity. The assumptions for the model are as follows:

(i) incompressible fluid with Boussinesq approximation;
(ii) thin shell approximation ($d \ll r$);
(iii) hydrostatic approximation (large horizontal length scale);
(iv) linear perturbation (mode velocities $\ll \Omega r$);
(v) effects of convective turbulence are represented as eddy viscosity/diffusivity;
(vi) ν, κ, and the Prandtl number Pr = constants; the "mean" density, temperature T, pressure p, ... etc. are taken to be constants with values chosen to be those at the middle of the solar convection zone (estimated by the mixing-length theory).

The linearized equations describing a slow, axisymmetric circulation in a rotating shell can then be written as:

$$\kappa \left(\partial_r^2 + \frac{1}{r^2 \sin u} \partial_\theta \sin\theta \, \partial_\theta \right) T' + \frac{\mu g}{R} W \delta\nabla = - \frac{q}{C_p} \tag{5}$$

$$\partial_r W + \frac{1}{r \sin\theta} \partial_\theta (\sin\theta \, V) = 0 \tag{6}$$

$$\partial_r (\ln p)' = \frac{\mu g}{R T^2} T' \quad (7)$$

$$\nu \left(\partial_r^2 + \frac{1}{r^2 \sin\theta} \partial_\theta \sin\theta \, \partial_\theta \right) U = 2 \, \Omega \sin\theta \, V \quad (8)$$

$$\nu \left(\partial_r^2 + \frac{1}{r^2 \sin\theta} \partial_\theta \sin\theta \, \partial_\theta \right) V = - 2 \, \Omega \sin\theta \, U + \frac{R T}{\mu r} \partial_\theta (\ln p)' \quad (9)$$

Here, the superscript ' denotes perturbation of a certain quantity, C_p is the specific heat under constant pressure, μ is the mean molecular weight, R is the gas constant, g is the gravitational acceleration, $\delta\nabla = (\partial \ln T / \partial \ln p) - (\partial \ln T / \partial \ln p)_{adiabatic}$ is the superadiabaticity, U is the zonal velocity, V is the meridional velocity, W is the radial velocity, and q is a heat source which is supposed to drive the linear system. This source may be created by a latitudinally dependent heat transfer or some other perturbation effects on the convective turbulence due to the rotation, but it really does not matter here because we are not interested in the passive response of the convection zone. What we are looking for is a resonance response which characterizes the natural preference of the system.

It turns out that in a convectively unstable atmosphere, resonances exist which depend on the superadiabaticity $\delta\nabla$. We shall show that the resonance mode with the largest horizontal scale can be used to match the observations. The search for this resonance is performed by varying $\delta\nabla$, starting from a large value (for example, 1.0) for inefficient convection, and gradually decreasing its value. This is allowed because we need not put all our trust in the value of the superadiabatic gradient predicted by the mixing length theory. Particularly, in the deeper layer of the convection zone, the convective velocity is small and the effects of rotation become very significant (Rossby number < 1); it is likely that the superadiabatic gradient is severely modified by rotation.

There is a useful constraint in varying the parameters. The product of the diffusivity generated by the convective turbulence with the superadiabaticity must be equal to the solar flux. So, once $\delta\nabla$ is given, κ and ν are fixed (Pr now assumed to be 1) and can be used in the above equations.

To solve Eqs. (5) - (9), the perturbed variables are first expanded in series of spherical harmonics:

$$q = \sin(\pi\zeta) \sum_l (-1)^{l/2} q_l \mathcal{P}_l \tag{10}$$

$$T' = \sin(\pi\zeta) \sum_l (-1)^{l/2} T'_l \mathcal{P}_l \tag{11}$$

$$(\ln p)' = \cos(\pi\zeta) \sum_l (-1)^{l/2+1} (\ln p)'_l \mathcal{P}_l \tag{12}$$

$$W = \sin(\pi\zeta) \sum_l (-1)^{l/2} W_l \mathcal{P}_l \tag{13}$$

$$U = \cos(\pi\zeta) \sum_l (-1)^{l/2} U_{l-1} \mathcal{C}_{l-1} \tag{14}$$

$$V = \cos(\pi\zeta) \sum_l (-1)^{l/2+1} V_l \mathcal{C}_l \tag{15}$$

where $l = 2, 4, \ldots\ldots,$

$\zeta = (r - r_b) / d$

$\mathcal{P}_l = [(2l+1)/(4\pi)]^{1/2} \quad P_l$

$\mathcal{C}_l = [(2l+1)/(4\pi\, l\,(l+1))]^{1/2} \quad P_l^1$

P_l and P_l^1 are Legendre and associated polynomials, and r_b is the radius at the bottom of the convection zone. In the expansion for U, the first term U_1 is called the shell component [$\mathcal{C}_1 \sim \sin\theta$], it represents a horizontally uniform shift in the rotational rate relative to the reference rate Ω. The second and higher order terms in the expansion describe horizontal differential rotation. Successive terms in the series contain higher and higher powers of $\cos\theta$. By substituting these expansions in Eqs. (5) - (9), we obtain a very large set of linear equations. For each l value, the conservation equations form a block:

$$\kappa \left(\frac{l(l+1)}{r^2} + \frac{\pi^2}{d^2} \right) T'_l - \frac{\mu g}{R} \delta\nabla W_l) = \frac{q_l}{C_p} \tag{16}$$

$$W_l = -\frac{d}{\pi r} [\, l(l+1)\,]^{1/2} V_l \tag{17}$$

$$(\ln p)'_l = \frac{\mu d g}{\pi R T^2} T'_l \tag{18}$$

$$\nu \left(\frac{l(l-1)}{r^2} + \frac{\pi^2}{d^2} \right) U_{l-1} = \Omega \left[f(l) V_l - f(l-1) V_{l-2} \right] \tag{19}$$

$$\nu \left(\frac{(4/3)\left[l(l+1) - 2\right]}{r^2} + \frac{\pi^2}{d^2} \right) V_l = \Omega \left[f(l) U_{l-1} - f(l+1) U_{l+1} \right] - \frac{\left[l(l+1) \right]^{1/2}}{r} \frac{R T}{\mu} (\ln p)'_l \tag{20}$$

where

$f(l) = 2\left[l(l+2) / (2l+1)(2l+3) \right]^{1/2} \simeq 1.$

These blocks are coupled in l to form a blocked-tridiagonal matrix equation which can easily be solved. For the modes we are interested in, the expansion series can be truncated at $l = 20$.

One can show that the linear system is characterized by the dimensionless parameter (Mayr et. al. 1984):

$$M_\Omega = - \frac{\delta\nabla\, g\, d^2\, P_r}{H_p\, r^2\, \Omega^2} \tag{21}$$

and a resonance occurs when $M_\Omega \simeq -1/L(L+1)$ for some integer L. The solution spectrum of such a resonance mode is quite flat up to the harmonic order L, but falls off rapidly beyond that. This justifies the series truncation. The first resonance occurs at a $\delta\nabla$ value of 2.6×10^{-5} which implies a ν value of 10^{12}. From this, we can determine the relative values of all the spectral components for Eqs. (10) - (15). Unfortunately, the amplitude of the resonance mode cannot be fixed by the linear approach. However, suppose that the solar differential rotation comes from such a resonance response with mild nonlinearity. Then, the low-order spectral structure of the nonlinear solution would stay close to that of the linear resonance. In this paper, we match U_3 [$C_3 \sim \sin\theta\,(5\cos^2\theta - 1)$] to the observed latitudinal differential rotation at the solar surface, and from that we can determine the absolute values of all the other components. Specifically, from the theoretical value of the ratio between the shell component and the largest differential rotation component ($U_1/U_3 = -1.46$), we predict that the radial differential rotation is negative and has a magnitude of about 1.38 degrees per day. In Fig. 4, the solid curve shows the observed solar rotation given by Howard and Harvey (1970); the plusses show the values produced by the model. The matching can be done so well because the predicted ratio of U_5/U_3 (= −0.118) is very close to that of the observed ratio (≈ −0.146). The dashed line shows the mean rotational rate at the surface; and the solid line shows the predicted mean rotational rate below the convection zone.

Only the first resonance ($|M_\Omega|$ largest) can match the solar differential rotation well, as the spectral distributions of other resonances extend to higher order harmonics. The next and higher order modes have no resemblence to the observed surface differential rotation.

The exact value of the predicted sub-rotation is of course not trustworthy because of the simplifying assumptions used in this model. However, there are several significant qualitative features of the model which may survive more rigorous modeling efforts:

(i) The mean rotational rate at the solar surface is smaller than that at the interior, and at the same time, a differential rotation which matches the observed surface rotation can also be maintained.

(ii) The difference in rotational rate between the surface and the interior is quite small.

(iii) Since $\nu/\Omega d^2 \sim 10^{-3} \ll 1$, the circulation is helio-strophic; the viscosity has little effect on the spectrum of the large scale resonance.

(iv) The predicted meridional velocity is very small, only about 10^{-3} times that of the zonal velocities.

(v) the temperature differential is small (zero in the model) at the solar surface. As long as the nonlinear effects associated with the large scale circulation can be considered as small, this is in fact a rather general consequence of the requirement that the solar surface is stress-free. This can be derived by operating ∂_θ on Eq. (7) and substituting $\partial_\theta \partial_r (\ln p)'$ into the

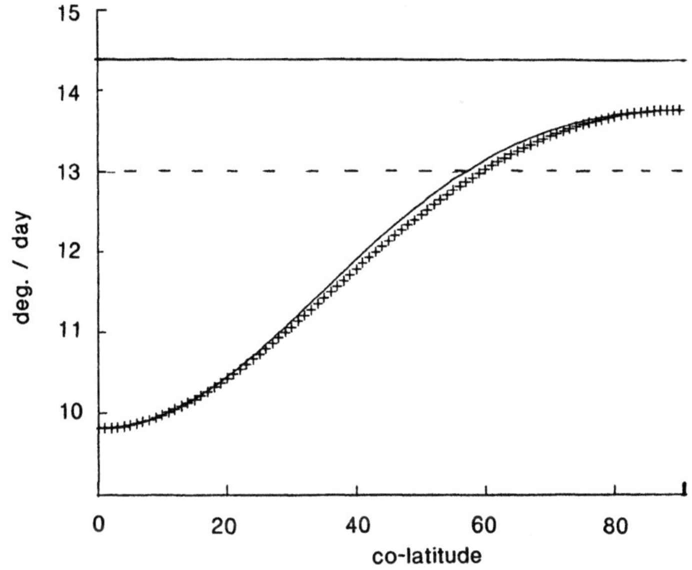

Fig. 4 Velocities of rotation

resultant of operating ∂_r on Eq. (9). Since $\partial_r U$ is zero at the surface and the left hand side of Eq. (9) is relatively small (helio-strophy), $\partial_\theta T'$ has to be almost zero. However, notice that $\partial_r T'$ is not zero at the surface; it primarily shows a P_2 variation across the latitudes.

Our model differs from most other linear models (e.g. Pidatella et. al. 1986) in that it identifies the global circulation in the solar convection zone as an axisymmetric resonance mode. Such resonances not only occur in a simple Boussinesq model, they have been noticed by Schmidt (1982) in a model that takes account of the vertical structure of the convection zone. Schmidt associated the resonances with discrete small Prandtl numbers. From Eq. (21), one can see that the same effect can be produced by fixing Pr but varying $\delta\nabla$ as we did.

This research is supported in part by the NSF (AST-8504399). SS thanks NASA for support. KLC would like to thank Drs. M. Stix and W. Unno for helpful conversations during the meeting.

REFERENCES

Chan, K. L. and Sofia, S. 1986, Ap. J., 307, 222.

Glatzmaier, G. A. 1985, Ap. J., 291, 300.

Howard, R. and Harvey, J. 1970, Solar Physics, 12, 23.

Mayr, H. G., Harris, I., Chan, K. L. 1984, Earth, Moon, and Planets, 30, 245.

Nakano, T., Fukuhima, T., Unno, W., Kondo, M. 1979, Publ. Astron. · apan, 31, 713.

Pidatella, R. M., Stix, M., Belvedere, G., and Paterno, L. 1986, Astron. Ap., 156, 22.

Schatten, K. H., Mayr, H. G., Levine, R. H. 1981, Solar Phys., 71, 169.

Schmidt, W. 1982, Geophus. As rophus. Fluid Dynamics, 21, 27.

HORIZONTAL REYNOLDS STRESS AND RADIAL ROTATION LAW OF THE SUN

G. Rüdiger
Sternwarte Babelsberg,
DDR-1502 Potsdam,
German Democratic Republic

I. Tuominen
Observatory, University of Helsinki
Tähtitorninmäki
SF-00130 Helsinki, Finland

ABSTRACT. Experiments with the design of the mean-field models for the differential solar rotation are reported, in which the rotational influence on the turbulence/convection is parametrizised in the sense of a series expansion of the Coriolis parameter Ω^*. The possibility of using only the one well-known free turbulence parameter is excluded by both observational and theoretical reasons. In particular, the observation of a positive horizontal Reynolds stress, $\langle u'_\theta u'_\phi \rangle \cos\theta \geq 0$, requires the introduction of additional terms, the number of which, in the next order, is four. While the parameters can be completely specified with the surface-dominated observations, the computational result is the inner rotation law. The models prove to have difficulties with the angular velocity along the rotation axis if, indeed, the horizontal Reynolds stress is adopted to be positive. In this case those smooth radial laws which are in accordance with the recent helioseismological results are not easy to produce.

1. INTRODUCTION

There is no fundamental doubt that the observable non-uniformity of the solar rotation has its source in such characteristics of the Reynolds stresses which are due to the rotational influence on anisotropic fields of turbulence and/or convection. Generally, the rotational influence can be weak or strong; it is linear in the rotation rate Ω only when this influence is sufficiently low. Slow and fast rotation are defined here in comparison to the scales of the velocity modes. The rotation can be considered as slow if the Coriolis parameter $\Omega^* = 2\ \tau_{corr}\Omega$ is smaller than unity. This parameter is certainly small in the sub-surface layers of the solar convection zone (SCZ) because of the short eddy lifetimes there. We conclude that the expressions for the turbulent angular momentum transport (TMT) in these layers are linear in the basic rotation rate $\Omega_\odot$.

Generally, the rotational influence on turbulence can be described with the Λ-effect formulation, where Λ_{ijk} is defined as that tensor which contains the entire influence of (uniform) rotation on the correlations

$$Q_{ij} = \langle u'_i(x,t) u'_j(x,t) \rangle = \ldots + \Lambda_{ijk}\Omega_k. \qquad (1)$$

The missing tensors are then restricted to involve either the gradients of Ω or those terms which exist already for $\Omega = 0$.

For slow rotation, the tensor Λ does not contain the angular velocity. If the vector $\hat{g}$ represents the anisotropy of the field, one can construct only one tensor with the correct symmetries, i.e.

B. R. Durney and S. Sofia (eds.), The Internal Solar Angular Velocity, 361–370.

$$\Lambda_{ijk} = \Lambda_V(\varepsilon_{ipk}\hat{g}_j + \varepsilon_{jpk}\hat{g}_i)\,\hat{g}_p \qquad (2)$$

- no other form is possible. The corresponding correlation tensor Q_{ij} possesses only one component,

$$Q_{r\phi} = \Lambda_V \sin\theta\Omega, \qquad (3)$$

so that the TMT terms must be written

$$Q_{r\phi} = -\nu_T r \sin\theta\partial\Omega/\partial r + \Lambda_V \sin\theta\Omega; \;\; Q_{\theta\phi} = -\nu_T \sin\theta\partial\Omega/\partial\theta \qquad (4)$$

(Lebedinski, 1941; Wasiutyński, 1946; Biermann, 1951). The gradient arises from the missing parts in Eq.(1). Notice that in (4b) there appears no term of the form $\Lambda_H \cos\theta$ with latitudinally uniform Λ_H. It is excluded by the shape of tensor (2) and also by regularity conditions of the Reynolds equation.

As the newly introduced eddy viscosity we shall always use the generally accepted value $5 \cdot 10^{12}$ cm^2/s.

2. $V^{(0)}$-THEORY

If the expressions (4) are applied to the conservation law of angular momentum, while neglecting all terms nonlinear in Ω the resulting rotation law depends on the depth only and not on latitude. However, this form of non-uniform rotation generates (Biermann-Kippenhahn-) circulation which as a secondary effect produces a latitudinal dependence (Zöllner, 1881). Such considerations are close to works based on the concept of "anisotropic viscosity" if we identify

$$\Lambda_V \equiv \nu_T V^{(0)} = 2\nu_T(s-1). \qquad (5)$$

When Kippenhahn (1963) and Köhler (1969) used s=1.2, i.e. $V^{(0)}$=0.4, they were able to produce a circulation which precisely generated the observed equatorial acceleration. For several reasons one must be careful, however. Firstly, the pole-equator difference in Ω is determined by the complete circulation pattern, including the bulk of the convection zone. If several cells exist then even relatively small changes of the cell structure can have dramatic consequences for the surface rotation law. Secondly, the system of equations which determines the mean field hydrodynamics in the SCZ reacts rather sensitively to its boundary conditions (Moss and Vilhu, 1983; Schmidt, 1982).

All calculations similar to the $V^{(0)}$-theory can be criticized, however, principally from two basic reasons. First, the turnover time of the eddies in the bulk of the convection zone is certainly too long in comparison to the rotation time in order to allow the use of only the linear terms in the basic rotation rate (c.f. Iroshnikov, 1969; Gilman, 1974; Rüdiger, 1977; Durney, 1983). Then it is unlikely that tensor (2) is the only one which contributes to the turbulent angular momentum transport. As we shall see below, it is very easy to construct higher-term tensors including the rotation rate in a higher power so that non-linear theories are possible.

3. THE HORIZONTAL REYNOLDS STRESS

Our second principal criticism of the models linear in Ω concerns the horizontal Reynolds stress. In this connection we have the general statement that all theories which produce the observed surface rotation law (e.g. via meridional flow), and which do not include higher-order terms in (4b), yield horizontal correlations in the mid-latitudes of the N-hemisphere, such as

$$\Omega_{\theta\phi}/\Omega_{\odot}\nu_T \cong -0.2 \tag{6}$$

Negative values, however, are *not* observed. We expect them for the direct surface layer (granulation), but the sunspot proper motions exhibit a *positive* correlation,

$$Q_{\theta\phi}/\Omega_{\odot}\nu_T \cong 1, \tag{7}$$

(Ward,1965; Gilman and Howard,1984). If this is accepted, a theory linear in Ω cannot be correct, since wrong sign of the horizontal correlation is then generated. The expression (4b) must be wrong at the depth of the SCZ where the sunspots are anchored. We must use a modification of (4) already in those layers, the depth of which is, unfortunately, not perfectly known:

$$Q_{\theta\phi} = -\nu_T \sin\theta \partial\Omega/\partial\theta + \nu_T \Omega H^{(1)} \cos\theta \sin^2\theta. \tag{8}$$

It is the Λ-tensor

$$\Lambda_{ijk} \sim (\hat{g}\Omega)(\Omega_i \varepsilon_{jpk} + \Omega_j \varepsilon_{ipk})\hat{g}_p$$

which provides the structure of the $Q_{\theta\phi}$-term in Eq.(8). The quantity $H^{(1)}$ must be positive and close to 1 in order to ensure the positivity of the horizontal Reynolds stress. As, however, $H^{(1)}$ goes as Ω^{*2}, and the Coriolis parameter is small in the surface layers, we must conclude that either

i) the outermost layer linear in Ω is much shallower than suggested by the mixing length theory, or

ii) sunspots carry information from very deep layers of the SCZ.

Both possibilities differ essentially with respect to the depth-dependence of the higher order coefficient $H^{(1)}$, and those analogous to it (see below). Case (i) implies a flat profile for $H^{(1)}(r)$ which does not dramatically decrease near the top of the SCZ. The position where the horizontal correlations are measured can be taken close to unity, $x_w = 1$. In the opposite case (ii), the higher turbulence modes are assumed to exist only near the bottom of the SCZ so that we write

$$H^{(1)} = (x_i/x)^{\lambda} H_i^{(1)}, \tag{9}$$

with (say) $\lambda = 8$. For the normalized radius x_i of the inner boundary of the SCZ we always use the value 0.7. With a polynomial series

$$\Omega/\Omega_{\odot} = \sum_{1,3,..} \omega_{n-1} P_n^{1} / \sin\theta, \tag{10}$$

one finds from (8) the relation

$$Q_{\theta\phi}/\nu_T\Omega_{\odot} = w\sin^2\theta\cos\theta, \text{ with } w = 15\omega_2 + (x_i/x)^{\lambda} H_i^{(1)} \tag{11}$$

if only the leading ω_2 term is considered. w and $\omega_2(1)$ have opposite signs. The observed value of ω_2 is -0.04 while that of w is 2, if measured from individual spots, and 5, if measured from sunspot groups (Gilman and Howard, 1984). Both empirical constraints shall be applied to both of the above mentioned two alternative cases, which might be treated with the specifications

$$\begin{array}{lll} \text{i)} & \lambda = 0 & x_w = 1 \\ \text{ii)} & \lambda = 8 & x_w = (1+x_i)/2 \quad \text{(say).} \end{array} \tag{12}$$

Though these numbers are arbitrary, we feel that they reflect the basic physical situation.

We have computed the response of the observed quantities to the coefficient $H^{(1)}$ in (8). The models have a density stratification and stress-free boundary conditions for the upper and lower boundary. The corresponding zeroth-order model (Ω=0) is described by the parameters

$$g^* = \mu g_{\odot} R_{\odot}/\gamma\Re T_{\odot}, \qquad f^* = \sigma R_{\odot} T_{\odot}^{3}/\rho_{\odot}\chi_T c_p \tag{13}$$

for which the values 1500 and 200 are used. For the Taylor number

$$T_a = \Omega^2 R^4/\nu_T^2 \tag{14}$$

the value $5 \cdot 10^5$ is used, which may be reasonable for the Sun. Faster rotation cannot be considered because an instability appears in the (linear) system of equations for a higher Taylor number. This instability is an interesting phenomenon in itself, but it will be discussed in another occasion. Another parameter coupling the pressure terms is the normalized surface temperature

$$T^* = \gamma\Re R_{\odot}^{2} T_{\odot}/\mu\nu_T^2 \tag{15}$$

for which we use the value $5 \cdot 10^7$.

The solutions for surface values of ω_2, mean latitudinal flow $\langle u_\theta \rangle$(in m/s), pole-equator temperature difference, and horizontal correlation w, are given in Table 1.

Table 1. Surface values of two models of different depth-dependance of $H^{(1)}$

model	λ	$\omega_2/H_i^{(1)}$	$\langle u_\theta \rangle/H_i^{(1)}$	$(T_p\text{-}T_e)/T_e H_i^{(1)}$	$w/H_i^{(1)}$
i)	0	-0.08	2.4	$0.9 \cdot 10^{-4}$	-0.24
ii)	8	-0.03	-0.3	$2.1 \cdot 10^{-4}$	-0.30

We see that $H^{(1)}$ must to be positive in order to generate an equatorial acceleration. In addition, no value of $H^{(1)}$ exists for which the product $\omega_2 w$ would be negative. The other physical observable quantities, i.e. the mean flow and the pole-equator temperature difference, are small enough to exclude further complications.

4. A COMPLETED MODEL

As we have seen, the solar mean-field hydrodynamics with only one parameter in the TMT expressions cannot provide the observed features. If really the observations of the $\langle u'_\theta u'_\phi \rangle$ correlations are accepted, then we are forced to introduce additional parameters. Indeed, the theory of rotating turbulences is incompatible with the idea that only two parameters, $V^{(0)}$ and $H^{(1)}$, are of relevance. One can construct another tensor of third order in Ω with the structure

$$\Lambda_{ijk} \sim (\hat{g}\Omega)^2(\hat{g}_i\varepsilon_{jpk}+\hat{g}_j\varepsilon_{ipk})\hat{g}_p$$

which suggests the appearance of a $V^{(1)}$ quantity in (4a),

$$Q_{r\phi} = \nu_T(-r\partial\Omega/\partial r + \Omega V^{(0)} + V^{(1)}\Omega\sin^2\theta)\sin\theta. \tag{16}$$

It possesses nearly the same power in generating equatorial acceleration as the $H^{(1)}$ quantity (Rüdiger, 1977). Similarly to $H^{(1)}$, it also goes as Ω^{*2} compared to the fundamental mode $V^{(0)}$, and hence exists only for not too slow rotation.

The TMT relations, however, do not contain the whole truth. There is also a rotational influence on the eddy heat flux,

$$F_i = \rho c_p \chi_{ij}\beta_j, \quad \text{with} \quad \beta = g/c_p - \nabla T, \tag{17}$$

which can be imagined by means of the simplified relation for the eddy conductivity tensor

$$\chi_{ij} \cong \tau_{corr} Q_{ij} \tag{18}$$

(Daly and Harlow, 1970), which at least fulfills the dimensional constraints. Obviously the meridional components of the eddy conductivity tensor, i.e. χ_{rr}, $\chi_{r\theta}$, $\chi_{\theta r}$ and $\chi_{\theta\theta}$ are of particular relevance. This means that just those velocity correlations which do not appear in the TMT expressions, must be known in rotating turbulences.

Let us introduce the formulation

$$\chi_{ij}/\chi_T = \delta_{ij} - \begin{pmatrix} VV^{(1)}\sin^2\theta & VH^{(1)}\sin\theta\cos\theta \\ HV^{(1)}\sin\theta\cos\theta & HH^{(1)}\sin^2\theta \end{pmatrix} \tag{19}$$

for the meridional components of the eddy conductivity tensor. It also reflects the concept of the so-called "latitude-dependent eddy conductivity" when special choices of the parameters are made, namely, when $VV^{(1)} = HH^{(1)}$ and $HV^{(1)} = VH^{(1)} = 0$ (e.g. in Pidatella et al. 1986). For a small pole-equator temperature difference, and neglecting the higher sine terms, we find

$$\begin{aligned} F_r &= \rho c_p \chi_T (\beta_r - VV^{(1)}\sin^2\theta\beta_r^{(0)}), \\ F_\theta &= \rho c_p \chi_T (\beta_\theta - HV^{(1)}\cos\theta\sin\theta\beta_r^{(0)}), \end{aligned} \tag{20}$$

from which we see that the quantity $HV^{(1)}$ normally dominates $HH^{(1)}$. One of the weak points of "latitude-dependent eddy conductivity" is the ignorance of this HV contribution. The inclusion of $VV^{(1)}$ as well as $HV^{(1)}$ into the theory leads to meridional flows and, as a consequence, to differential rotation.

To sum up, there are 5 mean-field parameters, $V^{(0)}$, $V^{(1)}$, $H^{(1)}$, $VV^{(1)}$, and $HV^{(1)}$, which produce differential rotation in depth and latitude. Two of them, $V^{(1)}$ and $H^{(1)}$, are able of direct generation of pole-equator differences of the angular velocity, while the other ones work via meridional flow. Yet there is also another circulation to take into account, the barocline flow. It occurs even if all the turbulence parameters are put equal to zero and is driven by a non-parallelity of density and pressure isoplanes due to the action of centrifugal force (of uniform rotation). Krause (1976), Mestel (1978), and Moss and Vilhu (1983), have discussed the effect. Although the surface flow in mid-latitudes,

$$\langle u_\theta \rangle \cong -a'_2 \quad [\mathrm{m/s}], \tag{21}$$

a being the normalized stream function, is not negligible in the barocline system, the differential rotation produced is only small.

5. THE OBSERVATIONS AT THE SOLAR SURFACE

Our aim is to determine what can be said, on the basis of the observed quantities at the solar surface, about the internal distribution of angular velocity in the solar convection zone. To this end, we construct models for the solar hydrodynamics which reflect our knowledge about the surface values and which predict the situation in the bulk of the convection zone. The solutions obtained can be discussed in relation to the angular velocity distribution which have been obtained by the analysis of the solar oscillations. The values for the turbulence parameters, which relate to the observed quantities, may also be compared to those given by turbulence theories.

The observed quantities which can be used are the following. The horizontal Reynolds stress was already discussed. If the proper motions of individual sunspots give the relevant value in the present context, we have $w(x_w) = 2$. In practice, in order to have reasonable models, we have to usually work with still smaller values. Using the first term of the differential rotation at the surface we have $\omega_2(1) = -0.04$. The observed meridional flow at the surface is at least small and we can use $a'_2(1)=0$. Values of the order of 10 (e.g. Snodgrass, 1984) have negligible influence. The pole-equator temperature difference has been observed to be very small, and hence $t_2 = 0$.

The observations of the solar oscillations give a rather flat profile of angular velocity at the equatorial plane and smaller or equal latitudinal differential rotation at the bottom of the convection zone than at the surface (Brown, 1985; reports in this workshop), as well as an inward increase near the surface (Hill, this workshop). This picture must be related to the observed difference between the sunspot and plasma rotation (e.g. Howard et al.,1984) and to the deceleration of the rotation of sunspot groups (e.g. Tuominen and Virtanen, this workshop). Both suggest that approximately $\omega'_e = -1$.

We will mainly be interested in the depth-dependence of the angular velocity and in the latitudinal differential rotation, both of which we will simply represent with radial profiles at the equatorial plane and at the axis

$$\begin{aligned} \omega_e &= 1 + \omega_0 - 1.5\,\omega_2 + 1.875\,\omega_4 \\ \omega_p &= 1 + \omega_0 + 6\,\omega_2 + 15\,\omega_4 \end{aligned} \tag{22}$$

The condition for the radial gradient has a connection to the turbulence parameters. The stress-free boundary conditions at the surface give

$$V^{(0)} = \omega'_0 + 6\,\omega'_2\,, \qquad V^{(1)} = -7.5\,\omega'_2 \tag{23}$$

and hence

$$\omega'_e = V^{(0)} + V^{(1)}, \qquad \omega'_p = V^{(0)}, \qquad \text{and} \qquad V^{(1)} = \omega'_e - \omega'_p \tag{24}$$

On the basis of turbulence theory (Rüdiger,1983; for $V^{(0)}$ Gough,1978, Hathaway,1984), the following conditions may be preferable: $V^{(0)}<0$ and $H^{(1)}>0$. The sign of $V^{(1)}$, however, is very sensitive to the eddy model used. Negative $V^{(0)}$ is in accordance with the observational constraint for the gradient of angular velocity.

6. THE MODELS

Computations have been done with stress-free boundary conditions for both surfaces. $V^{(0)}$ is always constant. The outer radiative boundary condition is in accordance with the Stefan-Boltzmann law, while for the inner one a spherically symmetric energy input is required. The equations are always truncated after n=5, so that we always use three orders for each perturbed variable.

In order to fix the five turbulence parameters, we have to use five observational conditions. The results for the models will be presented in the following compact way. The surface values of differential rotation and meridional flow for two models with differing depth-dependence of the first order turbulence parameters are given as

$$\begin{pmatrix} \omega_2(1) \\ a'_2(1) \end{pmatrix} = \begin{pmatrix} .001 \\ 15 \end{pmatrix} + \begin{pmatrix} .003 \\ -4.3 \end{pmatrix} V^{(0)} + A^{(\lambda)} T$$

where

$$T = (V^{(1)}, H^{(1)}, VV^{(1)}, HV^{(1)}),$$

$$A^{(0)} = \begin{pmatrix} -.02 & -.08 & -.8 & .7 \\ 5 & 2 & 1400 & 210 \end{pmatrix},$$

and

$$A^{(8)} = \begin{pmatrix} .0001 & -.03 & -.11 & .3 \\ 12 & -.3 & 240 & -100 \end{pmatrix}.$$

If now the turbulence parameter set is specified so that the above mentioned observations are reproduced, we obtain the values given in Table 2.

Table 2: Turbulence parameter set for 5-parameter models, i) $\lambda = 0$, ii) $\lambda = 8$

	$V^{(0)}$	$V_i^{(1)}$	$V_i^{(1)}$	$VV_i^{(1)}$	$HV_i^{(1)}$
i)	2.4	- 3.4	1.6	.007	.003
ii)	- 1.5	6.5	10	.11	1.2

Temperature and density pole-equator differences were small at the surface. The result shows that $V^{(0)}$ and $V^{(1)}$ have different signs. Uniform turbulence parameters lead to positive $V^{(0)}$ and depth-dependent $V^{(1)}$ ..... $HV^{(1)}$ to negative. In Fig.1 the generated rotational laws are depicted for the equator as well as for the poles. The pole-equator differences are not generally decreasing inwards. There is, however, another result. Case (i) yields the equatorial angular velocity with general inward increase while case (ii) leads to a maximum in the middle layers of the SCZ.

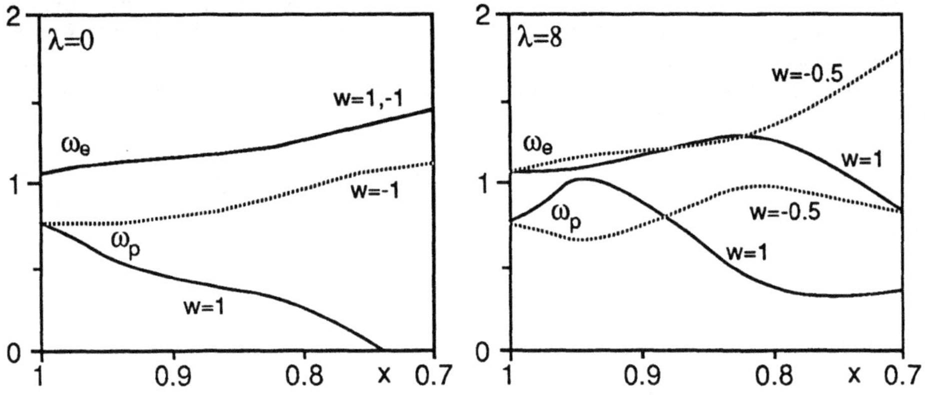

Figure 1. Rotational laws for the equatorial plane (ω_e) and for the axis (ω_p) with λ=0 (left) and λ=8 (right) with positive and negative horizontal stress (w).

However, these models have some unsatisfactory properties. In addition to the property that the values for ω_4 and a'_4 are not as small as is observed (only ω_2 is included in Fig.1), the pole-equator differences, especially of the angular velocity, are not decreasing inwards. Thus, even the most evident constraint that Ω must be positive everywhere will be violeted sometimes on the axis. In order to determine the reason for this property, the bottom values of angular velocity at the equatorial plane and at the axis as a function of w for the model with $\lambda = 0$ are given in Table 3.

Table 3: Influence of the horizontal stress on the differential rotation at the bottom of SCZ, a model with $\lambda = 0$

w	ω_e	ω_p	$V^{(0)}$	$V^{(1)}$	$H^{(1)}$
1	1.45	neg	+2.4	-3.4	+1.6
-0.5	1.42	0.82	+0.4	-1.4	+1.0
-1	1.41	1.12	-0.2	-0.7	-0.4
-1.5	1.40	1.40	-0.9	-0.1	-0.9

We see that decreasing w to negative values makes the bottom rotate rigidly and changes the sign of $V^{(0)}$. This result already demonstrates the dramatic consequences of the observed horizontal stress. Fig.1 gives also the rotational laws for negative w , with small $H^{(1)}$, i.e. near the "classical" Boussinesq-value for the horizontal stress. In the model with λ=0 the rotation at the axis behaves smoothly. On the contrary in the model with λ=8, at the bottom the differential rotation remains strong but at the equatorial plane the bottom rotation increases.

7. CONCLUSIONS

The models, of course, are crude compared to the real Sun, but we feel the following result to be significant: By ignoring the observation concerning the positive horizontal Reynolds stress, it is relatively easy to construct a model for solar differential rotation which reflects correctly the other observed properties at the solar surface and which is in accordance with the result obtained by the methods of helioseismology. In this case, the observed horizontal stress cannot have relevance with the present formulation of the problem, being e.g. a much larger scale phenomena than our turbulence parameters describe, $H^{(1)}$ being small in the solar conditions.

However, if we take account of the observation of the horizontal stress, the angular velocity distribution suggested by helioseismology cannot easily be produced. In this case, we must conclude that the present theory is too simple to describe correctly the solar conditions.

REFERENCES

Biermann, L.: 1951, *Z. Astrophysik* **28**, 304.
Brown, T. M.: 1985, *Nature* **317**, 591.
Daly, B.J. and Harlow, F.H.: 1970, *Physics of Fluids* **13**, 2634.
Durney, B.: 1983, *Astrophys. J.* **269**, 671.
Gilman, P.: 1974, *Ann. Rev. Astron. Astrophys.* **12**, 47.
Gilman, P., Howard R.: 1984, *Solar Phys.* **93**, 171.
Gough, D.O.:1978, in *Workshop on Solar Rotation*, University of Catania, eds. G.Belvedere and L. Paternò, p.337.
Hathaway, D.H.: 1984, *Astrophys. J.* **276**, 316.
Howard, R., Gilman, P.A., Gilman, P. I.: 1984, *Astrophys. J.* **283**, 373.
Kippenhahn, R.:1963, *Astrophys.J.* **137**,664.
Krause, F.: 1976, *Astron. Nachr.* **297**, 241.
Köhler, H.: 1966, *Mitt. Astron. Ges.* **21**, 91.
Köhler, H.: 1969, *'Differentielle Rotation als Folge anisotroper turbulenter Viskosität'*, Thesis, Göttingen.
Iroshnikov, R.S.: 1969, *Astron. Zh.* **46**, 97, (*Sov. Astr.* **13**, 73)
Lebedinsky, A.I.: 1941, *Astron. Zh.* **18**, 10.
Mestel, L.: 1978, in *Workshop on Solar Rotation*, University of Catania, eds. G. Belvedere, L. Paternò, p.364.
Moss, D. and Vilhu, O.: 1983, *Astron. Astrophys.* **119**, 47.
Pidatella, R.M., Stix, M., Belvedere, G., Paternò L.: 1986, *Astron. Astrophys.* **156**, 22.
Rüdiger, G.: 1977,*Solar Phys.* **51**, 257.
Rüdiger, G.: 1983, *Geophys. Astrophys. Fluid Dyn.* **25**, 213.
Schmidt, W.: 1982, *Geophys. Astrophys. Fluid Dyn.* **21**, 27.
Snodgrass, H.B.: 1984, *Solar Phys.* **94**, 13.
Ward, F.: 1965, *Astrophys. J.* **141**, 543.
Wasiutyński, J.: 1946, *Astrophys. Norveg.* **4**.
Zöllner, F.: 1881, 'Über das Rotationsgesetz der Sonne und der grossen Planeten', in *Wissenschaftliche Abhandlungen,* Bd. 4, Commissionsverlag von L. Staackmann, Leipzig.

SUBJECT INDEX

CZ, RC stand for convection zone and radiative core, respectively. *Magnetic fields*, rotation (effect on) means the effect of magnetic fields on rotation; (effect of) means the effect of rotation on magnetic fields.

GPSR Compliance

The European Union's (EU) General Product Safety Regulation (GPSR) is a set of rules that requires consumer products to be safe and our obligations to ensure this.

If you have any concerns about our products, you can contact us on ProductSafety@springernature.com

In case Publisher is established outside the EU, the EU authorized representative is:

Springer Nature Customer Service Center GmbH
Europaplatz 3
69115 Heidelberg, Germany

Batch number: 10815974

Printed by Printforce, the Netherlands